THE DEATH OF SCIENCE

THE DEATH OF SCIENCE

A COMPANION STUDY TO MARTÍN LÓPEZ CORREDOIRA'S
THE TWILIGHT OF THE SCIENTIFIC AGE

ANDREW HOLSTER

Universal-Publishers
Boca Raton

The Death of Science: A Companion Study to Martín López Corredoira's
The Twilight of the Scientific Age

Universal Publishers
Boca Raton, Florida • USA
2016

ISBN-10: 1-62734-076-9
ISBN-13: 978-1-62734-076-2

www.brownwalker.com

Publisher's Cataloging-in-Publication Data

Names: Holster, Andrew. | López Corredoira, M. (Martín), 1970- The twi-
light of the scientific age.
Title: The death of science : a companion study to Martín López Corredoira's
The twilight of the scientific age / Andrew Holster.
Description: Boca Raton, FL : Universal Publishers, 2016. | Includes biblio-
graphical references and index.
Identifiers: LCCN 2016935687 | ISBN 978-1-62734-076-2 (pbk.) | ISBN
978-1-62734-077-9 (PDF)
Subjects: LCSH: Science--Philosophy. | Science--History. | Science--Political
aspects. | Science--Methodology. | Peer review. | BISAC: SCIENCE /
History. | SCIENCE / Philosophy & Social Aspects. | SCIENCE / Re-
search & Methodology.
Classification: LCC Q174.8 .H65 2016 (print) | LCC Q174.8 (ebook) | DDC
500--dc23.

CONTENTS

INTRODUCTION

"In the temple of science are many mansions, ... and various indeed are they that dwell therein and the motives that have led them there. Many take to science out of a joyful sense of superior intellectual power; science is their own special sport to which they look for vivid experience and the satisfaction of ambition; many others are to be found in the temple who have offered the products of their brains on this altar for purely utilitarian purposes. Were an angel of the Lord to come and drive all the people belonging to these two categories out of the temple, it would be noticeably emptier but there would still be some men of past and present times left inside... If the types we have just expelled were the only types there were, the temple would never have existed any more than one can have a wood consisting of nothing but creepers." Albert Einstein, 1918. Address for Max Plank's 60th Birthday.

Modern science is in a state of unprecedented crisis. It is suffering from a chronic illness that has been advancing steadily since 1960's, finally accelerating to fatal proportions in the last decade. While technology runs rampant today, transforming our physical and social worlds beyond recognition, the creative vocation of *scientist* and the institutions of *science* that originally produced the platforms for this technology are in a death-spiral. This is testified by a growing flood of criticism from leading scientists in recent years, about the failures of scientific institutions, and their corruption by larger ideologies of power and wealth and bureaucracy that have come to rule our world. Martín López Corredoira's *The Twilight of the Scientific Age* (2013, Brown Walker) is a particularly striking critique, giving a vivid and scathing analysis of the state of modern science. López Corredoira is both a serious scientist and a philosopher, and it is this combination that gives his account a depth and resonance beyond the more conventional critiques of ordinary scientists simply lamenting their working conditions. I take López Corredoira's text as the starting point of this book, and explore key themes in greater detail, following his style of putting forward frank criticism, informed by personal examples. The result is a grim prognosis for the survival of science in its present institutionalised structures. These are not structures suited to science, but forced on it by the larger bureaucratised organisation of a state-controlled capitalist society. Science is fatally corrupted by the forces that flourish in this mode of organisation: ambition, greed, power politics, mass propaganda and intellectual mediocrity. López Corredoira concludes that science, in this modern institutionalised form, is on its death-bed. Here I analyse the process of its death. In the remainder of this Introduction I briefly recapitulate key themes and conclusions.

THE DEATH OF SCIENCE

López Corredoira's account is part sociology, part history and part philosophy. He recounts a breakdown in scientific culture, seen through examples of cultural syndromes at one level (mediocrity, bureaucracy, greed, competition); but also considered as the inevitable passage of a stage of civilisation in the larger panorama of history. He suggests that the most meaningful scientific research has now been completed in many core sciences, and the bulk of 'new research' being undertaken is largely derivative, a waste of time, and increasingly expensive. He sees a lack of scientific imagination, and a lack of scientific leadership, with the executive decisions of science ruled by bureaucratic mediocrity. He sees a conformist senior scientific community, and a powerful influence of scientific censorship and propaganda, working for the exclusion of heterodox or challenging or original ideas. He laments the flood of useless information drowning out wisdom or philosophy, and thinks science as we know it is in its twilight.

This will surely conflict with popularist perceptions of the success of science. Most people no doubt think science today is strong, mainly because they see new technology – shiny new computerised machinery and production systems, spawned on a vast scale. These are systems locking us into dependency on global networks of industry, finance, communications, energy, transport, food production, medicine; and harnessing our human lives to goals imposed by wealth and power elites. But while this technology looks like advanced science to outsiders, it is not science at all: it is merely the crude beginnings of an era of robotic automation. The *scientific* revolutions and discoveries that enabled this technology occurred some decades ago: it has just taken industry these decades to develop the machinery these discoveries make possible. Science itself is forgotten, and our scientific institutions are now converted to industrial-bureaucratic corporates, dominated by swarms of technologists and managers of the most mediocre scientific ability. In the meantime, the real scientists, the rare creative talents intent on science as a form of intellectual discovery, have left the building.

There is both an intellectual and a pragmatic dimension to this crisis. Intellectually, we see the corruption of the ideals of science, the exile of its most gifted personalities, and a fatally diminished intellectual culture. This may appear an abstract concern to most people, but it is a tragedy for the tiny minority of intellectuals who find themselves excluded from their natural vocation. And without the contribution of this tiny minority of creative genius, science as an intellectual and idealistic pursuit dies. This book, like López Corredoira's, is in fundamentally support of these intellectuals and ideals.

This corruption of ideals is also entwined with pragmatic failures of science. The question we must ask is this: can we afford to let the *intellectual* spirit of science die so soon after spawning such vast technological power? Our technologies are far from sustainable: they are still crude and transitional, and dependence on them in their current fragile state will be disastrous. The corporatisation of science transfers power over the scientific endeavour to a corporate management elite, with the vision of engaging a vast work-force of standardised, interchangeable scientific technicians in a 'science factory', controlled by prescribed rules and processes. But this excludes the *scientific intellect,* the critical role of the creative intellectual. Without the ongoing application of real *scientific intellect,* we are trapped in a half-built building that will eventually collapse on our heads.

More generally, it should be stressed that technological power does not solve any of the serious problems we face, unless used intelligently and benevolently. Our present system delivers tools of vast power, produced by the efforts of our rarest intellectuals, into the hands of spoilt children: the domineering personalities who rule business, politics and bureaucracy, egocentrically obsessed with their personal success, trapped in their sense of their own importance, ruled by simplistic visions of competition and superiority. Without *intelligent* guidance, advanced technology is more dangerous than helpful. The death of the scientific intellect recounted here leaves us without the critical intellectual function, in either scientific or public or corporate institutions, to address our real problems. These are real and urgent problems for the very survival of our civilisation entering a period of unprecedented globalisation and cultural turmoil.

The immediate source of failure is evident to scientists themselves: *the bureaucratic-corporate-capitalist system is a failure when applied to the creative intellectual domain of science.* Recent years have seen a growing flood of criticism from leading figures within science – including many aging Nobel laureates of the past, who can see how dramatically science has declined in quality in their own life-times. Science today is variously criticised from within its own ranks as being obsessively bureaucratised and institutionalised; commercialised and corrupted by greed; drowning in mediocrity; exhausted in imagination; trapped in authoritarian dogmas; bigoted and exclusive of heterodoxy; and failing to meet multiple challenges of our time. These are not exaggerations.

On the philosophical side, science is a system of thought that is meant to help us make sense of our lives as human beings, to help us to understand the natural world and our place in it. This was its explicit aim when it originated as *natural philosophy,* in the 17th – 19th centuries, before it became compartmentalised into the specialised technical silos of today. But here modern science has also become a dismal failure. The 'scientific philosophy' that gained dominance over science institutions in the $C20^{th}$, propagandised by leading Establishment figures as the 'scientific world view', has long appeared empty of meaning or wisdom. This modern 'scientific philosophy' is an incoherent kludge of Positivist-Materialist-Empiricist doctrines, long abandoned by serious philosophers themselves. But it remains a distinct integument of the scientific culture, providing a powerful ideological function, ensuring conformity and a sense of superiority and solidarity. It is an intractable intellectual failure of our time, and has offered nothing useful to real philosophy or real science for many decades. Although in the 1920-30's 'scientific philosophy' originally sparked valuable developments in logic and technical analysis, it soon became bogged down in scientistic dogmas. Transplanted from Europe to America, the second wave of C20th Positivism emerged as a doctrinal worship of scientific authority, and a contempt for any other systems of thought.

The *philosophy of science* proper, the academic forum where such questions are meant to be discussed, has now become an archaic tea ceremony; a specialist self-referring field that makes little contact with real philosophy or real science or real life. It is fixated on abstract dogmas (called "my position" by philosophers), and bogged down in trivialities. It primarily reflects a dismal failure to solve mid-20th Century problems of semantics. As a popularist ideology, 'scientific philosophy' is now strongly associated with atheist attacks on religions (militant atheists falsely claiming Science in their cause), and with technocratic contempt for traditional moral, social or religious philosophy. Within science, this official 'philosophy' represents a powerful force for conformity, legitimating hostility to heterodox ideas and original thinkers within science, and ridicule of challenging subjects on the borders of science.

The crisis in science itself stems intellectually, I believe, from the separation of the *philosophy* of science and the *philosophy* of specific sciences to a specialist subject outside the sciences themselves – starting in the late C19th, when Natural Philosophy was replaced by scientific specialisations. (Lord Kelvin, the great polymath C19th scientist, and incidentally a religious man, was *Professor of Natural Philosophy* through his long career, up to the end of the century: the last of his era.) Subsequently, science has been compartmentalised on an industrial model, akin to a factory production line, with little integration between sciences, and a huge rift between sciences and humanities. Natural scientists today do not think they need any understanding of the philosophy and history of their own subject, and are socialised in their training to be deeply contemptuous of it.

This is analogous to a police force being contemptuous of psychology and morality and human understanding, believing their job is simply to apply brute force to enforce their goals. This arrogance leads to mistrust and eventually hatred of the police in the general community, and they end up in the role of a paramilitary force trying to dominate a hostile civilian population. The prominent scientific leaders of today – the propagandists for the scientistic ideology – similarly believe that they have the brute 'scientific force' to override philosophical concerns, having no need to consider conceptual subtleties or heterodox theories or recognise a diversity of ideas. ("When I hear of Schrodinger's cat," says arch-Positivist Stephen Hawking, "I reach for my gun", echoing a notorious Nazi propagandist.)

But the conditioning role of philosophical paradigms in science does not disappear just because it is ignored and suppressed from conscious recognition: it is merely unconsciously replaced by simplistic dogmas, acting as doctrines of faith. Principles of *conceptual analysis, semantics and epistemology* that are routinely taught in core physical sciences of physics and chemistry as *scientific philosophy*, and saturate the subtext of science textbooks, are based on a startling ignorance of real philosophy. They represent a hopelessly incoherent kludge of the Positivist-Empiricist-Materialist ideology long discredited by serious philosophers. These are presented as *foundational doctrines*. Critical discussion of philosophical principles and issues is generally out of bounds in university science courses.

On the other hand, it is also quite understandable why scientists have become contemptuous of academic philosophy. The separation of 'scientific philosophy' from the scientific disciplines has equally led to its degradation, and it has now become an intellectual fantasy world. Intellectual standards in philosophy, it may be fairly said, are generally dismal. Academic 'philosophers of science' rarely have any working experience or competence as scientists. What they teach appears of no practical use to real scientists. As well as being scientifically incompetent, they also fail to address social issues of science, disdaining life outside their offices in the ivory tower. What do we pay them for? Serious scientists, whatever philosophical naivety they may have, are still generally hard-working teachers and researchers, and rightly contemptuous of such academic drones.

López Corredoira thinks the academic philosophy of science is so futile it should be abandoned. I sympathise, but we should recognise that there are *some* modern philosophers of great ability, and the philosophy of science and of specific sciences has achieved some results of fundamental importance. They are merely buried in the vast academic dross of philosophy. I think philosophy should be *transferred back into the sciences proper*. Science departments *should teach philosophy of science and the philosophy of their own science* as part of their core curriculum. Indeed, I think this is the *only* way academic sciences will be restored to health. But this is unlikely to happen. There is no longer any tradition of philosophy in the natural sciences, and simply no competent academics for such roles. For example, in my central area of interest, there is probably not a single competent *philosopher of physics* in NZ, in either philosophy or physics departments, across half a dozen traditional universities. And then what about the philosophy of biology, evolution, chemistry, information sciences, social sciences, etc? Where would you find people competent to teach the philosophy of all these subjects in science departments? Such people simply don't exist – certainly not in my country. Anyway, most science professors would strongly oppose this idea. For any realistic course must lead to science students questioning the naïve philosophies of their professors. Far from regarding the vacuum of philosophy in science as a problem, the scientific establishment is intent on *destroying philosophy*, in an act of intellectual genocide.

But I believe that a philosophical dimension of thought, including analysis of fundamental concepts, is critical to *doing science* – especially to discovery and creation and evaluation of new theories. If you cannot embrace conceptual novelty, you can only become a technician working within a fixed paradigm. This is what the core Science Establishment is now dedicated to. This is a profound change in the culture of science since the heyday of Lorentz, Plank, Curie, Einstein, Bohr, Schrodinger, Dirac, and their colleagues who created modern physics in the early C20th.

Science also has a critical role in reflecting back on philosophy – including social, metaphysical and ethical beliefs. A number of sciences profoundly influence our world-view: in particular, the sciences of physics and chemistry, geology and astronomy, biology and evolution, psychology and consciousness, information theory and semantics. Pragmatists and utilitarians may see science as merely a technology factory, producing product technologies, and not care about the 'ultimate truth' of scientific theories. They may care only about the practical use of science to make us richer, more powerful and more comfortable. For them, that is the only reality. But there is far more to it than this. Modern sciences profoundly inform our *metaphysical views* whether we know it or not. These are our beliefs about fundamental nature of the world and our selves, our meaning and purpose, our origins and fate. Here of course we have the classic modern 'War of World Views'. The 'scientific philosophy' is claimed (by the Scientific Atheists and anti-philosophers) to prove Materialist Reductionism, to deny the existence of a spiritual identity, to make human life an essentially meaningless accident, and morality a subjective delusion. This contradicts traditional religious and philosophical views, which embrace a transcendental reality of spirit or personal identity, and affirms meaning, purpose and morality as intrinsic and real aspects of existence. This has become a bitter divide between the science culture and the humanities.

However (although apparently unbeknownst to most scientists), *all the major programs in C20th 'scientific philosophy' have dismally failed – and failed to prove any of their metaphysical doctrines – Materialism, Positivism, Reductionism, etc.* The characteristic view of the scientistic philosophers is that *science is true.* Or, since scientific theories sometimes change, that *science is usually true, and when it is false it corrects itself, and leads to true theories.* But there has always been a chasm between reality and ideology here. After a century of work trying to *prove that science is true,* the scientistic philosophers have failed to prove any such thing. In fact, it became very evident a long time ago that *there is no guarantee, nor indeed even much likelihood, that any general scientific theories of our era are true, and certainly none are complete!* This applies particularly to very general and abstract theories, like fundamental theories of physics.

This may seem strange: almost everyone who is open to science recognises that science makes *progress* – and what can this progress be except progress towards *truth*? What can progress mean except finding *true theories*? And yet, most scientific theories in the past have proved to be *false*, and have been abandoned and replaced at a fundamental level. But despite this series of *false* theories, scientific understanding has clearly progressed. Indeed, many *false theories* – e.g. Newtonian mechanics and gravity – represent decisive advances in scientific knowledge. But if they are ultimately false, what do they really tell us about the nature of reality?

The first part of this book addresses this classic conundrum in the philosophy of science, an issue that also underpins López Corredoira's account. The answer given here is that science really progresses in giving *robust explanations*. They are robust precisely when they are robust against future theory change. But there is no reason at all to think that our current theories provide us with anything resembling the *ultimate truth about the metaphysics of the natural world*. This contradicts the claims of scientistic propagandists that science has now irrevocably established the fundamental nature of the world: e.g. that quantum mechanics and General Relativity truly identify the fundamental nature of space and time and matter, with no reasonable doubt possible. This attitude is dismissed here as arrogance, and almost certainly false.

But in this attitude we see two powerful destructive drives within modern science. On one hand, the belief that science has *reached the ultimate truth about fundamental questions* provides the rationale for suppression of heterodox thinkers – those who would question fundamental paradigms of present science (like quantum mechanics or relativity theory or materialist reductionism), who typically have their work suppressed and their careers destroyed by powerful establishment scientists, for the heresy of *questioning the truth of science, or scientific authority.* And the other hand, we see an insistence by the same establishment scientists – acting as philosophical interpreters of science – on drawing vast metaphysical certainties from shaky scientific theories and superficial reasoning – and treating sceptics about their metaphysical fantasies as anti-scientific heretics. These are two powerful negative syndromes characterising the social and philosophical dimension of modern science. The scientific community cannot blame *this* on outsiders: it is a form of intellectual fascism engendered from within science.

The discussion goes on to consider the wider social context of science. Although 'science' now saturates our culture at many levels, it is in a most peculiar position, as it has been throughout its brief history: it is almost entirely dependant on public funding for its existence. It produces no saleable products. Its only significant commercial product is *science education*, but that too is dependant on public funding - and on a public education system that forces its consumers (mainly school children) to take science courses. Most purchasers of science textbooks are not willing consumers. Barely one percent of adults takes any interest in science – only a small fraction of one percent become 'professional scientists', and most of these are no more than technicians, not research scientists. Science is deeply esoteric to the general public, and completely uninteresting to outsiders in its detail – including scientists from other fields.

Of course people do like to see 'science headlines' like the discovery of a new planet, or a new cancer drug, or a new species of hominid – but this is not an interest in *science as such*, it is an interest in strange and wonderful facts, or in technology that may be personally useful. The shallowness of public interest in science is seen by comparing with other cultural fields. While people today can typically name dozens or *hundreds* of musicians, actors, writers, artists, sports stars, politicians, and even business personalties, probably few could name more than *one* modern scientist off the top of their head – Super-Scientist Albert Einstein – with perhaps Stephen Hawking or David Attenborough as the other two widely recognised modern scientists (in the Anglo world at least, where they are TV celebrities).[1]

Science in its detail is *very* esoteric to ordinary people. I think academic scientists do not realise how esoteric it really is. Few adults understand what a single equation in physics means – not even Pythogoras' theorem, although they learned it in school. '$E = mc^2$' may be widely quoted, but it rolls off most tongues only as a noise: few people have any idea what it means. Few people can cross the abstract divide from *arithmetic* to *algebra*, where numbers are replaced by variables – even though everyone is supposed to learn this in high school too.

There is a good reason: science is actually a very *unnatural,* very *un-instinctive* mode of thought. It is not simply an extension of 'common sense', as Bronowski[2] and other science popularisers have tried to portray. On the contrary, it is distinctly tangential to 'common sense'. It is very abstract, and requires a peculiar discipline of thought: the discipline of *formulating explanatory hypotheses and refining explanatory judgements,* as well as the peculiar hobby of *inventing theoretical entities,* and the peculiar behaviour of *suppressing emotive reactions to beliefs.* Science is a distinctive and peculiar *cultural invention,* the most recent major cultural institution to develop in Western civilisation.

---

[1] An internet site lists 'the 50 most influential living scientists' as follows. How many have you heard of? What have they done? I know the work of barely a third of these myself – mainly the physicists and computer scientists. What if it was a list of the 50 most influential living musicians/sports stars/politicians instead? The list is:
Alain Aspect, David Baltimore, Allen Bard, Timothy Berners-Lee, John Tyler Bonner, Dennis Bray, Sydney Brenner, Pierre Chambon, Simon Conway Morris, Mildred Dresselhaus, Gerald M. Edelman, Ronald Evans, Anthony Fauci, Anthony Fire, Jean Fréchet, Margaret Geller, Jane Goodall, Alan Guth, Lene Vestergaard Hau, Stephen Hawking, Peter Higgs, Leroy Hood, Eric Kandel, Andrew Knoll, Charles Kao, Martin Karplus, Donald Knuth, Robert Marks II, Craig Mello, Luc Montagnier, Gordon Moore, Kary Mullis, C. Nüsslein-Volhard, Seiji Ogawa, Jeremiah Ostriker, Roger Penrose, Stanley Prusiner, Henry F. Schaefer III, Thomas Südhof, Jack Szostak, James Tour, Charles Townes, Harold Varmus, Craig Venter, James Watson, Steven Weinberg, George Whitesides, Edward Wilson, Edward Witten, Shinya Yamanaka. See: http://www.thebestschools.org/features/50-influential-scientists-world-today/ For people complaining that Mr. Spock from Star Trek is missing, sadly he died before the list was compiled.
[2] Bronowski (1978) is an interesting book however, much better than modern versions by the crop of recent scientific propagandists.

People are interested in *results* of science – in 'scientific facts' and discoveries – which is to say, in *implications* of science. But this is quite different to scientific enquiry itself, which means an interest in the nitty-gritty of *scientific explanations and evidence*. It is like being interested in the content of a program on TV – '*The News*' say - without any interest in how the TV works – or how *The News* is produced. And why should we be interested in how the TV works? Most of us are never going to try fix one, or look inside one ourselves. The vast majority of people are likewise never going to *do any science themselves*. They are never going to *use science methods or enquiry to investigate anything*. They are never realistically going to *exercise independent scientific judgement* to evaluate anything. Nonetheless it is important, in a world of propaganda and half-truths, to understand something about how science is produced – just as it is important to understand how *The News* is produced.

I think people are really interested in *philosophy* rather than science – and science has its human interest only because it informs our philosophical outlook. By philosophy I mean a natural curiosity about the nature of the world; a desire to understand ourselves, our meanings and purposes; extending to an interest in theoretical ideas, and to rational processes behind our beliefs. The natural intellectual genius of the human race is really for the great smorgasbord of *philosophy*, not the thin gruel of science. It is driven by our need to self-consciously rationalise our own belief systems. We have complex beliefs systems, fraught with fallacies, incomplete knowledge, contradictions, change and instability. We are *self-conscious* of our own rational processes and beliefs, and there is a huge mental churn required to keep this in order. Although much of this rationalising work no doubt takes place unconsciously, and in dreams, we also consciously obsess about it. People constantly *philosophise* about their beliefs, discussing judgements of behaviour and morality, puzzling over implications of facts, trying to make sense of discordant information, and so forth. We are trying to understand the *meanings* of things.

This process is partly centred around an intuitive sense of *explanation* – we want to know why things happen as they do, what their meanings and implications are. But this is not *scientific explanation*. It is far more intuitive and psychological, calling on agencies and propensities, referring to symbolic meanings and motives and values, making intuitive observations of patterns, triggering off emotive reactions, and so on. The step from this to abstract philosophy, as a systematic rationalisation of ideas, developed independently in many cultures, long before Western Philosophy evolved a branch called 'natural philosophy'. The latter is based on a specific mode of *causal explanation* that eventually became 'modern science', but it forms only a small part of philosophy as a whole. This 'scientific mode of rationality' cannot possibly replace our general capacity to think philosophically.

In this respect, ordinary people are much more philosophically thoughtful and sophisticated than academics typically realise. People are constantly philosophising about the meanings and explanations of things. Not in an academic mode, but in an intuitive way. The radio talk-back host, mulling over topical questions and inviting responses from the public, is philosophising. The worker discussing frustrations of dealing with management, and trying to decide how to respond to the system, is philosophising. Academic philosophy, at the level of mediocrity generally taught to undergraduates, destroys rather than enhances people's intuitive abilities to think about meaningful philosophical questions.

On the other hand, people are much less *scientific* than scientists expect – including most academics and professionals. Very few people, *including few science graduates,* can make meaningful independent scientific judgements on their own. It is not because of a lack of intelligence: it is because science is counter-intuitive, it draws on a very complex web of background beliefs, and it requires a very specific mental discipline. As a result, 'scientific belief' - among science students as much as ordinary people - is overwhelmingly based on *authority,* not on *evidence* (as claimed by scientistic propagandists). Indeed, the most valuable talent in a world awash with false information and propaganda is to be able to identify the reliability of sources of information – and modern science is now full of propaganda to convince us that *science is the only reliable source.* This is rationalised by the idea that *scientists themselves base their beliefs directly on explicit evidence,* but this is entirely a myth. Scientists base their beliefs overwhelmingly on what they read in textbooks – trusting that the textbook is based on evidence somewhere down the line. When the line breaks down, as it has today, science too is degraded to dogma. We are urged to trust science, but when there is a scientific controversy, *whose science* do we trust? The 'science' with the best propaganda. Science is a *faith-based discipline,* embodying *faith in the (Text) Book and the High Priest ('expert opinion').* Exercising your own independent judgement in a scientific discipline will get you in big trouble if it conflicts with Scientific Authority.

The meaningful aspect of science does not lie in its empirical data as such, but in its power to inform our personal systems of beliefs, or philosophy. The modern scientistic propagandists, preaching a Materialist-Atheistic ideology, exploit this philosophical dimension of interest in the same way as religious demagogues, just with a different aim: the aim of *destroying* belief in meaning. In fact, most individual sciences *per se,* as specialised enquiries, have no intrinsic human interest at all. This reflects the very peculiar position of science as *the subject almost no one is actually interested in.*

Music, literature, drama and art are all creative intellectual activities too: but they can all support themselves, at least partially, in a capitalist world, by selling their products. This is because people *like them.* Pure science cannot support itself commercially from its products at all. Its products are *scientific theories, observational data, and explanations* – and there is no popular or commercial market *at all* for scientific theories, for scientific knowledge, as such. (Just try to sell a scientific theory to someone! "How would you like to buy my theory that: $T = 2\hbar^2/m_e m_p^2 Gc$? Isn't that lovely? No? Not even a dollar? Oh why not? You don't have any *use* for it?") There is no market outside the circles of other scientists – and scientists cannot pay for science simply by buying it off each other in a circular economy.

Iconic modern science projects – like the LHC, the Hubble telescope, the Apollo missions – are of course vastly too expensive to be possible except as large national or international projects. But even small-scale, routine laboratory-type science has become an expensive activity, and the cost of pure science in most fields today is beyond anyone not funded by governments or corporates. 'Independent scientists' who typically work as consultants generally make their incomes from routine commercial science – gathering and analysing information of specific commercial interest to some business. Such work uses scientific technology and methods, but it is usually of no scientific interest as such: only of short-term practical interest, as specific information. (In fact, if you do discover something of genuine scientific interest in the course of such work, it is unlikely your employer will have any interest in it. It is more likely to make the project manager angry that you have gone 'off task': for they are only interested in what they have commissioned you to report on.) Thus *pure science* is in a very awkward position in our society economically, and it has no natural place among more traditional activities and professions.

Professional science as we know it is a very recent – by far the most recent generic profession. There have been doctors and lawyers, engineers and architects, mathematicians and accountants, bureaucrats and planners, merchants and bankers, philosophers and priests, soldiers and sailors, artists and musicians, prostitutes and tax collectors, across many millennia and many civilisations. But natural science has barely existed for 400 years in the West, and is only about a hundred years old as a professional career path for significant numbers of people. It appeared only sporadically in other cultures before that - and only as an anomaly, an expression of a certain rare and eccentric personality. It remains the *natural vocation* of only a rare type of personality.

Until the later C19th, science was very much a self-appointed *vocation*, largely restricted to hobbyists with independent means, and a few independent researchers who could find wealthy patrons. The modern organisation of specialist sciences in professionally-taught university faculties only emerged in the later C19th. For the first few decades of the C20th, science was limited in scope, and survived on a relative pittance compared to the vast budgets of today. It was really only when physics became a critical prop for military technology, with the nuclear arms race, and subsequently a nationalist competition in the cold-war space race, that science started getting the huge levels of public funding it enjoys today. In the same era, politicians began to recognise the value of science as the foundation for advanced technology, and realised that public investment in science would pay off economically down the line – although it might be ten or twenty or thirty years later. Commercial economics has no place for projects of such long-term value – it is based on short-term values, trading of goods with immediate demand.

So science does not fit within the natural categories of the commercial world, or capitalist economics, where *profit from sales of products and services* is the driving value for an enterprise. Yet it has been forced to fit, because capitalist economics is the overwhelming principle of modern social organisation. But the true value of pure scientific knowledge is impossible to measure economically. It is both ephemeral, and uncertain, and largely belongs to the future.

A symptom of this is an artificial construct of *value and productivity* within modern science. 'Scientific productivity' is measured by a surrogate: *numbers of peer-reviewed papers published.* Science has consequently become a competition to publish as many papers as possible (with something like one and a half *million* peer reviewed papers published per year – and presently doubling about every 15 years – which implies some ten million unpublished papers *submitted* per year, to about 25,000 officially indexed peer-reviewed journals – a number also steadily expanding[3]). Yet most published papers have no apparent value – about 90% are never referenced again – while a tiny number have a large impact, with hundreds of citations[4]. This *peer-reviewed-paper-productivity* measure, despite its huge distortions of scientific *value*, has been the fundamental organising principle for the scientific-academic bureaucracy for decades now. It has led to a serious crisis of quality and to manipulative power-politics.

The failure of peer review is now recognised as a major crisis in science, and a lot of recent material has appeared on this issue. López Corredoira also has an interesting discussion of this. In Part 2, I put forward a series of examples from my own work, from some 15 years ago, that illustrates the severe lack of objectivity and the abuse of peer reviews in the game of academic power politics. These examples are also chosen to illustrate the themes addressed in the Part 1: the powerful role of metaphysical paradigms in science, projected as ideological contests. The examples I give here concern the metaphysics of time in modern physics, and attempts to suppress realism about *time flow.* Following López Corredoira, I also propose a reformation of the system of journal publications, to address the problem of bias and manipulation by reviewers.

Another major feature of modern science is the pressure on *pure science* (producing knowledge of intangible value, with no capitalist measure) to become *technology science* (producing tangible products). Today, corporates as well as governments fund extensive 'scientific research and development', primarily aimed at *product technology.* Of course there must be a domain for this kind of applied science. But nowadays, 'pure science research' is rarely done without also being motivated by potential product opportunities. As such, this

---

[3] E.g. see Morrison *at alia* (2014).
[4] And many that have no immediate impact must have a high real value, in terms of quality and originality, but are lost in a swamp of mediocrity and trivia.

research activity loses the primary intellectual values that drive pure science. The spirit of science dies. My own country, a second-world scientific nation, provides a dismal example of this corruption of science to profit-making commercialism, with the conversion of public scientific institutes to business enterprises, under the spell of a neo-liberal capitalist ideology. Pure science research institutes in New Zealand were converted in the 1990s to 'CRIs' – crown research institutes – specifically tasked with making profits. The scientific value of their subsequent work appears to have diminished to practically zero. In fact, I think they now have a negative value, stifling private competition from outside the Government sector, and suppressing the function of providing public-good science, in favour of trying to establish their own competitive advantage. Their performance is so poor that senior academics and independent scientists now repeatedly call for various CRIs to be disbanded. The ongoing *pretence* of having scientific institutions is more harmful than having nothing, as well as more expensive. These institutions have proved impossible to reform. If we want a science sector in NZ, it must be started again from scratch.

Science of course has now become a key Public Institution in our society, with huge public funding and infrastructure. It has developed a mythological status, being treated as a kind of god: an official *Oracle* that society calls on to answer certain kinds of questions. "*Scientists say that ...*" is the stock phrase in news bulletins when reassurance of objective knowledge is needed. But there is increasing popular doubt about its reliability, effectiveness and value. One telling sign is that where science in the past had this largely unchallenged role of *Oracle*, of representing scientific *conformity*, of answering controversial questions with authoritative evidence, we now see conformity breaking down, and science impotent to resolve numerous controversies. It now provides little intellectual leadership on difficult topics. This is an inevitable result of its degradation into mediocrity.

The authority of *Science as Oracle* has diminished most among the most highly educated classes, with well-informed people today routinely sceptical and cynical of 'scientific experts'. They appear as just more talking heads on TV, with a message to sell from their respective institutions. We know now that 'scientific facts' frequently change, just like 'legal facts', and 'scientific conclusions' depend on who the scientists are – which side of a debate they are on, which industry they work for. There are also many people today, well-informed without being scientists, with intense personal interests in niche subjects – ranging from health to energy to farming and genetic engineering to UFOs and parapsychology – who have grown contemptuous of the views of the 'Scientific Establishment' in their own areas of special interest.

Such 'amateurs' who question scientific orthodoxy are dismissed as cranks by academic scientists; but their weight of numbers grows; and 'professional science' is often proved incompetent in such areas – and prone to corruption when commercial profit is involved. What should be even more alarming to the Scientific Establishment, many of the best creative and original professional scientists *feel the same way about their scientific colleagues too*. The best heterodox scientists often see the Establishment scientists (who typically attack their new ideas) as dull, mediocre, conformist and dogmatic. The best creative scientists do not trust the bastions of modern science either! There is no doubt that thousands of the most creative young students, who would have become the premier scientists in an earlier age, have withdrawn from science in recent decades, because they find its culture so stifling and conformist. Thus we see the authority and prestige of science under threat on multiple fronts.

Science – or a scientific methodology – is also supposed to be applied within the business functions of corporates and bureaucracies – tasked as they are with critical planning and research functions. This is a joke. And not the funny kind of joke. The death of science is nowhere more evident than in the research, analysis and planning functions of government departments. Here we find 'science' reduced to its utterly lowest quality – where it is no longer recognisable as science at all. In Part 3 I illustrate this, again with examples from personal experience. In the NZ Ministry of Education I observed 'education research' reduced to a state of farce, a farce perpetuated decade after decade. There is nothing unique about this example: this failure is universal across government bureaucracies.

These also act as public *Oracles*, controlling official information and peddling government propaganda, as well as controlling state funding of wider research communities. They dominate every sector of society – education, justice, health, science, environment, welfare, finance, business, transport, housing, farming, security, defence; every major institution in every developed country is under the control of such monolithic government departments. Their *scientific capability* can only be described as farcical. Here we find the death sentence on science completed. If the future success and survival of our society and civilisation is dependent on the scientific competence of government institutions, we are doomed.

I also try to relate these syndromes to the socio-psycho dynamics of these institutions. Why are such bureaucracies such dismal failures? It is not just one or two, but *all of them, practically without exception*. It is not just a phenomenon in New Zealand: similar complaints about incompetence, malevolence, and invasive encroachment of public bureaucracies on personal lives seem to be heard from most developed countries. And it is not just government departments: large corporate bureaucracies may not be quite as *hopelessly* incompetent as government departments, but they are similarly fraught with bureaucracy and ham-strung by intellectual mediocrity. The banking and finance sector, for example, is an intellectual vacuum at the highest executive level, and completely incapable of addressing the systemic problems of the global financial system. Of course its executives don't care as long as they continue to make vast fortunes.

I attempt to identify some core mechanisms at work here. One is to relate the hierarchical power structures that define these organisations as *cultural ecologies*, to the characteristic adaptive behaviours of their members. The power-roles defining the power hierarchies are the definitive common feature of all such organisations. I argue that these power structures are instinctively exploited by power cliques of management, working for their own ends. The dynamics of competition in the environment they have designed for themselves inevitably leads to the degradation of these organisations, through a severe degradation of talent. They are doomed to failure by the *mediocrity and exploitative attitudes* of their management.

A second theme is the dominance of a specific cultural paradigm, reflecting a certain kind of shared *social metaphysics* characterising our age. This is an intense drive to a *rule-based* model of social behaviour. This *rule-based* mentality is not new: it is seen throughout the long history of institutional tyranny, from the Spanish Inquisition to Nazi Germany. It is a characteristic of a distinctively fascist vision of morality that many people seem preconditioned to adopt. It has always been the dream of politicians, moralists and bureaucrats to have centralised control of people's behaviour in total detail; the world of Orwell's *1984*. In the past this was prevented by natural technological limitations; but it has taken on a new virulence today, empowered by the use of computerised information-and-decision-making technology, to exhaustively encode rules of behaviour, and to monitor and control behaviour, and administer punishments. I note that the same *rule-based* metaphysics underpins a number of distinctive modern scientific paradigms: those that characterise the new tyrannies within science.

Scientific institutions have been overtaken by the same general syndrome as public bureaucracies – forced on them by the bureaucratic masters who fund and control them. They are obsessed with 'management processes' and 'business models' and 'methodological rules' and neo-liberal commercial reforms – obsessions destructive of the pursuit of science. The *creative individual* at the heart of the scientific enterprise has been trampled to death in this managerial scrum. "But *Modern Business Science* shows that this is the only rational way to organise large-scale enterprises!" the neo-liberal management scientist tells us. "We must control and measure our *Inputs* and *Outputs* and use our economic theory to *Optimise Outcomes!*" In their vision, scientists are "Inputs" to an industrial machine, just as factory workers are "Inputs" to an industrial machine – a raw resource just like other material resources. Human capital is interchangeable and replaceable, productivity has a one-dimensional measure, value is reduced to money, and the System runs by pre-defined Rules imposed by executive decision-makers.

The first thing these neo-liberal business executives fail to recognise is that they themselves and their peers – the managerial elite and the 'business scientist' gurus – are actually the most mediocre and incompetent, relative to their roles, among *all the worker hierarchies*, and setting *themselves* in charge of such organisations dooms them to failure from the start. In any case, although this industrial-type business model succeeds up to a point in mass-production factories (albeit denigrating the conditions of the enslaved human work force), it simply does not work at all in science. Science has no paint-by-numbers *rule-based methodology*. It cannot be *automated by business process*. It cannot be done by people who *do not have a creative scientific intellectual drive*. The institutionalised model of science cannot identify genuine scientific talent, cannot give it the creative freedom to work, and cannot retain it.

This raises the question of whether there is a practical prospect of rejuvenating science in its current forms. Should we try to save the failing institutions of modern science? Of course the Establishment assumes that we must keep fiddling with reforms to preserve their System. The System is their God. They would be children lost in the wilderness without it. But I believe it is too late to save this bureaucratised incarnation of science, and futile to keep supporting its failed institutions. It should be allowed to die. I believe it is time to allow our corrupted science industry to collapse, as a bankrupt institution, and trust the scientific spirit – by which I mean the spirit of *natural philosophy* rather than that of *scientific technology* – rejuvenates real science anew. Science will one day be resurrected, at a grass-roots level, through the efforts of those with a natural vocation to be scientists. But the time has come when real scientists should now abandon our failed scientific institutions, cease supporting the bureaucratic tyranny that sweeps across our world, cease working as technology slaves for state-corporate management; and find new places to work, new roles to define themselves, and new ways to fulfil their natural vocation.

Creative scientists were once called *natural philosophers,* and sought a deeper understanding of nature and our place in it, in an integrated view combining the technical sciences and the social sciences, metaphysics and social morality. This quest for understanding has been destroyed in the modern era of science. I think it is time to reinvent that occupation, and for real scientists to return to that role once again. In the last part of this book I make some comments about the challenges for independent scientists, who wish to withdraw from the present system, and work instead for a future resurrection of science. However this book is not about the resurrection of science, but about its death.

The final part of this book is a set of Appendices, containing exhibits to illustrate certain key points. These contain technical detail that goes beyond the style of the main text. They correspond to specific points made in the main text. I briefly explain the rationale for these.

- Appendix 1 is an extract from a brilliant monograph of philosophical observations about physics by the great pioneering particle physicist, T.D. Lee, in 1988, made not in the tradition of academic philosophy of science, but by applying his natural intuition. He makes many lucid observations. Relates to the example of *time reversal and time symmetry* in Part 2, and to the value of genuine philosophical reflection within science.

- Appendix 2 is an extract from an article by Einstein, in 1920, *'The Aether and Relativity'*. Einstein in his mature years took the concept of an Aether seriously, and even affirms its reality in a certain sense in the context of the General Theory of Relativity – contrary to popularist ridicule of the concept in physics. Relates to the example of the Ether in Part 1, and temporal metaphysics in Part 2.

- Appendix 3 gives some technical background to the issue of time reversibility of quantum mechanics, relating to the first peer review example in Part 2. Requires basic quantum mechanics.

- Appendix 4 illustrates what is meant by a *realist semantic interpretation*, illustrated first with the example of Pythagoras' theorem, and then with the metric equation of Special Relativity. This is for scientists who are mystified by what philosophers mean by 'semantic analysis'. Relates to the case in Part 1 for realist philosophical analysis to be taught in science.

- Appendix 5 goes a step further into logic, and presents an important proof in semantics, a *reductio ad absurdum* of the core Positivist principle of meaning. This is to illustrate that technical philosophy is a real subject, with analysis backed by formal techniques of proof.

- Appendix 6 goes another step again, and presents core concepts of *possible world semantics*, illustrating the real metaphysical dimension of debates over foundational concepts of science. Metaphysics is encapsulated in the construction of the *logical space*, required for the formal representation of concepts. Scientific theories implicitly postulate theories of the *logical space*.

- Appendix 7 presents various extracts on the Alvarez controversy, over the astroid-impact theory of the extinction of dinosaurs. This is a classic illustration of how scientific debates are often fraught with controversy and spite. Relates to example of explanations in Part 1, illustrating that evidence for well-accepted explanations is not necessarily straightforward.

- Appendix 8 provides material on the evolution of peer review, illustrating the open review system, and giving a further example of conflicting peer reviews about an interesting heterodox theory in physics.

- Appendix 9 provides material on the recent phenomenon of 'predatory publishers', and issues about the commercial corruption of the scientific and academic journal industry.

- Appendix 10 provides further material on the evolution of the academic journal industry, and the trend to open access journals.

- Appendix 11 provides material on a new system or surveillance on researchers, imposed through a universal identification scheme called *ORCID*. This prevents independent researchers, without approved *institu-*

tional affiliations (with academic, government or corporate institutions) from even *submitting* scientific papers.

- Appendix 12 is an extract from NZ writer Nicky Hager's 2002 book *Seeds of Distrust,* about a notorious incident of political manipulation of science in NZ. A model of investigative reporting into political corruption of science, containing many insightful observations.
- Appendix 13 presents arch-heretic Rupert Sheldrake's *Ten Dogmas of Science,* and the virulent personal attacks on him, started by Sir John Maddox, the arch-conservative editor of *Nature,* and continued by many other scientistic propagandists from sceptic and debunker groups.

PART 1. THE TWILIGHT OF SCIENTIFIC PHILOSOPHY

López Corredoira's perspective.

López Corredoira sees a corruption of science in a larger framework, as part of a more general degradation inherent in modern society: a loss of individuality and humanity to the gods of money and power, bureaucracy and institutionalisation. I begin with an extended quote that sets the general framework for his point of view.[5]

> "In general one has the impression when reading about the history of civilization that human beings were behind the dynamics of the societies... Nowadays, however, one has the impression that individuals are just simple marionettes whose strings are pulled by some abstract and superior entity. I am not talking about the Christian god but about that almighty god of modern capitalist times: money. Money is the great boss of our society. It governs the decisions of individuals and has much more power than the different nation-states or other human organizations...
>   [I am talking about] the generation of structures in our society which have become automatic in some sense, and thus have become self-sustaining autonomous structures. Once implemented, they may work independently of the will of human individuals. The economy, our present

---

[5] I quote extensively from López Corredoira's text, so that the reader may have the material in front of them, without having to constantly refer back to his book. This is no substitute for reading his book. I deal primarily with only the central third of his book, the critique of modern scientific institutions. The first third of his book tells the stories of the early historical development of some sciences he knows well, and these are very interesting – López Corredoira has a real empathy and detailed knowledge of the scientific subject matter and the cultural setting of the history he talks about, and this is of a much greater quality than the kind of conventional potted histories of science typical of most scientists who turn to presenting the history of their subjects – Hawking being a typical desultory example in physics. The last third of the book reviews some of López Corredoira's favourite 'classic philosophers', and I find this material fascinating, partly because these are philosophers rarely encountered in modern Anglo-American philosophy, and partly because López Corredoira's interpretation is fresh and original. His larger perspective on culture and philosophy is not adequately conveyed by the material I have selected here, and is not meant to be. I also note that I have not included all López Corredoira's references in the references to this paper, and where he makes references, I give the author's name and date, but not page and chapter numbers, so to trace these references the reader must go to his text. Thanks go to Martín López Corredoira for permission to quote from his material.

financial system, appears in my opinion as the most important of the autonomous structures to have emerged in our civilization. A monster,
Frankenstein's Creature, which once created, emancipates itself from its
creators and turn against the interests of individuals, pushing them to slavery. We could live in a society in which individuals worked much less than
they do now, but capital and its interest in unstoppable growth, pushes
individuals to produce more and more, far beyond their needs. People ask
for more and more work in a time when the work force is more dispensa-
ble than ever, given the high industrialization of our society; an absurd situation according to some thinkers, including myself.

We are destroying our environment; our towns are becoming uglier
and uglier, with so many factories, cars, concrete, etc.; our forest, seas, riv-
ers, etc., are becoming contaminated; but we cannot stop the advance of
the destruction because money pushes people to follow the career path of
progress, technology and destruction, and few people want to be outside
that system, few people want to live without money in their pockets.

In my opinion, knowledge has become another autonomous structure
within our society, although it is less important than money. In this new
system of gods, money will be like Zeus, the major god, while the rest of
the autonomous structures will be minor gods, depending always on the
great god of money. Indeed, we must recognize that most of the people
working as scientists in our society regard their research as a job, in order
to earn a salary, and in most of the cases the obtaining of that salary and its
associated status are the only things which keep them in their boring and
passionless research fields. (p.101-102).

What makes López Corredoira's critique most interesting and valuable is that
he brings this large claim to life by exploring its details through illustrations
from the real practice of science, giving a psycho-social account of how
individuals respond to the cultural framework imposed on them. The framework is fixed by the larger 'autonomous structures' of bureaucracy, hierarchical power and money that govern the organisation of science, as they
govern all modern social institutions. But the modern scientific bureaucracy is
implemented in its detail by the adaptive behaviour of individual scientists,
locked in a struggle for careers. The result is the triumph of mediocrity and
self-serving pettiness in the social elite, with the genuine creative heterodox
thinkers, who created science in the first place, being driven from the field.
Over the last half-century or so, orthodox scientific institutions have increasingly been left in the hands of a feudal mafia of power-hungry gate-keepers, a
controlling conformist mediocrity of incompetent researchers. Yes: evolving
exactly into the form of a government bureaucracy or capitalist corporation.
The result is tyranny.

"The evolution of science towards a tyranny of knowledge is unavoidable, because of the logic of the construction of knowledge itself, both because the knowledge is becoming more and more solid, and also because the dominant groups within the dominant ideas in science are becoming more and more powerful. The distinction between power and truth in science is difficult. p.117

López Corredoira emphasises the confusion of truth with power, and the gulf between the perception and the reality of success. Although *science* has descended into failure over the preceding decades, *scientific institutions* have become far more powerful, fuelled by perceptions of success controlled through media propaganda.

"The point is that there are strong economical interests surrounding the business of science, and the publicity in the mass media is used to create the illusion of great science, supported by huge investments of public money. The spirit of science is already in decline, close to death; but the body of science, the hierarchy around the business of research, is still very much alive and will not decline simply because of a lack of interesting results. P.97.

He laments the corresponding degradation of science to technical specialisations, tasked with gathering information, but lacking any kind of wisdom. This reflects the business model of an industrial production line imposed on science.

"Scientists have been specializing for quite a long time, but it is now a question of micro-specialization… It is a case of converting the scientific process into an industrialized mass-production system. p.103.

"Science has abandoned wisdom and become a mere technical profession. It is supposed that humanity has become wiser because of the greater amount of knowledge gained, but it is not so. In general, scientists and philosophers in the past were much wiser, even with less knowledge. p.104.

Despite his outspoken criticism of scientific institutions, he is a scientific traditionalist at heart, with tolerant but relatively conservative views about what science means. He speaks for the value of science, and believes it is the proper method to real knowledge about the natural world.

"Scientific knowledge is something good; I think I have expressed this idea with sufficient clarity in Chapter 2. But anything in excess is bad. Water is good for living beings, but an excess of water drowns them. … Science is not being eroded by the limits of knowledge but by unlimited knowledge.

Certainly, there is a limit to the number of really interesting questions that can be answered, a limit which we have already surpassed, but society will not stop its activity just because of that limit; rather, I think society will stop the obsessive collection of knowledge when it feels lost among the information, when it gets fed up with getting drunk after so many science parties, when it falls down exhausted and wonders whether we are servants of the structures we have created or we control them, whether man was made for science or science made for man, as Unamuno said. p.105.

He is careful to contrast his position with simple anti-science views:

"A more aggressive attack on present-day science comes from some authors who talk not only about the end of science but also about whether we should do anything to save it because of the way it has behaved in the last few decades (e.g., Oblomoff, Ségalat). p.100.

"In my opinion, science is still of great value and should not be confused with present-day scientific institutions, in the same sense that religion should not be confused with the church. I think we must not level the charge against science but against the capitalist system which currently supports it. P.100.

Here he conforms to a popular sentiment among scientists themselves. There is an outpouring of criticism by scientists of all walks against the bureaucratisation of their culture. This general phenomenon can probably be traced back to the cold war and the rise of corporate capitalism in the US. Following the Second World War, scientific institutions in the US became dominated by the twin organisational model of corporate elitism and government bureaucracy. Their elite bureaucrats fulfilled the Capitalist vision of science as a technology factory, providing slaves for the military-industrial estate.[6] It is obvious to everyone engaged in real science that this capitalist business model for scientific organisation does not work. I will come back to this question at the end of this study: *what should or could our response be to the corruption and degradation of science?* Should we try to 'fix' or 'save' its failed institutions? Or as more radical critics hold, should we leave it behind, as we left the political institutions of feudal aristocracy behind, and find a new way? But before approaching that essentially political question, I review López Corredoira's views, and give some pertinent examples of my own, both in support, and in some cases in contrast.

---

[6] Initially large numbers of physicists and engineers were conscripted into the US effort to create a nuclear arsenal and advanced weaponry, justified by the politics of the cold war, and evolving subsequently into the present Orwellian state of permanent 'war on terrorism'.

Heterodoxy and orthodoxy.

López Corredoira recognises the key, traditional goal of scientific *research* is the exploration of new knowledge, and the key problem lies in making judgements of the *quality of novel ideas*, in fields with apparently diminishing returns. What new research should be funded and supported? What new ideas should be investigated? The research enterprise itself depends on *the potential for new or revolutionary ideas to be found*. But as time goes on, the most obvious veins of research are explored, and new ideas are harder to come by. It is a bit like a gold mine. The first prospectors strike it rich and find nuggets in easy reach; but when the first strikes have been panned, you need to switch to industrialised methods to extract what is left. And when there is no gold in the ground left to find at all, it doesn't matter how big you make your industrial gold mine, you will only find microscopic grains. The spirit of gold mining changes too. The first prospectors are individualists, exploring new frontiers. The first gold miners are adventurers, prepared to undertake hazardous journeys with serious risks and endure harsh conditions. They are individualists, and entrepreneurs, working for themselves. Later the mine becomes an industrialised factory, run by production-line wage-workers, requiring expensive heavy equipment, needing large capital investment, with the enterprise and the profits controlled by executives and accountants from comfortable offices.

Gold miners are always keen to determine if there is likely to be gold in a certain piece of ground. But scientists have no way of telling if there is any scientific gold to find in a certain area or not. In this respect, *scientific discovery* – the process of *finding* ideas, theories, concepts, possibilities - is not a part of the 'scientific method'. Orthodox philosophy of science (positivism, empiricism, Popperianism, etc), which proclaims the triumph of the 'empirical method', says nothing about the *creation* of scientific ideas – or the *pre-scientific judgement* of the quality of new ideas. The conventional account of the 'scientific method' is an arm-chair rationalisation, designed to *justify scientific knowledge*. It is primarily a banal account of a method to test a given empirical hypothesis. But it assumes hypotheses, theories, ideas, are given: plucked out of thin air. This is where the 'scientific method' has a yawning chasm: it has no method for either the creation of ideas, or for the judgement of value[7].[8]

---

[7] In fact the 'foundationalist' tradition of philosophy of science, based on empiricism, fails even at its very first hurdle, the simplest task of providing a method for judging the truth of empirical theories. Intended as a rigorous justification of the *certainty and superiority of science as a form of knowledge*, orthodox philosophy of science has succeeded in proving precisely the opposite: there is no certainty inherent in the 'scientific method'. There is no guarantee that any general scientific *theories* are true. In practise, *there is no generalised scientific method*. There are certain general principles – objectivity, impartiality, rational inference, valuing evidence over authority - applied in different

Given that science has the task of *extending knowledge* (rather than just accumulating more facts to support present theories), then science must explore *new ideas*. Genuinely new ideas are notoriously difficult to judge – and impossible to judge for the vast majority of 'scientific technicians' churned out of our academic institutions. For new ideas, by their very nature, are discovered by scientists with rare talents and by rare accidents. They are ideas hiding behind the facades of conventional paradigms, hiding in the gaps of the textbook explanations and exam exercises that define 'scientific reality' for academics. They require novel thoughts, concepts, models, formalisms and solutions. Ordinary science students struggle to understand and recite even the conventional text-book problems that *have been solved for them*. If they cannot verify even these by themselves without following and trusting a text-book, they have no capacity at all to make judgements of *new solutions* at the forefront of their field. How many scientists mature from the state of *science students* to develop any kind of mature scientific judgement, when confronted with novelty? The answer is obvious enough from history and experience: *very few*. The present scientific ideology and mediocrity that dominates scientific institutions often reduces this to: *none*. I think most NZ science institutions

---

ways according to context. The notion that there is a well-defined 'scientific method' as a set of *rules* students can learn in science school is a myth.

[8] I have recently worked on two different scientific projects. One (the 'dandelion experiment') was a minor experimental study of effects of electromagnetic radiation on water, done by wilting and reviving dandelions in water subject to different treatments. It took about three months to get conclusive results. Another is the development of a unified foundational theory of physics, which involves proving various equations and developing predictions. I have worked on this on and off for 25 years. (WEBREF Holster 2015 (c)). It has taken years to get solutions in some areas. Almost everyone thinks the dandelion experiment is good science: it makes perfect sense as a classical type of controlled experiment. The result (WEBREF Holster 2015 (a)) helps confirm observations others have made in other contexts. It contains no conceptual novelty. Conversely, almost nobody thinks the unified theory is valuable – no academic physicist or mathematician will consider it. Of course it is beyond almost everyone's technical expertise, unlike the dandelion experiment – but people make immediate judgements anyway. This reflects preconceptions about its likely *success*. But the *value* of a unified foundational theory, although obviously far greater, is ignored. It is obvious that people think I should stick to the nice simple 'model science', and regard the prospect of finding a unified theory as delusional.

The point is that almost no one registers any difference in the *significance or value* of these two research ideas. Rather, people make immediate judgements about whether it *looks plausible methodologically* to them, quite regardless of its significance or triviality. People cannot judge the unified theory; but think it is implausible to discover such a thing, and will be wrong, and hence 'bad science'. There is a complete disregard for any *value scale*. The unified theory *is* likely to be wrong – but then it is a failed attempt at climbing a mountain – while the dandelion experiment is most probably right, but it is a Sunday excursion up a gentle slope.

actually have *no scientists* capable of making meaningful judgements of novel scientific ideas.

López Corredoira laments that the *mediocrity* of modern science – built into the logic of its modern mass-production culture – swamps out genuinely new, ambitious, or 'revolutionary' ideas, when they do appear, in favour of safe, trivial, pointless research projects, based on old and failed ideas. He gives the following hilariously sarcastic image.

"There are many naïve persons, scientists or non-scientists, around the world who still believe that science is an open process in which the best ideas are quickly recognized and accepted, while the wrong ideas are immediately discarded. This kind of individual thinks that achievement in science nowadays depends on intelligence, on genius. They think that someone could be working hard in a laboratory, or developing some theoretical idea and, if they were to make a revolutionary discovery, they would open the door of the room in which they was cloistered and shout along the corridors "Eureka! Eureka!"; then, colleagues would approach and say: "Have you made a revolutionary discovery? Come in! Come in! We were waiting for you ..." and the genius would have the chance to show their new discoveries and their colleagues would open their mouths, surprised by the new idea, recognizing its merit, and carrying the genius on their shoulders while shouting "Torero, torero,...!" ("Bullfighter" in Spanish). This was never the way in which general ideas were accepted. p.108.

"The reality is that nobody is waiting for revolutionary ideas, they are not welcome, now less than ever, and the difficulties that professional researchers have when they want to challenge dominant ideas (e.g., Campanario & Martin) are enough to dissuade them in their enterprise or cause to be rejected as outsiders by the system. p.111.

The same point has been reiterated over and again in recent times by many original scientists – including Nobel laureates - who have made significant breakthrough discoveries. Many testify to being attacked and excluded by their colleagues, for years or decades, precisely for being original and challenging. It is not so surprising, for at the end of the day, they are *competing with their orthodox academic colleagues.* Academics are notoriously ruthless when their self-interest and self-image is at stake. This is *precisely* what it means for science to adopt a *corporate bureaucratic culture.* Corporate executives and government bureaucrats are the model of ruthlessness in defending their self-interest – pragmatism overrides truth - and the institutionalisation of academics in a corporate culture produces the same result.

This distinction between the thankless mission of developing *new and challenging or revolutionary ideas,* and the often pointless but well-funded grind of the routine science industry, is a crux of the problem. On one hand, López

Corredoira is somewhat conservative, in expressing scepticism that there *are* many 'revolutionary ideas' out there to find. He suspects that, like an exhausted gold field, much of our significant scientific knowledge has already been found. In this respect, research science is in a death-spiral, devoting more and more resources to find more and more trivial knowledge, and leaving behind a growing slag-heap of useless information, like the vast charnels of consumerist waste our society cranks out in a hopeless quest to improve the quality of life through consumption - instead of recognising that there are natural limits to consumption, and coming to terms with the meaning of what we have. On this point I will disagree somewhat though: I think there is far more real scientific knowledge to discover – indeed, that our scientific world view will undergo a dramatic transformation before it discerns what the world of nature really encompasses. I will return to this.

On the other hand, he recognises that insofar as there *are* really new or revolutionary discoveries to be made, modern science suppresses real attempts to develop significant knowledge. He presents this with a contrast between 'orthodox' with 'heterodox' science.

"Science is not a direct means for reaching the truth. Science works with hypotheses rather than with truths. This fact, although recognized, is usually forgotten. It gives rise to the creation of certain key groups within science which think that their hypotheses are indubitably solid truths, and think that the hypotheses of other minority groups are just extravagant or crackpot ideas. These are usually referred to as the orthodox and heterodox positions in a given field. p.107.

"On the one hand, heterodox scientists are possessed by a feeling of being an unappreciated genius, have too much "ego", normally working alone/individually or in very small groups, creative, intelligent, nonconformist. A vast majority of them are men. Their dream is to create a new paradigm in science, something which completely changes our view of the field of science in which they are working. For instance, there are many of them who are trying to show that Einstein was wrong, maybe because he is the symbol of genius and defeating his theory would mean that they are greater than Einstein. Most of them are crackpots. On the other hand, orthodox scientists, who constitute the majority of the community, are dominated by groupthink and snowball effect, following a leader's opinion as in the story of the emperor's new clothes, are good workers performing monotonous tasks without ideas of their own in large groups, are specialists in a small field which they know very well, conformist, domestic. Their dream is to get a permanent position at a university or research centre, to be the leader of a project, to be a recognized science administrator. Most of them are like sheep, some of them with the vocation of shepherds as well. Luminet compares these people doing "normal science" to craftsmen, and

compares those scientists who pursue a revolutionary science to imagina-
tive artists. p. 109 – 110.

López Corredoira's description here is something of a stereotype, and he
does not tell us what 'crackpots' are.[9] However, there is some basic truth in
this description of heterodoxy and orthodoxy within science. In ordinary life,
people come with a spectrum of orthodox and heterodox beliefs and impuls-
es - there is no simple distinction of personality types on this basis (although
openness is one major psychological trait, it interacts with half a dozen other
major traits). But in the community of science a clear and critical distinction
has formed: a great divide between 'us' (the professional scientists: people
with titles and offices) and 'them' (the ignorant public, including amateur
'crackpots'). It echoes throughout the rhetoric of scientific propagandists
attacking heterodox thinkers. (*"Sheldrake[10] is not a scientist, he is a crackpot, a
pseudo scientist, any real scientist will tell you that his ideas are scientific nonsense…"*). In
their private beliefs, not all orthodox scientists are so prejudiced; but for
those with official power roles, things are indeed very black and white. Scientific culture forces conformity, designed to make an orthodox 'class solidarity'
the norm, and heterodoxy and creative individuality the enemy. Conformity
takes comfort in a shared mediocrity.

There is also deeper kind of personality trait related to the difference between science as an exploratory activity of *discovery*, and science as an academic activity of *mastering a domain of pre-existing knowledge*. It is parallel to the
difference between explorers and tourists. Both are travellers, and there are
lots of people today for whom 'travel' is a major hobby. But most of us travel
as tourists, visiting places and cultures that are novel and exciting to us
personally, but are actually well-known, thoroughly explored, and essentially
safe. Travelling to experience other cultures is a good thing to do, but it is
quite different to the traditional role of *explorers*: visiting unknown and dangerous lands, driven by an excitement of discovering *new and unexplored* frontiers. The real explorers in history typically endured great hardship and real
danger. But the age of the intrepid explorer is essentially over on the Earth's
surface, because it has been systematically explored. Nowadays, travelling has
become the popular middle-class hobby of tourism. And few modern tourists
would have been explorers in the old days. Conversely, few genuine explorers

---

[9] The main thing he leaves out is that, if most of the heterodox scientists are crackpots, then most of orthodox scientists are equally crackpots. Calling people 'crackpots' is not very helpful, or even meaningful – it is like calling people 'dishonest'.
There are all shades of dishonesty. Everyone is dishonest when it comes down to it.
Everyone is a crackpot when you scratch the surface too. If there is one thing
humans are universally deluded about it is the *sanity* of our own belief systems.

[10] An iconic example of a heterodox scientist of great talent who has been widely
attacked by the science establishment. See (Sheldrake, 2012).

are interested in tourism. They seek other challenges. There is really no role left for true *explorers*, in the geographical sense. Their age has gone.

I think the scenario painted in the Twilight of Science is analogous to this. Science is seen as having already explored most of the dangerous and exciting places. Scope for new *discoveries* is increasingly limited for modern scientists, and most are now professional 'scientific experts', tourist guides through textbooks who take their pride in mastering existing knowledge. The spirit of science has changed in the process, and this makes science unattractive for the explorers: the people wired for novelty and discovery, and typically some kind of heroic individuality. Thus science changes its nature *when it has been fully explored,* and scientists of the present age are now quite unlike the explorers of the past in spirit.

The scientific culture changes its fundamental values when it becomes dominated by this different type of personality, just as frontier outposts change their personality when they become settlements, and then towns, and then urban environments. In the first phase of 'civilisation', the original jungle or prairie is rendered into farms by pioneers – those able to tame nature from its raw and untamed state. In the next phase, the farms are taken off the pioneers by carpetbaggers, lawyers and bankers – those able to dominate the system of legal ownership. And then the whole environment is subsequently *incorporated* into the estates of the wealthy and powerful: who have learned how to dominate the entire social jungle. Colonial towns often retain a sentimental nostalgia celebrating the deeds of the pioneers: but this is a false image, designed for a sense of social solidarity, and themes for tourist attractions.

López Corredoira illustrates how this cultural shaping in modern corporatised science works, through a number of concepts, such as the *snowball effect, groupthink, supervedettes.*

"The snowball effect … is a feedback loop: the more successful a line of research is, the more money and scientists are dedicated to working on it, and the greater the number of experiments on observations that can be explained ad hoc, such as in Ptolemaic geocentric astronomy; this leads to the theory being considered more successful. In some cases, the system supports conservative views, but there are also cases of speculative lines of research that have been converted into large enterprises. For instance, in theoretical physics, string theory has absorbed a lot of people and funds, as well as marginalizing and deprecating other approaches… p.85-86.

Groupthink is a widely recognised phenomenon in social psychology:

"In a sociological analysis, Janis categorizes the symptoms of groupthink as: 1) An illusion of invulnerability, shared by most or all the members, which creates excessive optimum and encourages the taking of extreme risks. 2) An unquestioned belief in the group's inherent morality, allowing the members to ignore the ethical or moral consequences of their decisions. 3) Collective efforts at rationalization in order to discount warnings or other information that might lead the members to reconsider their assumptions before they recommit themselves to their past policy decisions. 4) Stereotyped views of enemy leaders as too deviant to warrant genuine attempts to negotiate, or as too weak and stupid to counter risky attempts made at defeating their purposes. 4) Self-censorship of deviations from the apparent group consensus, reflecting each member's inclination to minimize to himself the importance of his doubts and counterarguments. 5) A shared illusion of unanimity concerning judgments conforming to the majority view (partly resulting from self-censorship of deviations, augmented by the false assumption that silence means consent). 6) Direct pressure on any member who expresses strong arguments against any of the group's stereotypes, illusions, or commitments, making clear that this type of dissent is contrary to what is expected of all loyal members. 7) The emergence of selfappointed mindguards—members who protect the group from adverse information that might shatter their shared complacency about the effectiveness and morality of their decisions (Dolsenhe). P. 110.

These are essentially mechanisms of *adaptive cultural evolution*. They show how personality traits are shaped to sustain a cultural environment in a self-supporting feed-back loop. In fact there is an obvious *adaptive evolution* at work in institutions that everyone knows. People have different natural personality types, and institutional structure reflects this. There is a distinct sub-set with a primary *will to power*, who seek powerful positions and status, and become managers and executives. They intrinsically understand each others' motivations as egocentric, which is why they can form a coherent power elite. Those with a primary *intellectual drive* seeking interesting intellectual challenges – challenges of exploration and discovery - become researchers or technologists. They intrinsically understand each other's motivations as creative performance. The latter have no social power and the former have no intellectual vision.

As workers in bureaucracies often say: anyone who *wants* to be the manager should be disqualified for that reason. Managers are the people who wave their hands furiously when a position of authority is in the offering, saying: *Pick me! Pick me! I'm no good at technical stuff, my talent is for making decisions! I am a natural leader!* This of course reflects the concept of the leader as the person in authority.

This forms a stable eco-system when the organisational structure can sustain the division of roles. This is the primary function of the bureaucratic power structure: to allow the strata of power-seekers, who have limited appetite for the real tasks of the organisation, to dominate and control those with the talent to work. It is exactly the analogue of the feudal power structure, and at a deep level, the essential structure of a slave society. Heterodox intellectuals have no natural place within such systems: they are frontier explorers stuck in the urban jungle. They have no chance working outside the power system either.

> "Without fame, without money and without the recommendation of, or support from, a prestigious team of researchers, even the best of scientists, working in the most important fields, would be not listened to. "An individual with few resources, achieving what we could not do with billions of Euros. This is a scandal, and we cannot allow it". This is the message of the real capitalist society where money exerts its power. A new Einstein working in a patent office would be a scandal.
>
> Unzicker (2010, ch.1) thinks that in physics there are no longer idealistic individual thinkers, only large organizations, political interests and the rules of the science market. The role of the genius is being filled by the average technocrat. P. 86.

Now here is a crux of the problem. Heterodox science contains the subset of creative geniuses capable of making real progress. Orthodox science contains few of these. But to orthodox scientists the geniuses appear indistinguishable from 'crackpots', as saints may appear indistinguishable from lunatics to ordinary 'sane' people. López Corredoira himself says of heterodox scientists that *"most of them are crackpots"*. He assumes he himself can identify 'crackpots' too. I am sure he can - up to a point – but can he – or anyone else – *systematically* tell crackpot ideas from radical discoveries of real value?

On this point, I do not like the term 'crackpots'. Many people with *wildly* unconventional ideas are sane and creative; and generally speaking I think heterodox thinkers have the same spectrum of *rationality* as orthodox conformists. The latter are just as much 'crackpots', it is just that their 'crackpot ideas' are not *original*, but popular irrationalities that are too familiar to detect. The fact is that *everyone is irrational and a-rational* in large degree. Our primary judgements are overwhelmingly made first by reference to authority, and when we try to *think through consequences for ourselves*, we overwhelmingly think by 'intuition' and feeling, not by 'logical rational calculation'. I also don't see much difference in the egoism stakes – although I think what is involved is perhaps a drive to *individualism*, as much than simple egoism. Anyway, heterodox and orthodox scientists are *both* inclined to intense egoism. Egoism is a universal condition of the human race. Barring a few saints or prophets perhaps, everyone is at the center of their own world view, whether we recognise it or not. What is required is tolerance and respect and empathy for the ideas of others *despite our egoism*.[11]

---

[11] The expression of egoism is at the heart of all our social issues, because it is at the heart of human personality. The furnace of the human ego carried us through tens of

Nonetheless, there is this phenomenon that scientists are effectively divided, by personality and culture, into 'heterodox' and 'orthodox' thinkers, somewhat as López Corredoira describes. The unsuccessful or misguided heterodox thinkers who go off in pursuit of theories with no realistic chance of creating successful science vastly outnumbers the small group of gifted – or lucky - heterodox thinkers who make genuine scientific discoveries.[12] The central point is that the orthodox scientists cannot distinguish the latter from the former - and do not *want* to. Thus we have one psycho-social dynamic that pushes modern science to its deathbed: it excludes its primary creative explorers. These people are rare, and without them science dies.

Philosophy of science and philosophy in sciences.

López Corredoira talks not only about science but about philosophy at length, and sees problems besetting science originating in part as philosophical failures. He is keen to distinguish his approach from that of contemporary philosophy academics.

"It is probable that some will identify the present book as a philosophical criticism, a charge made by many philosophers against science. Certainly, I think that the way of thinking in this book is philosophical. However, it must not be confused with the type of presentations made by self-claimed professional philosophers. Philosophy of science done by scientists or philosopher-scientists is not the same thing as philosophy of science done by pure philosophers without direct contact with the profession of scientific researcher. In my opinion, when talking about science or nature, listening to active scientists who produce their own philosophical reflections is the best option; and when talking about Philosophy with a capital P I prefer to listen to the great philosophers, that is, to the important classical philosophers rather than the mediocre specialized academicians of our own epoch. We may wonder whether the present-

---

thousands of years of life-and-death struggle in the wilderness, before we developed the tools to bring our present domination of nature and physical security. But now, egoism appears as a central tragic flaw of human nature. It threatens our destruction from within, as humans have become their own most dangerous enemy. How we address this will be the key to our longer-term survival as a civilisation. The only thing I am sure of is that it is not going to be addressed by technology or science, and it is not going to be addressed by business models or economic theories. Most likely it will be resolved by force and violence, as it usually has in the past. Which is to say: it will not be addressed.

[12] But note heterodox thinkers are not only about 'creating science': they may also be about creating interesting ideas, opening pre-scientific concepts for exploration, challenging orthodox preconceptions for the sake of it, and simply being autonomous human beings.

day philosophers of science may help science to be better, and my answer is negative; paying attention to them is a waste of time. P. 149.

He is scathingly critical of the modern academic philosophy of science, but still supportive of the need for a mode of philosophical reflection *in science*. I will address this point in some detail. To me, a central cause of the failure of modern sciences is precisely the failure of *philosophy in sciences* – the exclusion of philosophical thought from science, in favour of technocratic specialisation. Indeed, this is a prime cause behind the exclusion of heterodox thinkers, who are naturally philosophical, sceptical rather than authoritarian, able to question conceptual assumptions. For philosophy requires a different type of thinking to technical learning or 'puzzle solving'. Science used to be called *natural philosophy* before this specialisation came to dominate, and in earlier eras, physicists were free to discuss issues of philosophical complexity. Over the last 50 years especially, the quality of philosophy *in physics* has been extremely degraded – killed deliberately in fact from the 1940's by an orthodox ideology. Today you find ready expressions of contempt for philosophy among physicists. Few have any idea of what philosophy means.

López Corredoira recognises the difference between the general *philosophy of science* and *philosophy in or of specific sciences,* e.g. the philosophy of physics, the philosophy of biology, etc. The philosophy of science is a generalised study of 'scientific method', typified by foundational programs justifying *why science works, why science progresses, how the scientific method guarantees truth, how scientific theories have meaning,* etc. These are typified by programs like Positivism, Instrumentalism, Empiricism, Pragmatism, Popperianism, Kuhnianism, Structuralism, etc, consolidating in the early-mid C20th, now fragmented into a thousand philosophical programs ('isms'), ranging from extreme *scientistic* philosophies insisting on science as the only valid means to *any* kind of knowledge, to equally radical critiques of scientific 'truth' as subjective, culturally relative, or merely a pragmatic form of behaviour with no meaning or truth at all.[13]

---

[13] There is a much longer general history of course, relating to *epistemology,* which goes back to Aristotle as the major classical figure in Western scientific tradition, with Bacon and Descartes leading a conversion of philosophy from metaphysics and ontology to epistemology in the 17th century, at the explicit birth of modern science. Hume and Kant are the most prominent C18th philosophers who brought epistemology to the forefront. The modern tradition of empiricist philosophy of science starts most definitively with John Stuart Mill's inductive empiricism, and Mach's positivism, in the C19th, which subsequently had a resurgence in the 'right wing' *foundationalist* movements of positivist-empiricist-materialism in the early C20th. The excesses of this led to the Popperian revolt, and then Kuhn and Lakatos, who brought some critical sociological perspective, and then Feyeraband's more extreme 'anarchistic' view. Subsequently, on the 'left wing', we have structuralism and post-structuralism and post-modernism, branching into Marxist theories and post-colonialist theories

The foundational problems, of scientific method, epistemology and semantics, have most conclusively *not* been solved, and there is *no* consensus about anything in this field, except within ideological cliques. Like modern philosophy generally, it is a mess (and a paradigmatic example of the extremes of mediocrity that science itself does *not* want to descend into). For all the different programs or *isms* (*positivism, instrumentalism, operationalism, Popperianism, rationalism, empiricism, relativism, constructivism, etc etc ... there are hundreds*) are intimately connected with larger metaphysical systems and semantic theories, and no one agrees on these.

To the outsider, the philosophy of science looks like a mess, and it is a mess. This is not to say there had been no progress in the philosophy of science over 100 years. In fact, there are a lot of real results – but they are mostly negative, especially showing the *failure of all the foundational programs*. There are practically no positive solutions to the fundamental problems, and the best work is buried in a morass of triviality and nonsense. If you think modern science produces trivia, try looking at the philosophy of science! Of course, scientists shudder to think their own subjects could descend to this state of mediocrity – and yet there is work of great genius scattered in this mud too.

López Corredoira's attitude that "paying attention to [philosophers of science] is a waste of time" is a little too harsh. We need to extract the real insights of philosophy of science, because they have been generations – centuries actually - in the making, they have been painful to establish, and they will *not* be reinvented from scratch except by going back over this painful process. We see this in the repeated rise and fall and rise and fall of variations of *positivism*, for example – a philosophy dominating in the C19th, with anti-realism about atoms, overthrown in the early C20th with the atomic realism of Plank and Einstein and Rutherford, etc, then re-established in the early-mid-C20th through instrumentalist interpretations of quantum mechanics, then partially overthrown again in later C20th, and now with multiple divisions and no consensus.

Philosophy within specific sciences does not have to try to solve these grandiose methodological problems: it concentrates on analysing specific concepts and assumptions, the metaphysical preconceptions underlying specific theories of a science, providing logical analysis of fraught topics. E.g.

---

and feminist theories, turning the philosophy of science into radically *politicised theories*. Political ideologists think everything revolves around politics, and they naturally present philosophy of science as a political agenda. This is the full circle to the death of science for sure, which was created as the ideal of *separating objective evidence for beliefs from political, religious, moralistic and emotive ideologies*. The reactionary back-lash to this then appears in the equally extreme scientistic ideologies of today – witness sceptic and debunker sites full of abusive diatribes against 'pseudo-science'. The subject is now radically polarised.

the *philosophy of physics* is typically about things like the possibility of a realist interpretation of quantum mechanics, the completeness of quantum mechanics, the meaning of fundamental probabilities, the reality of space-time and time flow, the nature of reversibility and time symmetry, the meaning of physical symmetries, etc. It is not distinct from physics proper: it is an extended critical analysis of the conceptual foundations of theories, and leads to reformulations of theories and to new theories.

E.g. the famous EPR (Einstein-Podolsky-Rosen) paper on the incompleteness of quantum mechanics, Bell's theorems, and subsequently the Aspect experiments represents a famous episode in the *philosophy* of physics, leading from conceptual analysis and philosophical debate (featuring Einstein and Bohr) to a more rigorous theory (Bell, Kochen-Specker), and thence to a certain degree of resolution by experiment (Aspect). The larger debate is still alive, but many of the possibilities have been closed off, and the concepts sharpened. The many worlds interpretation of quantum mechanics of Everrett and the deterministic interpretation of de Broglie-Bohm are related philosophical developments, addressing the same conceptual problem.

I believe this kind of *critical philosophy of physics* is at the heart of theoretical creativity in physics. It is what thinkers like Lorentz, Plank, Bohr, Einstein, Schrodinger, Dirac, Bohm (to name some instinctive *realists* who were central in the early development of modern physics), excelled at naturally and intuitively. Now physicists may scoff at the idea that this requires *philosophical thinking* that is anything special in comparison to 'scientific thinking'. But in the same way, some physicists have scoffed at the idea that mathematics is anything special that physicists could not have simply invented for themselves. This is simply arrogance. Pure mathematicians think in a different way to most ordinary physicists, and without mathematical theories at hand, physics would never have advanced. Natural philosophers also think in a different way to most ordinary physicists, and without philosophical conceptions at hand, ordinary physics would never have advanced. Natural mathematicians and natural philosophers were once found within the ranks of physicists and scientists. The great problem now is that physics has excluded those who think philosophically from its midst, and it has been left with a truly impoverished sense of philosophy, lacking insight into conceptual analysis, logic, semantics.

To return to López Corredoira's own distinction, for a great example of philosophy *in* physics, which is practical and accessible and enlightening *for physicists themselves*, without getting into dense abstractions of analysis, see (Lee, 1988), *Symmetries, Asymmetries, and the World of Particles.* I have included a section of his observations in Appendix 1. I think this is the kind of philosophy in physics that López Corredoira also admires, and it should be compulsory reading for physicists. However, it does not address other issues that philosophers are preoccupied with, and these are real issues, central to López Corredoira's account, commented on next.

Scientific realism and truth.

I will now try to illustrate in some detail how these problems in both the general *philosophy of science* and the specific *philosophy of physics* dovetail into López Corredoira's account. López Corredoira is a kind of scientific traditionalist and is in some respects quite conservative in his philosophical views, and this comes out in his positive vision of scientific knowledge, as well as his negative evaluation of the philosophy of science. He is a *realist* in the sense that he *believes science makes claims about a real world, and its propositions are objectively true or false, and science progresses by improving the truth of its claims, and makes realistic advances in discovering truth.*

> "I must clarify explicitly that my view of science in general is realist rather than constructivist/relativist. I may accept some degree of constructivism in a few speculative areas in science, but in general I think that science is talking [to] us about truth in nature, not merely a truth based on social consensus. I should clarify my position because a postmodern philosopher might interpret my criticism of the system of science as an attempt to defend constructivism. One might wonder why, in spite of all the problems that scientific institutions possesses, all the corruption and biases, we should trust that science is talking about truth. My answer is that science in general, all through its history, has shown the robustness of many scientific ideas as solid absolute truths about nature, independent of the social context. There is no doubt from my side that science talks about truths. Nonetheless, science is a slow process and it is quite possible that wrong ideas may dominate science for a long time. P. 151-2.

But justifying this kind of view is exactly the problem that the philosophy of science has failed to solve for two centuries! How does science give any guarantee of *truth* about anything? If it could be shown it has a method that, properly enacted, does indeed guarantee a progression to *truth*, then the realist position would be justified. But no one has been able to do this. In my view, it is impossible. There is no such method available to human beings and human society, and there is no such guarantee.

Rather than *guaranteeing* a progression to truth, this is an *ideal of science*. It is a good ideal and the right ideal, but it may not succeed in the end, and it will at best only succeed partially. It is like democracy, which is supposed to bring freedom and justice, but fails repeatedly to achieve its ideals. There are political philosophers who try to prove that democracy has some internal logic that guarantees its ideals: but there is no such guarantee. It depends on the quality of its practice. In a corrupted society, democracy guarantees nothing, and in a corrupted institution, science guarantees nothing. The institutions of NZ society I live in are deeply corrupt, corrupted by mediocrity and incompetence and self-interest of the managerial elite, and the practice of democracy dramatically fails its ideals for large numbers of people. And in all the various institutions in NZ I have studied in and worked in, the practice of science is corrupt, and pathetically fails its ideals.

On the positive side, we can only say that the present institutions are better than some alternatives – but they are still failures. Whether there are better alternatives again, whether these systems can be reformed or revolutionised, is an open question I return to at the end of this essay.

López Corredoira says in the quote above that:

"My answer is that science in general, all through its history, has shown the robustness of many scientific ideas as solid absolute truths about nature, independent of the social context. P. 152.

The key problem is that physics (more than any other science) proposes *general laws of nature* – Newtonian mechanics and gravitational theory for instance, or Special Relativity or General Relativity or quantum mechanics, etc. These general laws correspond to even more general metaphysical frameworks, conceptions of the fundamental nature and contents of reality. E.g. modern physics assumes the theatre of all reality is a *space-time manifold*. But such general laws and their corresponding ontologies can never be proved to be *universally true*. And when such laws are overthrown, the entire metaphysical framework of the theories is overthrown. Even the very entities they refer to appear to disappear from reality. Today physicists think that Newtonian point particles, and classical (indivisible, point-like) atoms are not real – they are merely idealisations or approximations to quantum particles, ghostly wave functions that are absolutely *unlike* classical material particles or atoms.

The intuitive and characteristic move is to say that earlier theories are *approximately true*, and science progresses through approximations to truth, but never establishes the absolute truth of general theories. This was Popper's main idea: to replace the ideal of *absolute truth of general theories* with one of a progression through approximately true theories. López Corredoira is unjustly critical of Popper[14], and does not give him credit for his real achievement in philosophy of science, which was to help overthrow the dominant Positivist philosophy of his age (which is fundamentally wrong in every respect, and

---

[14] I think there is a frustration with *Popperianism*, which has become an ideology taught in many philosophy of science courses, reflected by the slogan that 'scientific theories must be falsifiable'. This has spread to many social science and even business studies courses, which now often try to teach some 'scientific philosophy', and usually give desultory versions of Popperianism – often naively mixed with Positivism and Empiricism - as the accepted theory. But Popper himself was a more interesting and diverse thinker than the desultory legacy of 'Popperianism' suggests.

inculcates a totalitarian philosophy), and try to replace it with a modified form of realism. López Corredoira is most impatient when he discusses mainstream philosophy of science and semantics, and I think he does not entirely appreciate either the real difficulties or real achievements in this subject, as I will briefly try to illustrate.

It is very easy to slip into thinking that this move from *truth* to *approximate truth* is just a simple matter of degree, and realism is retained. To illustrate, consider the claim: *"All natural numbers are divisible by 2".* The immediate response may be: *"No, that is only 50% true, because every second number is not divisible by 2."* But it is not 50% true: it is simply *false.* Mathematical statements are not partially true and partially false. What if we try: *"All natural numbers are divisible by 2 or 3".* Well now there are fewer exceptions – only $1/3$ of examples fail instead $\frac{1}{2}$ - and we might say "that is *more approximately* true". What if we improve it with: *"All natural numbers are divisible by 2 or 3 or 5".* Now there are fewer exceptions again, and we might say it is more approximately true again. But what does *approximately true mean?* Given that none of these statements is actually *true at all?*

Taken literally, these statements are not 'approximately true' at all: they are simply *false.* For they are *universal generalisations: they start with "All cases of …".* What does it mean for a universal generalisation to be false? That there is at least *one* exception. However there is some intuitive sense in which the second statement is more 'approximately true' than the first statement – there are *less exceptions* - but this is only a *relative concept.* I.e. the propositions taken literally, in themselves, are not 'approximately true': they are simply true or false. The concept of 'approximately true' only applies to *relative comparisons.* And it does not apply to the *universal generalisation as such.* It involves a reformulation of the statements to refer to *the number of true or false cases of the sub-statement within the universal generalisation.*

If this sounds a little nit-picking, it is deliberate: it is a crucial part of the philosophical method to specify claims *precisely and accurately.* If we allow ourselves to uncritically say that these statements are 'approximately true', instead of identifying that they are *false,* we immediately get into a bog of intuitive thinking that leads to nonsense. If we want to bring out some sense in which they are 'approximately true', we must formulate this precisely. It is not the *universal generalisations that are approximately true* at all: as I keep saying, *those statements are false.* They have no intrinsic property of 'a degree of approximate truth' *at all.* It is only some construction around the inner clause of the universal generalisations that may be said to have a measure of 'approximate truth'. I make this point as an example of the 'nit-picking' concern over *stating things with literal precision* that philosophical or logical analysis really requires. Physicists are used to a rough-and-ready intuitive mode of thought, and become very impatient with the method of precise logical analysis. It takes a long time to teach physics students to grasp the mode of thought that real logical analysis requires.

Popper's idea was that general scientific theories – universal generalisations - progress by becoming more *approximately true*.[15] It should be emphasised that this threatens to throw out the concept of *truth* as such. It threatens the idea that we can infer any general properties of nature from such theories. E.g. Newtonian mechanics is *approximately true* – in terms of predicting certain phenomenon with approximate accuracy – but it is actually, literally, *false*, and it does not tell us the fundamental nature of space or matter. The statement that "Newtonian mechanics is approximately true" tells us something about empirical predictions – but the Newtonian laws *per se* do not tell us truths about the fundamental entities, e.g. the symmetries of space and time and forces and motion etc. Analogously, if the statement that *"All natural numbers are divisible by 2 or 3 or 5"* was true, it would tell us something very profound about the natural numbers – but it is false, and it tells us *absolutely nothing true* about the properties of the natural numbers. It does not tell us that *"most numbers are divisible by 2 or 3 or 5"* – that is a completely different statement.

In the mathematical example, we might say that these false approximations are ultimately converging to a real truth, i.e. that *"All natural numbers (except 1) are divisible by a prime number"*. But we *never reach this true statement* by continuing the progression of including more prime numbers in the statement. And in foundational physics, the real hope is not that we will find closer and closer approximations in an infinite series: it is that we will *find the ultimate true theory of the universe*. And this will tell us the fundamental truths of nature – everything about the fundamental nature of reality itself, for a naturalist or materialist.

So Popper's theory of a progression of 'approximate truth' first of all makes truth *relative* in the sense that we can only *compare the 'approximate truth' of a theory against another theory*. And secondly it removes the idea that we can ever find real truth, as far as general theories describing the general nature of physical reality are concerned. And as long as general theories remain uncer-

---

[15] Popper of course is most famous for *falsificationalism*, (which he sometimes calls *negativism*), the theory that scientific theories must be capable of being falsified by evidence, which was a reaction against Positivism, which holds that science progresses to *true* theories. Popper realised that general theories, like fundamental theories of physics, cannot be proved true. Popper proposed that being *potentially falsifiable, by possible evidence*, is essential to distinguish scientific theories from 'pseudo-scientific' theories. To save scientific progress, he proposed that scientific theories approach truth though approximations. But his philosophy of science is completely inadequate, and its dogmatisation into part of the scientistic ideology of our day has turned into yet another destructive force against real science. Has any modern movement in philosophy of science been of any direct benefit to real science? The only one I can think of is *realism*, which is not a *philosophy of science* per se, but a more general philosophical stance that provides some antidote to the anti-realism rampant as positivism in physics, behaviourism in psychology, capitalism in economists, formalism in logic and linguistics, etc.

tain or mere approximations, what *metaphysical lessons* – truths about the nature of reality - can we draw from them?

Popper's theory also suffered from a fatal logical error. To be more than just a suggestion of a theory, it requires a precise method for defining *degrees of approximate truth* – defining when one theory is "more approximately true" than a rival theory. Popper proposed such a method, and defined it precisely. Good work, Popper. At face value it looked sensible, but it is logically inconsistent, essentially because there are many different ways of counting the cases that conform to an 'approximately true theory'. This inconsistency was shown in a proof by Pavel Tichy, about whom I say more below. Tichy presented his critique in a seminar attended by Popper himself, and (legend has it) he concluded with a statement to the effect that "therefore Popper's theory of verisimilitude (approximation to truth) is of no value".

Sir Karl apparently sat silent for some minutes, while everyone held their breath waiting for the great man to respond, and he finally said: "Tichy's demonstration of the failure of my theory is sound, but I disagree with his conclusion that it has no value. If I had not presented my wrong theory, Tichy could not have presented his beautiful disproof of it."

This is an example of what I mean by the statement earlier that the philosophy of science *does have valuable results* – and results of genius – but they are primarily *negative*, showing the failure of multiple attempts, in the C19th and C20th, to give foundational solutions to the problems of scientific epistemology and semantics.[16] Now there are positive results from philosophy too, but as illustrated next, real achievements of the first order are often lost in the great swamp of trivia that academic philosophy churns over.

The tragic example of Pavel Tichy.

Pavel Tichy, who disproved Popper's theory, is a tragic example of a brilliant heterodox thinker. He was a Czech political refugee to NZ, and spent most of his mature working life at Otago University in Dunedin. He was the greatest philosophical logician to work in NZ – indeed one of the handful of great logician-semanticists of the modern era, in company with Frege, Russell, Goedel, Tarski, Turing, Church, Carnap, Montague. He created a formal system of logical semantics in the 1970s-80s called TIL (transparent intensional logic) that represents *the* fundamental breakthroughs needed to solve a host of problems that have plagued the foundations of logic and formal semantics since Frege created the subject in the late C19th, and continue to

---

[16] But see (Oddie, 1986) as an attempt to rescue Popper's logic of 'approximate truth'. This is a valuable insight into the logic, although I do not think this is the right model for scientific theories myself. Like López Corredoira I am more interested in real scientific theories and what they really tell us.

plague it to this day. He committed suicide in 1994, aged 54 and still at the height of his powers.

Practically no one in NZ philosophy even remembers him today. Among NZ philosophers, I think only a couple of his graduate students (including myself), and a couple of his professional colleagues at Otago University ever appreciated his ground-breaking work in his life time.[17] The problem is that his theories are too technically demanding for practically any of the 50-odd professional philosophers in NZ – even though many *teach courses on the same problems of semantics that he solved, focussing on any number of inferior theorists from the earlier C20th, tying their students up in knots, and continuing to hash up their own naïve theories of the same subject.* He has a handful of dedicated supporters in the Czech Republic, Slovakia and one or two Easterly European countries, among whom he is considered a legend. But the wider Anglo-American community of logicians and philosophers of science – a large domain within Western philosophy – has almost completely ignored his work.

This is a perfect example of the death of science[18] that López Corredoira presents: a rare heterodox genius, ignored by an extremely mediocre academic community preoccupied with their own trivia, and too incompetent to *recognise ground-breaking solutions to fundamental problems when presented with them.* Philosophy is often accused of this preoccupation with trivia.[19] But it is in the most developed branches of *natural science* that there is now the same problem.

As a footnote to this, in 2003, Tichy's supporters initiated a project to publish a collection of his published papers. To complete this, they needed to raise $20,000 for publication costs. I wrote a letter to the Vice Chancellor at Otago University, supporting their application for a grant towards this. I described Tichy as one of the seminal figures of C20th philosophy and logic, the most outstanding philosopher-logician to work in NZ, and his work as of enduring

---

[17] His disproof of Popper's theory is quite widely known, but it is an isolated result and not part of his major theories.

[18] What Tichy primarily contributed to is a *science*: a scientific theory of formal semantics, observed empirically in natural language and conceptual thought. It is accurate to the empirical phenomenon, even though they are largely observed introspectively, by reflecting on how our own conceptual constructions in thought relate to our linguistic constructions in language. Tichy began studying mathematics and subsequently moved to philosophy, and his ideas are central to many philosophical issues, but they are still fundamentally *scientific*, in being demonstrable solutions to real problems, rather than the endless chatter about meta-problems that most philosophers engage in.

[19] Not always fairly of course. Sometimes problems that seem like trivia to outsiders are actually of central importance when you get to know a subject properly. The accusation is nonetheless true in large part.

quality and importance. I emphasised the importance of having an accessible collection of his papers for scholarly research.

The curt reply I received back declined funding on the grounds that Otago University had already funded a graduate student scholarship into TIL (Tichy's main theory), as well as funding a couple of visits to Otago sometime past by Pavel Materna, professor of logic and philosophy at Charles University in Prague, and Tichy's closest supporter and research colleague over his lifetime. The letter concluded with a message that if I was looking for funding for myself, I should go through appropriate university research grant channels. I gained the distinct impression I was not welcome to do so, although I had no intention of this in any case.

Some time later, this position was apparently reversed, and some funding was granted. The *Collected Papers of Pavel Tichy's Collected Papers in Logic and Philosophy*, edited by Vladimir Svoboda, Bjorn Jesperson and Colin Cheyne, was published in 2004 jointly by Filosophia, The Institute of Philosophy at the Academy of Sciences of the Czech Republic, and Otago University Press. It is 901 pages and contains 46 papers. That is not a great many papers by academic standards: competitive academics today publish many hundreds or even a thousand papers. But these are all papers of quality and originality, filled with dense, complex and lucid analyses, original theoretical developments.

The response from the VC reflects the attitude of bureaucratic administration that López Corredoira objects to so strongly. As an academic bureaucrat, removed from the subjects he oversees as a business CEO, the VC's problem is balancing funding priorities. He regarded awarding a scholarship for a graduate student as equivalent to the value of publishing the life's work of a rare genius. He even saw the fact that the student scholarship was *on the subject of the theory of the genius* as a reason to deny further funding! The student scholarship represented a similar funding amount, so therefore, for the CEO, it represented a similar *value* donated to philosophy. There is also no appreciation that Tichy himself is the one who has *given the real value to philosophy*, through long years of hard and lonely work. There is instead the attitude that Tichy (or his supporters in his name) are asking for some kind of hand-out for his benefit.

The VC might also say that, in retrospect, funding was not justified after all, because despite Tichy's *Collected Papers* being published, it did not succeed in bringing any advantages for Otago University, Tichy remains as obscure as ever, and the reputation of OU has hardly been enhanced. Did this publication bring in any new students, or raise the profile of OU? Scholarly work pursuing Tichy's TIL theory is primarily limited to about 20 papers, mainly by Marie Duzi, Bjorn Jesperson and Pavel Materna (working in the Czech Republic), published in a volume *Procedural Semantics for Hyperintentional Logic*, 2010. However the story of Tichy's *achievement* is not over yet. We should remember that Frege's theory of logic in the C19th is now considered the great watershed in the history of logic, and every logician and semanticist in the world knows who he is.[20] But Frege was just as obscure as Tichy, almost totally ignored in his own life time. He was rescued from obscurity by Bertrand Russell in the early C20[th], and it was a couple more decades before he became widely known. He is now at the center of a vast industry of academic philosophical trivia, an industry of anachronistic misinterpretations of his theories – something Tichy tried to redress too.

As a final footnote to this story, at the same time I wrote a paper, intended as an introductory article to Tichy for the *Internet Encyclopedia of Philosophy* (for which I had just written an article on time in relativity theory (Holster, 2003, B)). The IEP had no article on Tichy. I should add these articles are free contributions, and I was invited by the editor, Brad Dowden, to propose further article topics to contribute. The peer reviewer for IEP, undoubtedly a good academic in Montague logic, had never heard of Tichy, and was unsympathetic to his *objectual* or *realist* approach. He came from a competing philosophy of logical semantics. He did not think Tichy deserved an article in the Encyclopedia. He rejected publication, not because of the quality of my article, but because he *did not like the ideas or approach that it effectively advertised for a competing school of thought.*[21] The article is now posted at *(WebRef Holster (k))*.

---

[20] Although few physicists have heard of Frege, thinking logic means Boolean logic, i.e. the trivial logic of propositional connectives: *and, or* and *not*. Paul Davies, a *supervedette* physicist and prolific philosophical interpreter of modern physics, has written an entire book, *The Hole in the Zero,* on the origins of logic in the C19[th], attributing it to Boole, without a single mention of either Frege or the predicate calculus. His ignorance of semantics and logical analysis is especially obvious in his many articles on his pet subject, the philosophy of time in physics, where he is taken as a leading authority. Stephen Hawking and many other popular philosophical interpreters of physic *who have never bothered to study modern philosophy or logic* show similar ignorance of philosophical logic and analysis. Hawking and many other naively positivistic physicists openly deride philosophy without knowing anything about it.

[21] A particular point of contention for the reviewer was that Tichy independently discovered *intentional logic* simultaneously with Richard Montague, but he published his system a few months later. Montague, (who was murdered by one of his students shortly afterwards, preventing him from making further advances in the field), is taken as one of seminal figures of C20[th] logic for this discovery. Tichy's independent discovery is well documented, but workers in the school of Montague Logic are either unaware of this, or opposed to Tichy's alternative treatment. Tichy's system is to my mind superior, having a superior formalism, and a consistently objectual semantic approach, that let him go on to develop the much more advanced version of hyper-intensional logic of TIL. In any case, in response to this claim of simultaneous invention, the reviewer resented the idea that Montague might share the credit

The IEP still has no article on Tichy. The quality of his theories goes unrecognised.

López Corredoira's fundamental problem.

The foregoing is partly to set the context for what as I see as the fundamental problem in López Corredoira's account: the problem of *judging heterodox ideas* is left unsolved. He is quite aware of this.

> "Who decides what is or is not an absolute truth in science? That is the key question. On the one hand, conservative sectors which have the power to control science will always claim that their ideas are undeniably absolute truths; on the other hand, the few voices of the revolutionary creators of new ideas, mixed with the many voices of crackpots, will cast doubt on anything claimed as absolute truth. Think, for instance, about extreme ideas, such as creationism, which claims that the earth is only six thousand years old and denies the evolution of the species. The fact is that there is no absolute criterion for separating absolute truths from false hypotheses, although there are some quite clear cases. P. 115.

López Corredoira implicitly and explicitly makes such judgements himself – "there are some quite clear cases". You can't even be a scientist without making such judgements *for yourself*. But the problem of judging *other people's* heterodox ideas is that *they all look like crackpot ideas to the outsider*. How does he judge? He emphasises that there is a problem with the stability of scientific knowledge – wrong theories accepted for long periods, true theories rejected for long periods - but still thinks that science achieves certain *absolute objective truths*.

> "False theories may last for centuries or even more than a thousand years and still be accepted without any doubts. In Chapter 2, I gave some examples of this, Ptolemaic geocentric astronomy being one example. There have also been many other theories which were important for a long time, but which were eventually demonstrated to be wrong: the phlogiston theory, the caloric theory, Newtonian optics, the proposal of the existence of an "ether", etc. The opposite thing is also true. There are ideas which have been rejected, and forgotten for a long time, until they are later recovered and become successful ways of explaining phenomena, examples being the heliocentric theory, tectonic plate theory, etc. P.107.

---

for priority of the discovery of intensional logic. For him or her, it was about a competition for power and influence.

Now how does he know that these specific ideas – the phlogiston theory, the caloric theory, Newtonian optics, the 'ether' theory – are really wrong? López Corredoira thinks science eventually accepts good theories, and conclusively dismisses bad ones.

"Science has achieved some truths which will possibly last forever, and no revolution will be able to deny them. Atomic theory will last forever, I guess; I think Archimedes' principle of hydrostatics in incompressible fluids will persist forever; the bacteriological origin of some illnesses will be certain forever; etc. This explains why science is a more conservative system than other human affairs. Only in some fields where the truth is not so immediately clear—for instance, the origin of life, cosmology, the final components of matter, etc.—is there the possibility for some kind of revolution, and this is the place where conservative scientists try to defend their power against the possible revolutions of progressive ideas. p.113.

Here he seems to suggest that in most fields "the truth is immediately clear", and that "only in some fields where the truth is not so immediately clear ... is there the possibility for some kind of revolution". But this is the problem: heterodox scientists will strongly disagree about what "truth is immediately clear". Who decides and by what criterion?[22]

Now López Corredoira considers himself a materialist, reductionist, empiricist and atheist – so conservative in some ways about the limits of science. He believes the world is purely physical.

"We live with illusions in order to cheat our fears. At present, we live with the illusion that a longer life with a lot of medicine and surgical operations is better than a shorter life without visiting physicians. However, we are going to die, and there is no life after death; that is the most important truth, for which science cannot produce anything more. P.169.

He rejects 'metaphysics' (I think meaning 'speculative metaphysics') as a form of objective knowledge – although without dismissing its value as a human activity. He is passionate about the value of philosophy as a part of *cultural* heritage – but he means traditional philosophy, and he is dismissive of 'modern philosophy' as having a role within science proper. He is a reductionist, careful to emphasise for example that his realist language for abstract entities

---

[22] See Hankins (1985) for a good short detailed account of the development of science in the formative period from roughly 1600 – 1800, showing how diverse scientific opinions and concepts developed. This kind of diversity of scientific opinion and controversy is not tolerated today.

like 'knowledge' do not commit him to any metaphysical beliefs that abstract entities exist in themselves.

> "Indeed, the consideration of "knowledge" as an entity is an artefact but so is the concept of "person", and many others which are useful in shaping our vision of the world. In fact, the only entity in nature is nature itself, in the form of matter-energy, but we can carry out an exercise in thought by isolating some particular structures and treating them as if they were autonomous entities, self-organized structures which make their own decisions. In my opinion, free will is an illusion, and all organisms or beings are simply fragments of nature, driven by its laws.

His metaphysical belief in materialism (which he sometimes states as *naturalism*)[23] already sets powerful limits for him about *what ideas can be taken seriously in science*. For instance, he is dismissive of research into spirituality or psi or paranormal phenomenon, and thinks this is essentially an area for cranks, since he believes these things have been scientifically ruled out with practical certainty.

---

[23] I note that in various conversations, López Corredoira rejects that materialism is a 'metaphysical' belief, while to me it is. But 'metaphysics' has different meanings in different traditions. To some it means belief in abstract, ethereal, or unobservable entities or substances – e.g. Platonism about abstract objects. In this sense, materialism, affirming the existence of concrete physical entities, is not metaphysical. To some it means the Kantian class of *synthetic a priori* beliefs - that have no empirical proof – typically claimed of belief in God (although what about logic and geometry and mathematics? Are they empirical?) In this sense, *materialism*, postulating a purely material world to which we have empirical access, is not seen as metaphysical. I use the term in the more customary sense in modern Anglo philosophy, meaning *general claims about the fundamental nature of reality or ontology (the entities that exist), that go beyond any direct observational proof*. Thus, Cartesian dualism (with mind and matter as fundamental), materialism (with only matter as fundamental) and idealism (with only mind as fundamental) are generally contrasted as three classic metaphysical theories of substance. Claims like *every event has a deterministic cause*, or the *every event is determined by laws of nature applied to the present physical state*, are similarly metaphysical in this sense, because they are so very general that even if true, they cannot be empirically verified. This does not mean they do not have empirical consequences of course. It should be emphasised in any case that materialism is not simply the claim that the material world is real – which is just physical realism - but the much stronger claim that it encompasses all of reality, and that every real phenomenon has a material cause and explanation.

Materialism usually means a more specific belief in a reductive account of *consciousness*, and denial of any dimension of spirit, soul, mind, etc, that transcends physical existence. I do not see how science can claim any proof for this when there are no *real causal explanations* for the phenomenon that convince people of these spiritual beliefs – explanations that would be robust against plausible theory changes.

"Wallace would move towards spiritualism (even geniuses get lost in stupid ideas; Newton also dedicated an important part of his life to alchemy or theology), and he maintained that natural selection cannot account for mathematical, artistic, or musical genius, as well as metaphysical musings, and wit and humour. Wallace eventually said that something in "the unseen universe of Spirit" had interceded at least three times in history. The first was the creation of life from inorganic matter. The second was the introduction of consciousness in the higher animals. And the third was the generation of the higher mental faculties in mankind. He also believed that the *raison d'être* of the universe was the development of the human spirit. These views greatly disturbed Darwin, who argued that spiritual appeals were not necessary and that sexual selection could easily explain apparently non-adaptive mental phenomena. P. 45-46.

López Corredoira thinks we *already essentially have the permanent scientific truth* about the nature of consciousness and mind and spirit: i.e. whatever we refer to by these terms are really material entities or material processes.[24] But here we already have a contentious judgement. On this point, and other points of scientific judgement, I disagree with him – and there is no broad scientific consensus. To me, consciousness has never been explained by materialist science, indeed there is no prospect in sight for its explanation in current physical science; and to me, psi is a perfectly valid area for scientific research, whatever you think its prospects of success are, because the host of psi phenomenon have *never been satisfactorily explained by natural science,* and certainly not in a way capable of achieving a broad consensus. From his faith in evolution to give a materialist explanation of the origin of life he also says:

"Ideas like 'intelligent design' or similar are not science. P.46.

And yet to me this is an open question. For a popular point these days: how do we know we were not 'designed' in part by an alien species that deliberately colonised earth with life - since we seem to be on the cusp of being able to send genetically pre-designed organisms to colonise other planets ourselves? If we ever achieve a high-level mastery of genetic design, how many end-features would we pre-design to evolve in a scenario of transferred evolution? Of course, for many scientists, 'intelligent design' means something more specific: the direct intelligent design of humans by God, with a connotation of *creationism.* But in reality, many shades of 'design' and many types of 'designers' are possible. Scientific evidence for 'intelligent design' should revolve around isolating features of genetic code and environment that could not plausibly have just appeared by *chance,* and that point to a deliberate intention.

---

[24] Although he does not think we yet have scientific explanations of how consciousness is created by the brain, and this remains an open question.

Intelligent design is an identifiable feature of our own technology (e.g. a clock). What we can subsequently infer about the *designer* – whether God or alien or human – is another question, and preconceptions about this should not prevent us examining various structures, environments, or indeed the world as a whole, for evidence of intelligent design. Stephen Meyer's (2009) *The Signature in the Cell* is an extended argument that *the hypothesis of intelligent design is an open scientific question* – and Meyer gives a sophisticated treatment of central issues in the philosophy of science, especially the nature of explanation.

Now if we plan to liberalise the sciences to more readily encompass *heterodox ideas,* who is going to make the judgement calls on which heterodox ideas have some scientific quality, and which are 'crackpot' theories? This is the fundamental problem – the problem that philosophy of science has never been able to solve, and barely even tries to address: that *the pre-scientific judgement of ideas* (before they are subject to close scientific scrutiny, evaluation, or experiment) *is not covered by the scientific method.* The orthodox solution to this is peer-review by experts: but of course the peer review of heterodox ideas by orthodox scientists is precisely the original problem. It falls back on *subjective judgement, not objective evaluation.*

López Corredoira has a number of suggestions to redress this problem. His fundamental impulse is to *give judgement and decision making back to the individual scientists,* and take it back from the bureaucratic system of 'expert peer reviewers' and power hierarchies of 'scientific authorities'.[25] He starts with a proposal to liberalise publications.

> "We have too many people doing science, and few of them who have the power do not want to be mixed up with the vast hordes of creators. It is all a question of the will to power. Therefore, they prefer to maintain the barriers which separate the good science from the bad science, with all the consequent biases introduced in this selection: Lots of mediocre misguided ideas are published in important journals, whereas many bad papers and a few genuine ideas are left aside unpublished or published in minor publications which almost nobody reads.
>
> My suggestion would be to allow the important media to publish all kind of contributions, at least in an electronic version on the internet, which is cheaper. For instance, arXiv.org could stop the filtering of papers and allow the posting of all kinds of publications related to the sci-

---

[25] And the question from this point of view is whether Newton or Wallace should be *allowed to publish* their ideas about spirituality and so on. For me, and I expect for López Corredoira, the history of science and ideas would simply be that much poorer if their ideas on these subjects were repressed. For more totalitarian scientists of our day, these are simply bad ideas, examples of 'bad thinking', and should be repressed in the interests of making science appear as a uniform progression of 'good ideas'.

entific topics they cover, even papers from amateurs. Some scientists think this would be terrible because important knowledge would be drowned by huge numbers of papers by amateurs, but this is not true: It can be shown that before 2004, the date at which the filtering of papers was introduced on arXiv.org, the number of papers produced by crank amateurs was a small portion of the total; they do not disturb the flow of information produced by professional scientists, and they introduce some fun touching on questions which are not usually dealt with by paid researchers. P.186-187.

arXiv.org is the major preprint site for physics, and he is really only suggesting it return to its former, pre-2004 model of being *self-regulating*. (And he is right: its filtering process has only led to its manipulation by editors to censor ideas.) But he has a more radical proposal in mind.

> "For major journals, instead of a peer review system to ensure quality, I would suggest that anybody with a Ph.D. in any scientific discipline may publish; something similar is also proposed by Gillies. It might be supposed that a Doctor of Philosophy already has sufficient maturity to know what he is doing[26], although the journals could ask experts to recommend revisions of the papers, but only to improve them with constructive criticism, not to reject them unless the author decides to withdraw the paper; or they could allow experts to post their comments on the published papers (Ietto-Gillies). P. 187.

This doesn't seem realistic to me – won't academics just send rafts of papers to the highest-ranked journal in the field? Anyway, most major journals are privately owned *commercial enterprises*[27], and the owners certainly would not consider this viable. It undermines their whole function and purpose and rationale. Far more appealing I think would be to provide *alternative free on-line self-regulating pre-publication journals in all fields*. We have free preprint sites such as *arXiv.org* for physics, and *philsci-archive* for philosophy of science. There is a problem that these two sites in particular are controlled by private interest

---

[26] No! Having a PhD is no assurance of maturity or good judgement! Rather, I think, it is a reasonable criterion for someone to have a *right* to publish their views, even if they are wrong. But the drive in modern science journals is to make research publications equivalent to '*text-books*': to sustain the delusory idea that what is published is *verified scientific truth*. In mathematics, you generally cannot publish a result unless a *proof* is verified – but this is an entirely different story, because first there are strict (although not infallible) criteria for sound mathematical proofs; and second, the role of the mathematics journal is precisely to give some guarantee that claimed proofs have been verified. Physics *research publications* cannot have the same standards of finality or authority as physics *textbooks*.

[27] See Appendix 9,10,11 for some further detail on journal publishing.

groups, with distinct ideologies, and they are prone to abuse – not in general perhaps, but *specifically in the context of publishing heterodox ideas in the specialist fields of the editors.*

General sites provided by an independent authority, with censorship only of abusive or offensive material, or material entirely outside the scope of the subject (e.g. publishing political diatribes or novels as science preprints) would have real value. The *philpapers.org* site is a good model for this – it does not have the censorship of *philsci-archive*, which is an instrument of the Pittsburg University Department of Philosophy and History of Science, in conjunction with the Philosophy of Science journal, and used to control the subject to their agendas. I will return to this example later in Part 2.

López Corredoira then makes a number of further suggestions in regard to funding and resource-sharing mechanisms. They are interesting ideas, but I do not see most them as practical – although probably he does not either: they illustrate the *need* for reforms, not the practicality of such reforms. My favourite one however is the following radical proposal.

"The assigning of positions, either Ph.D. grants or postdocs or permanent positions, is more difficult to solve, given that there are more applications than positions. This should be addressed with an appropriate education system, and a new direction in the job of researcher. Indeed, many of the present-day researchers do not have a true vocation for science and natural philosophy; they are mostly technocrats who might be redirected to jobs as engineers or computer experts or something similar. A reduction in the salaries of researchers would also reduce very significantly the number of people who want to do research; perhaps, the State could offer a free house and food and a small salary for other expenses, to encourage an austere way of life, dedicated to study and thinking rather than linking business and conference-tourism with science. Once this regulation is done, I am sure that the number of people who would want to do research would be much lower, and with a higher passion for science; there would be a place for all of them without need of a competitive selection system. There would also be fewer papers and conferences, less knowledge in which to be drowned, which would also be a positive outcome.

I like that suggestion a lot, especially in the NZ context.[28] However, the point I want to make here is that none of López Corredoira's suggestions addresses

---

[28] Perhaps there could be a policy that *anyone who completes a PhD, or writes a scholarly or creative book of serious intent and quality,* has an entitlement to a living allowance sufficient for housing, food and basic living expenses, to continue working independently. This is especially relevant in a scientifically second-world country like NZ, where there are few opportunities for intellectuals. Giving an entitlement sufficient to

the fundamental problem of *how to judge the value of heterodox ideas themselves:* how to tell the ideas of 'cranks' and 'crackpots' from those of geniuses; how to tell which current scientific laws or theories are 'truths that will last forever', and which will be subject to revision or revolution in the future; how to tell which heterodox ideas *have enough value to be worth pursuing.* If he could solve that problem, he would solve the fundamental problem of philosophy of science. Without solving this fundamental problem, the options to find a methodological solution to the problem of organizing scientific research to appropriately value heterodox ideas are limited.

But I do not believe that problem has a methodological, i.e. rule-based, solution – it inevitably comes back to individual human judgment: the application of 'refined intuition'.[29] However, I do think there is an alternative concept to the simplistic realist goal of achieving *truth* that helps analyze our intuitions about scientific *robustness and progress*, which I make a brief detour to explain next.

Explanation as the key quality criterion.

I posed the question earlier: how does López Corredoira know that *"the phlogiston theory, the caloric theory, Newtonian optics, the proposal of the existence of an 'ether', etc"* have been conclusively excluded from science – and will not be revived at some later date to haunt us? Alternatively, how does he know that theories of *"the origin of life, cosmology, the final components of matter, etc"* are still open to revision? I think the answer lies in the ability to judge *the quality of explanation achieved in these areas,* not the ability to judge the 'objective truth' of proposed general laws of nature or theories.

---

survive on, say twice the NZ unemployment benefit, to anyone who completes a PhD or meets a criterion like writing a serious book, would provide support for an intellectual life, with a status above 'unemployed dole bludger'. It would allow intellectuals some *freedom* to operate outside a degraded bureaucratic and academic system. In scientifically second-world countries like NZ, where the bulk of intellectual talent leaves for developed countries like Australia that offer career opportunities, this would help encourage an intellectual culture at a grass-roots level. López Corredoira is talking about policies for developed countries, with first-world research and development sectors, which is a somewhat different prospect, but I think essentially the same principle applies: the principle that intellectuals primarily need *freedom* to pursue their own *intellectual* goals, according to their own judgement – instead of forcing them into conformity to bureaucratic systems, competing for funding controlled by self-serving academics, which rewards those with social power networks and presentation skills to impress expectations of managers.

[29] C.f. Feyerabend's view that science in practice is *anarchic,* and not a rational progression based on a well-defined method at all; see Feyerabend (1975) for his classic critique.

In the case of phlogiston, for example, it was proposed to explain the nature of fire. The accepted explanation of fire is now that it is a chemical process, essentially of oxidisation. Now our present explanation of fire may not be entirely right in all its details, there may be important things still missing from the account, but it is surely correct in its basic identification of cause and effect. Fire results when some unstable compounds react chemically with oxygen, forming new compounds, and releasing heat. The heat comes from the molecular bonding energies. This general explanation is so well-tested experimentally it is scientifically conclusive. As a result, the whole purpose of the phlogiston hypothesis is redundant. Now phlogiston has never been detected, but how do we know it might not one day be revived as a substance? What if the chemical theory is one day found to be wrong? Well, that might happen – in fact it did happen! The original classical chemical theory was wrong – a lot wrong - it was replaced by better classical theories and eventually the quantum theory - but *the essential causal explanation for fire still did not change. Phlogiston was not reinstated.*

The chemical explanation is a *causal explanation*. It identifies the causes of fire. It lets the fire investigator say: "this particular fire started *because* there was fuel and oxygen present and a source of ignition (activation energy) and consequently the compound in the fuel oxidised into the new compound ...". Now the *causal* nature is related to the fact that this supports *counterfactuals*. "If (counterfactually) the situation was different and there was no fuel, or no oxygen, or no ignition source, then there would have been no fire."

It is widely recognised in the philosophy of science that *laws of nature* must support *counterfactual reasoning*. This goes hand in hand with the fact that good explanations must equally support counterfactuals. Note that the explanation is *not just a set of facts: "There was fuel, there was oxygen, there was ignition. There was fire."* It is also not a *logical derivation* (as in Hempel's positivist theory of explanation, widely taught in philosophy of science): i.e. it is not a derivation like:

> *There was fuel, there was oxygen, there was ignition. [Facts]*
> *When there is fuel, oxygen and ignition there is always fire. [Universal law]*
> *Therefore: There was fire. [Logical conclusion]*

It is a statement of *causation*: this is what the term 'because' signifies. And *causal inference is not captured by propositional logic.*

Explanations can be stated as *propositions* of course: "there was fire *because* there was fuel, oxygen and ignition". This states a proposition, and it may be true or false in a particular case. They can be quantified to give general scientific truths: e.g. *"all the planets, asteroids, comets, etc, in the solar system rotate around the sun because they are attracted by its gravity."* But this is not a simple quantified proposition, like laws of physics, or a simple fact, like statements of initial conditions. It contains a special logical operator: *Because*. This acts as a *counter-*

factual implication. Few outside the circles of philosophical logic know this, but the logic of counterfactuals – which is essential for the logic of causal explanations – has proved extremely troublesome, ever since the time of John Stuart Mill in the C19th, and still has no accepted formalisation (as quantificational logic has)[30]. The term *because* has no accepted formal logical analysis. Yet statements of *causation* are surely the most durable and objective scientific truths we have.

Now there is indeed an insurmountable problem, exactly as Popper and many others have argued, in conclusively justifying the universal truth of general (universalised) laws or theories. You may believe that quantum mechanics or GTR (the general theory of relativity) gives us exactly true universal laws of nature – but is it possible to *prove* that? How do you know they will not be replaced in the future, found to have exceptions, to be approximations, just as classical physics turned out be an approximation? Well *we just can't know* – as Popper holds, we can only verify that general theories are *false*, when they are found to fail: we can never conclusively prove they are true. In fact, all the evidence is that there must be a 'unified theory' that goes beyond these two partial theories – since they contradict each other - and it is an open question whether either QM or GTR will turn out to be *exactly true* in the new framework theory, or whether they will be reduced to approximations within a larger theory, as happened with the classical theories of mechanics, gravity and electromagnetism. And it is not simply a matter of proving factual observations: these large theories *bring their own theoretical ontologies,* of unobservable hypothetical entities, and when they are overthrown, their whole *ontology* goes out of the window.

Personally I believe QM and GTR are both approximations – which means their fundamental theoretical ontologies are wrong. Most physicists seem to believe they are exactly true – that we live in an historically privileged era, when our fundamental theories are finally absolute, unlike every other era in history. I think this is delusionary. But it has a huge effect on how you look for a unified theory. If you constrain the theory to render GTR *in exactly its present form* or to render QM *in exactly its present form,* then you place a huge constraint on the possible theories you can consider. If you allow theories

---

[30] See (Tichy 1984) for an excellent paper on counterfactual logic. I emphasise that 'logic' is not the simple matter that most scientists think. Physicists are rarely aware of more than 'propositional logic' (Boolean logic), and quantificational logic (universal quantifiers, introduced by Frege), and most are probably not even aware of the latter. But these two fragments of logic are just the tip of the iceberg. Modern logic includes counterfactual logics, intensional logics, hyper-intensional logics, modal logics, temporal logics, as well as model theory and lambda calculus and recursive function theory, which are also forms of logic. I have argued in (WEBREF Holster (n),(o)) that it is impossible to formulate a formally adequate logic for physics without at least using an intensional logic such as the intensional component of Pavel Tichy's TIL.

that generate GTR and QM as approximate theories (as happened in the change from classical to modern physics), then there are far more possibilities. Most orthodox physicists seem committed as a scientific dogma to retaining GTR and QM in *exactly* their present forms, and they consequently snub attempts to formulate a theory that renders them as approximations. They think that *this* is delusional, and refuse to contemplate it. Hence we find a very real and insurmountable barrier to heterodox unified theories of physics. Any theory that revises the foundational framework of QM and GTR steps outside the orthodox paradigm, and is rejected from consideration in the *pre-scientific evaluation*, before it can be scientifically evaluated at all.[31]

But in any case, returning to the original question: I agree with López Corredoira that *the phlogiston theory, the caloric theory, and Newtonian optics* are all wrong and have been made permanently redundant concepts in science – because conclusive and quite comprehensive *explanations* have been given, using alternative theories, for the phenomenon they are proposed to explain. And *these theories have no role other than to explain these phenomenon.* There is no other reason to consider them as viable theories.

I also agree with him that theories of *the origin of life, cosmology, and the final components of matter* are still wide open for revision – because the explanations offered are far from being conclusive or comprehensive. To see what I mean, try to complete the following six sentences:

---

[31] And this is precisely what I have found in submitting theories such as (Holster Webref C, G) to physics journals. They are rejected *before review* with comments like: *"We do not consider papers of this kind for review in physics."* What kind? The kind that explicitly propose foundational alternatives to GTR or QM. *"But what about all the 'alternative theories' physicists propose?"* you may wonder. *"E.g. string theory, quantum gravity, many worlds quantum mechanics, etc, etc? Physicists are always talking about how creative they are, how they are constantly proposing new ideas. Don't these get reviewed?"* Well, closer inspection shows two things. First, practically all the 'alternative theories' that are taken seriously *still take QM and/or GTR as the foundational theories.* They are variations or elaborations of these, they still assume that the fundamental laws – e.g. relativistic covariance and quantum uncertainty relations - are *exactly correct.* Theories that step outside *this* boundary are too radical. And second, the 'alternative theories' invariably have a history of repression in their early stages anyway – it usually takes decades for them to become acceptable, with their originators ignored or disparaged. When they gain official recognition from powerful gatekeepers, and gain official funding and research programs, they take on a life of their own – and start their own specialist journals. But for novel theories to reach this stage became very difficult in the late C20th, as orthodox interest groups increasingly dominate publication processes.

> *The dinosaurs died out because* _____ [32]
> *The Earth rotates around the Sun because* _____ [33]

---

[32] The theory of dinosaur extinction is an iconic episode in modern science, with a unique hold in pop culture too, and well worth study. When Luis and Walter Alvarez *et alia* published their (1980) theory that an asteroid impact caused the extinction of the dinosaurs, it was sensationalised by the press, but ridiculed by many scientific experts, and a bitter controversy continued for decades – and still persists. The asteroid theory was finally 'officially' sanctified by an international panel of 41 scientists in 2010 – but is still not universally accepted. Much scientific knowledge claimed about dinosaurs (anatomy, metabolism, ecology and life-style) remains speculative, and represents 'scientific fiction' rather than fact. (E.g. it was only recently confirmed that birds are descended from dinosaurs; it is still unknown whether there were warm-blooded dinosaurs.) Benton (1997) observes that: *"From the start, the [asteroid theory] debate mixed science and personalities, hype and hypotheses."*

Few leading personalities in the original controversy comes out with credit for their behaviour. Luis Alvarez (at the time the USA's most famous Nobel physicist) was ultimately vindicated by the Scientific Establishment, but his initial theory lacked detail about the process of extinction. It appears to be some 8 years before anyone proposed that extinction was due primarily to a global fire-storm (Wolbach *et alia*, 1988), rather than the subsequent global winter; but this too has become controversial; e.g. Robertson *et alia* (2013), Webref Atkinson (2015). In fact, the extinction process is complex and details are not well understood. What we can confirm now however is the counterfactual: *If the asteroid collision had not occurred, then other things being equal, the dinosaurs would not have become extinct at that time.* I.e. the asteroid strike was the prime material cause of subsequent processes.

In any case, there were real reasons for doubt initially, and Alvarez was not justified to ridicule his critics. But palaeontologists were wrong to react with ridicule of his theory in the first place. See Appendix 7 for some extended extracts from reviews of the controversy.

If you think palaeontology is past this stage, think again: there are multiple ongoing controversies. For a very heterodox example, see David Thrussell's (2015) interview of Graham Hancock, over a parallel kind of controversy, Hancock's theory that an earlier advanced civilization was wiped out by a global cataclysm that he now attributes to two meteor strikes, some 11,600 and 12,800 years ago respectively. Hancock has proposed a variety of hypotheses about this (not all equally plausible it may be said), and has been widely ridiculed by scientists. But evidence for these recent impacts now seems conclusive (since around 2007), and his hypothesis of an ancient civilization is surely plausible. It requires scientific research, not polemic and abuse to resolve these questions.

[33] Galileo was famously subject to persecution from the establishment scientists of his day for proposing that the Earth moves around the Sun. Scientistic atheists of course attribute this to persecution by religious authorities. But in fact it was the equivalent of *establishment scientists* of the day, experts on the Ptolemaic theory, 'owned' by the global corporate Church, who fronted the hostile response from the authorities – just as it is their counterparts, *establishment scientists* now 'owned' by the global corporate Science, who front hostile responses to novel or challenging theories in our day.

Life appeared on Earth because _____ [34]
The universe has its cosmology because _____
The electron and proton exist because _____
Humans have consciousness because _____ [35]

I would say science knows the answer to the first two with effective certainty – not absolute certainty, but as much as certainty has any meaning for human science. But science does not have any conclusive answers to the last four questions at all.

Note also that science discovered the answer to the second question in Newton's time: *the sun's gravity attracts it...* This answer remains true even though we have changed from the Newtonian theory of gravity to Einstein's theory, i.e. GTR. This shows that *theoretical explanations can be robust against changes of foundational theories.* And even if the detailed theory of gravity changes again, the answer to this will still be the same.

―――――――――――――

[34] For a long time in the C20th, and even to this day, many scientists were adamant that random molecular chemistry spontaneously gave rise to simple 'self-reproducing' forms, which then randomly evolved into complex forms, producing life on Earth. Alternative possibilities were dismissed by evolutionists for decades. But today it is recognised that this 'random molecular chemistry' explanation lacks any detailed mechanism, has no direct evidence, and other possible explanations are recognised – such as Fred Hoyle's proposal that life was transferred from asteroids or comets, that primitive life originates in interstellar nebulae rather than on planets, and the possibility life was deliberately transferred here by extra-terrestrial civilizations. (Which must be plausible if it is plausible that humans will transfer life to other planets). Such possibilities were long ridiculed by establishment scientists.

[35] Again, many scientists will claim they know the answer: "... because the human brain is of sufficient complexity to generate structures that generate intelligent behaviour." But this is not a scientific explanation: it is just a hand-waving indication about *where to look for an explanation.* No one can specify precise features of organic neuron clusters that generate consciousness, rather than just generating intelligent-looking behavioural responses. Many philosophers have spent decades trying to prove this is not a 'real scientific question', just a semantic confusion over the meaning of 'consciousness'. There is a growing appreciation in the last ten years however, in both scientific and philosophical circles, that the generation of consciousness is an open empirical question, and may require a fundamental paradigm change – e.g. is it fundamentally dependant on quantum effects, or on some unknown aspect of physical reality? Or as positivists assumed, merely on functional information processing? Renewed interest is partly due to the fact that new brain imaging techniques give scientists tools to try to correlate conscious perceptions with physical states of the brain. The long-held Positivist-Behaviourist dogma, that 'conscious perception' is 'scientifically meaningless' because it is subjective or private, i.e. not open to 'objective scientific observation', is now widely discredited among realist philosophers and scientists – although it remains wide-spread in scientistic circles.

This is why I think scientific knowledge can, in certain cases, be effectively certain, and science makes definitive progress over time. Even though general (universalised) scientific laws are never certain, this stable core of scientific knowledge is also theoretical, and is more than just an accumulation of simple facts. This stable core of scientific knowledge lies in *generalised causal explanations*. All real scientists have a powerful intuition that science makes progress. It cannot be confirmed as a definite progression to *true theories*. I think it can only be confirmed as a definite progression to *good explanations*.[36]

From this point of view, a key question to ask about any scientific claim (beyond simple facts of observation) is: *what is its explanatory value?* This is what I ask about the claim of Materialism, for example, that tells us that *consciousness is purely a material phenomenon*. What explanatory value does this have? In fact: none that I can see. "Your thoughts are generated purely by the activity of the atoms in your brain". This may or may not be true: but what explanatory value does it have? How does it explain my consciousness to me? It doesn't. It doesn't do *anything* to give a causal explanation of *why we are conscious*. It is just a hand-waving claim that *whatever explanation is given it will be materialistic*. To me, this is merely a metaphysical claim (in a negative sense), and a claim that no one in the present state of science can verify – there simply being no good causal explanation for consciousness at all yet.

In fact many scientists assume something stronger: a *functionalist* explanation of consciousness. The assumption is that if you map all the 'logical connections' between the firing of neurons in a brain (the 'neural net') into another representation (e.g. a digital representation on a computer), the result will be a *simulation of consciousness* – an artificial consciousness. There is absolutely no evidence for this – yet it is widely assumed by materialist physicists and computer scientists that this *must* represent consciousness. Here science turns into science fantasy. But this passes as sound scientific sense in many circles; while sceptics who question this dogma will typically be scorned as unscientific, anti-materialistic mystics.

Before going back to López Corredoira's discussion, and some examples of heterodox science, I pause to introduce some other general concepts that are central to understanding the philosophical debate.

---

[36] Of course, philosophers may now complain that *explanation* is a poorly analysed concept, or that the criterion for *good explanations* is unknown, or even that *explanation* is an intrinsically subjective concept, relating to a psychological feeling of understanding, not an objective quality like truth. Many will insist that *explanation* itself must in turn be 'reduced' to other concepts – as positivists like Hemple tried unsuccessfully to do. Well, it is true that the analysis of *explanation* remains primitive. Even the logic of *abduction*, inference to the best explanation, has barely developed any theoretical rigor. But all that means is that philosophers have failed to do their job. It still remains true that providing good explanations is the central, verifiable feature of scientific progress; and I believe good explanations are objective.

Kantianism versus empiricism.

I will briefly recount another central controversy in the philosophy of science and epistemology, which has a long history. This is useful to gain a perspective on the deep flaws in the empiricist philosophy - which leaves modern scientists with a completely misleading theory of science. The fundamental point is that *scientific theories are not simply derived by a logical process of 'induction' from atomic empirical observations. Theories are intrinsically based in the capacity of the human mind to apply conceptual systems, and scientific theories are limited to the kinds of models or conceptual structures the human mind is prefabricated to visualise.*

The simplistic empiricist-inductivist philosophy takes it as an axiom that *all empirical knowledge – knowledge of the external physical world - comes ultimately from empirical perceptions, or sequences of sensory impressions.* Such perceptions are limited to specific *facts,* and it is thought that we make generalisations from such facts, discovering patterns in phenomenon, encoded in general laws, to eventually give larger scientific theories. This empiricist philosophy was elaborated largely by British empiricists, with the iconic figures of Locke and Hume in the C18th, John Stuart Mill in the C19th, followed notably by Bertrand Russell in the early C20th – before being overrun by the more extreme Logical Positivist (also called Logical Empiricist) philosophy after the 1920s. Other European philosophy remained more open to rationalist and metaphysical traditions, reflecting a very Anglo *versus* Continental philosophy divide.

The primitive empiricist theory is often referred to as the 'tabla rasa' theory – that the human mind is a 'blank slate', on which *facts* are written by perception. Generalisations are then determined by *logical operations,* abstraction and generalisation. (We 'abstract' common features of individual facts, and 'generalise' them as patterns or rules). Russell and other early C20th empiricists were most concerned to develop logical theories justifying theoretical scientific knowledge as a structure built on top of empirical 'atoms'.

But this *tabla rasa* theory is utterly misleading. It has no *scientific evidence* at all. It is a metaphysical arm-chair rationalisation, itself based on 'logical introspection', or the projection of a conceptual vision. (When philosophers say: "Things *must* be this way, because it is the only way that makes *rational sense*", they are invariably in the grip of a metaphysical vision, and they are telling us about the limits of their imagination.) The scientific fact is that the human mind is not a 'blank slate': it is born with highly structured perceptual and conceptual structures *pre-formed.* These already determine its capacity to perceive and think and project conceptual models and represent them in language. In evolutionary terms, the human mind is already a highly-evolved perceptual device, attuned to its empirical environment. Thus it is perfectly logical that it has *inherited knowledge* – indeed it is *inevitable* that certain types of representational models are *innate* in our mental structures.

We cannot remember specific factual experiences from our ancestors – but we have inherited more or less specific structures to *represent and process knowledge*. Given that these have evolved to be accurate to represent the external world (an assumption of empiricism), they already represent a lot of pre-conceived assumptions about the world. These are preconceptions about the way the world is logically structured.

We see this in geometry, a classical example famous from Plato's time. Geometry appears to defy empiricism. For it is an *a priori* science, worked out from fundamental principles (axioms) that appear given to us by our own rational faculties. Euclidean geometry, explicitly based on logical reasoning from axioms, has the qualities of mathematics: and yet it quite accurately describes the geometry of the physical world. For straight lines and areas and volumes and so on, defined by empirical objects (rules and strings and light-beams and material surfaces), behave in accordance with mathematical geometry. The Egyptian pyramids and Greek temples were built using geometric principles, and they work. Thus it appears that some empirical knowledge is provided by our innate 'logical' ability to conceive of geometry.

Now of course you might object that geometry has since been generalised to non-Euclidean geometries, famously by Riemann in the C19th. And it is then an open question which variation of geometry is actually *exactly* empirically accurate. So physical geometry is not fully determined by *a priori* mathematics after all.

But this is not a real objection to the point. First, *non-Euclidean geometries* were also worked out *a priori*, as logical refinements of Euclidean geometry (generalising the 'parallel axiom' of Euclid). It is not as if we discovered non-Euclidean geometries empirically, by finding that Euclidean geometry is not quite accurate.[37] And second, the critical point: our ability to formulate theories of physical geometry is dependent on our innate ability to conceive of these mathematical geometries in the first place.

Now I agree that this innate ability does not *determine* truths about empirical geometry – rather, it is *necessary* for us even to begin conceiving empirical theories. If we could not consciously conceive geometric structures and reason about them mentally, we could never create scientific theories about physical geometry, *no matter how many 'perceptual facts'* about material objects we collected.

In fact, *we cannot create a <u>language</u> to represent perceptual facts about geometry without first having innate conceptual structures to host them.* The very idea that human perceptions are themselves some kind of 'atomic language' for 'empirical facts' to be imprinted on the 'tabla rasa' of the mind is a fantasy. Modern psychology shows that perceptions are highly constructed entities. Percep-

---

[37] Einstein's *General Theory of Relativity* established non-Euclidean geometry as an empirical theory; but Einstein needed the mathematical model of Riemannian geometry, established over half a century earlier, to discover his theory.

tions may be sparked by sensory inputs (light beams for instance), but what we experience and report in our perceptions are *constructed images*. We construct an imagistic world appearing to us as a world of 'solid objects', not a field of light-beams. We have no direct contact with external 'physical reality' in this sense. What we 'see' is an internal theatre, constructed by our minds. It is already rendered in a kind of mental language. We fail to see the vast bulk of detail in our environment – the eye only perceives a small band of wavelengths, the ear only hears a small range of frequencies. When it comes to secondary perceptions – of secondary physical qualities like colour and temperature – and even more so of human behaviour, social facts, qualities of other minds – our constructions of 'facts' are extremely subjective, and culturally dependant.

This failure of empiricism should be obvious to any scientist today, because our modern scientific theories are so explicitly dependant on something we bring from our minds: *mathematical models*. Modern theories of physics are expressed as complex mathematical models – so complex they are only just within the capacity of the best mathematical intellects to visualise. What if reality is a little more complex than any humans can visualise? Then no matter how much more 'data' we collect, we will never conceptualise the right theory. What if the nature of reality is of a different *kind* to the mathematical models we can visualise? Those models are suited to atomistic reductionist structures – a projection of the mind. What if the real structures are fundamentally different?

"How could they be?" You may demand. Well there are many inklings. One example is the discovery of *fractal structures*, like the Mandelbrot set, and closely associated *chaotic structures*, which lie outside of conventional dynamics to describe. Many scientists have become entranced by the possibility that reality may be based on fractal structures, fundamentally different to the atomistic models now pursued in main-stream physics.[38]

Another example is the experience of mystics, who report perceptions of 'reality' as having distinct 'holistic', 'timeless', 'interpenetrating' qualities that are *inexpressible* in normal language. *"Oh nonsense, they are fantasying! There is no scientific evidence! Religious buffoons!"* the scientistic empiricists say. But they are not listening. These mystical *experiences* are real and common; and this is exactly what we might expect if the human mind has evolved to have some deeper innate perception of the nature of reality, that goes beyond reductionist models of mathematical physics, or of normal language to express. This is to say, our very *paradigm* of physical ontology used in normal science may prove inadequate to describe the essence of reality – it may describe only superficial patterns that we are adapted as physical creatures to live with.

---

[38] And the theory of fractals and chaos is a classic episode in C20th science and mathematics, with the usual resistance and miscomprehension by the Establishment in its first decade. See Gleick (1987) for a classic account.

And the idea that reality may be more complex than we can comprehend mathematically is strongly supported in mathematics itself: for it is clear we can conceptualise only the *tiniest fragments of mathematical structure*. The real number line (or geometric continuum), apparently simple to our intuitions, actually has a second-order infinity of points: this infinity is so huge and its properties so unimaginable that we cannot hope to understand more than a tiny fragment of its properties in our finite and discrete languages.

Another example, of special interest to me, is the *dimensionality of space*. We are hard-wired to visualise three-dimensional spatial images. This appears to be evolutionarily selected by our environment: for in normal life, we move and act in three dimensions. Euclidean geometry not only encodes the assumption of a *'flat'* geometry (which Riemann generalised to *curved* geometries), but the assumption of *three dimensional geometry*. The most revolutionary realistic proposal of modern physics is that space has more than three dimensions. We cannot visualise hyper-dimensional structures mentally: we can only work with them using mathematics, at the cusp of human ability. If we (or rather: *if at least a few people*) did not have this mathematical ability, we could not even ask this question about reality from an *empirical* perspective. Science is constrained by our *innate mental abilities*: it is not a boundless logical power, constrained only by data collection.

All the applied sciences now require substantial mathematical-statistical modelling to make progress, to express theories. But they have outstripped the mathematical abilities of the vast bulk of their own practitioners. Few applied scientists in practice today can work with the mathematics required to comprehend their empirical data in meaningful *theoretical* structures – or to understand the work of the few who have created the most advanced mathematic models. This is a practical crisis in science, and empiricism comes to a shuddering halt: for scientists can and do keep endlessly collecting *data*, without having the capacity to make sense of it.

Empiricist philosophy fails to justify scientific progress, or to serve as any foundation for scientific method, because it ignores the fact that human science involves bringing the *human mind, with its pre-determined capacity for mental visualisations and for imposing its own rationality on the world,* to the party. The capacities and limitations of the human mind are intrinsic to scientific discovery and judgement. There is no automated logical method of scientific discovery apart from this. A philosophy of science that ignores and overlooks this fundamental fact is too crude to be taken seriously.

C18th philosophers, as exemplified by Immanuel Kant, were well aware of this, and put it in terms of a distinction between *phenomena* and *noumena* – a distinction going back in essence to the earliest philosophers in all traditions; for philosophy was born with the conscious realisation that *reality is different to appearance:* that the world presented to our senses is transcended by another world, hidden from our perception. In Kantian terms, 'phenomena' are what we perceive and measure empirically, but they are only the *appearance* of things. 'Noumena' ('things-in-themselves') refers to the 'ultimate reality' underlying appearances: the *being* of things.

The question for science is whether it can tell us only about *phenomena*, or whether it can penetrate to the *noumena*, and tell us how reality is ultimately represented. If you believe science identifies fundamental causes, fundamental irreducible essences of things, then you believe it brings us metaphysical truths about *noumena*. If you think this is beyond it, then you think it beings knowledge only of *phenomena*. The C19th Positivists – the most famous figure being Mach, who had an early influence on Einstein - thought that scientific knowledge is limited to phenomena. Scientific laws and generalisations, and their *theoretical ontologies* (like imperceptible 'atoms'), in their view, are merely systems of rules telling us about regularities among phenomena: ultimately, *connections of regularity among empirical observations.* They denied the reality of atoms and fields and forces and so forth – calling them 'intervening variables'. Their *terms* – the key terms in most scientific equations - have no *realistic reference* to anything in nature. They are to be understood instead through an *anti-realist semantics:* as meaning their 'observational' content. Thus the *huge* debate in C19th philosophy of science was about the reality of *atoms*, with most taking the positivist view that they are unreal and unverifiable. This view was not overthrown until the early C20th – Einstein's (1905) theory of Brownian motion was a turning point in establishing *atomic realism*.

The victory of *realism* over *positivism* was short-lived however: the *logical positivists* or *logical empiricists* soon appeared with a new brand of anti-realism, in the 1920's, and swamped scientific philosophy through the medium of quantum theory. This left an enduring stamp on C20th scientific philosophy: the dominance of the *empiricist-inductivist-positivist-materialist* philosophy. The dogma of 'inductivism' is a major strand in this philosophy of scientific method, which I briefly comment on next.

One of the persisting delusions of the empiricist ideology lies in a parallel belief in *induction*. Induction refers to a (mythological) method conceived of as:

(i) collecting factual observations, and
(ii) subsequently discovering patterns in them, and
(iii) generalising these to formulate general empirical laws, and
(iv) generalised these further to create higher-level theories.

'Theories' here mean comprehensive explanatory frameworks, with theoretical ontologies, covering large domains, as opposed to individual 'empirical laws of regularity'. You will see 'induction' referred to constantly, like 'scientific method' itself, when scientistic philosophers start pontificating about epistemology.

But 'induction' is not a real method: scientific discovery and inference does not resemble this pattern, except in very limited cases. There are cases where discoveries have been made by deliberate use of an 'inductive' method – where scientists deliberately collect data, and subsequently find patterns that turn out to represent significant 'laws'.[39] Certainly collecting data and looking for patterns is a valid technique *in certain middle-stages of a scientific investigation*. But it is only a very partial and inadequate method in general. Referring to the points above:

(i)   You have to start with some reason for collecting data, for recording certain variables from the vast number of possible variables; this is usually only when concepts and hypotheses have already matured enough to deserve careful *testing;*

(ii) there is no logical reliable way of discovering 'patterns' from observational data – sure, sometimes you find significant simple patterns; but usually patterns of *causality* remain hidden because they are multi-layered, multi-causal, they require a deeper framework to conceptualise; the science of pattern recognition, as pursued in AI studies, is still extremely primitive, and has no general methods that could possibly be used as a general basis for *scientific discovery and theory creation;*

(iii) science rarely progresses by formulating *law-like generalisations;* far more commonly it progresses by discovering *explanations,* primarily *causal explanations;* and these are made by intuitively identifying *causes,* often from single phenomenon, not by subsuming effects under 'inductive generalisations';

(iv) high-level theories – which are *explanatory theories,* explaining complex phenomenon through simpler ontologies of entities - are *never* discovered inductively: they are creations of the human mind. They involve creations of theoretical languages and ontologies.

Perhaps the area where 'induction' is most hyped today is in Artificial Intelligence (AI) studies – the use of automated computer programs to discover patterns in data. This is a very interesting field of study, no doubt about that, and it is of practical importance. But so far it is very naïve, and has no prospects for the kind of success in creating 'thinking machines', that can outstrip humans in scientific ability, that its proponents boast of.

Computers are effectively programmed in *ad hoc* ways that use low-level pattern recognition as a basis for taking actions – e.g. flying machines, responding to their environments. But these programs are developed and tested by humans, using *human intuition,* and hard-programmed into the

---

[39] We may think of Kepler's laws of planetary motion, and of ideal gas laws.

machines. There is nothing even remotely resembling the flexible conceptual ability of human beings to *think*, to construct novel languages, novel concepts – which we do by growing an organic neuronal network. This is seen easily enough because AI programmers have not made a dent in the fundamental problem of programming machines to interpret and process natural language *semantics*. The domain of AI is dominated by enthusiastic technocrats fixated on a paradigm of *syntactic programming* – i.e. manipulation of symbols – without realising that syntactic models must capture *semantics* to succeed. But perhaps this failure is a good thing at this point in history: we should all shudder at the prospects for harm of successful AI programs under the control of our present generation of technocrats and bureaucrats.

The most famous critique of empiricism, and the *major* philosophical tradition for most Western philosophers outside the Anglo-empiricist world, goes back to the German philosopher Kant in the C18th, mentioned above. Most European scientists of previous generations – such as Einstein – were familiar with the Kantian view, and this helped them keep a perspective on the Anglo-empiricist dogmas. I will not try to introduce Kant's philosophy here, it is far too involved for any summary. But one main legacy, known to practically all philosophers today, lies in a distinction he made between the four-fold classification of knowledge illustrated below. I present this here, as it is extremely helpful for understanding the philosophical discussions.

Table 1. Kantian classification of propositions.

Kantian classifica-tions	Analytic	Synthetic
A priori	Analytic a priori	Synthetic a priori
A posteriori	Analytic a posteriori	Synthetic a posteriori

- The *analytic/synthetic* distinction is about *semantic content*: does a proposition refer to facts about external reality *(synthetic)* or merely to logical or definitional facts about the meanings of terms *(analytic)*.
- The *a priori/a posterioiri* distinction is about *source of knowledge:* is a proposition known from external empirical evidence *(a posteriori)*, or from rational intuition or reasoning *(a priori)*.

This 2 X 2 classification generates four possible categories of propositions. The first two are allowed by both empiricists and rationalists.

- *Analytic a priori knowledge* is non-empirical, but tells us nothing substantive about the world. E.g. logical tautologies, like: *If P and Q are true then P is true.* Truths by definition, like: *All bachelors are unmarried.* These follow from the *semantics of the terms* alone.

- *Synthetic a posteriori knowledge* is empirical, and tells us substantive facts about the world. E.g. ordinary perceptual observations: *the sky is blue.* Empirical scientific observations: *The earth is the third planet from the sun.*

Empiricists insist that these are the *only* two classes of genuine propositional knowledge. They wish to identify: *a priori = analytic = vacuous,* and *synthetic = a posteriori = empirical.* But there is also the third combination:

- *Synthetic a priori knowledge* is non-empirical (being known by reason alone, not observation), but still tells us something substantive. For Kant, mathematics: $5 + 7 = 12$ and geometry: *the shortest distance between two points is a straight line* are examples of this. Also, for Kant, principles of ethics are in this category.

It is this class of the *synthetic a priori* that is contentious, at the center of long philosophical disputes. Empiricists want to rule it out, but Kant embraces it. It allows the possibility of metaphysics, in this specific sense: such truths would be available to us from our rationality alone, without empirical observation. *Rationalists* generally believe that there are such *a priori* truths.

The empiricists deal with apparently *synthetic a priori* propositions by re-assigning them to one of the other two categories, or rejecting them as 'meaningless metaphysics'. E.g. most empiricists take *mathematics and geometry* to be *analytic a priori,* denying it has any content – variously holding it is 'logical knowledge' (Russell - logicism), or 'true by definition' (Russell – conditionalism), or merely represents a 'syntactic game of symbols' (Hilbert – formalism – maths is not propositional at all). However John Stuart Mill, the leading empiricist of the C19th, took the view that mathematics is empirical, and thus *synthetic a posteriori.* Propositions about ethics, aesthetics, etc, are rejected as objective knowledge by empiricists on the basis of *materialist reductionism,* being dismissed as merely 'subjective belief'.

The traditional war in epistemology is between rationalism and empiricism. Plato is the most famous Rationalist, and mathematical Platonists (including Frege and Tichy) believe mathematics is real knowledge about a realm of abstract or non-physical entities, i.e. mathematical objects like numbers.

My point is not to embrace the Kantian classification as such, or hold that there is a genuine class of *synthetic a priori knowledge.*[40] Rather, it is that *the*

---

[40] The classification is criticised by Lakatos and Quine and others as not being absolute at all: for them, beliefs come in holistic structures, with degrees between

class of synthetic a posteriori (i.e. purely empirical) propositions is certainly not adequate as a foundation for science – for scientific knowledge is really an inextricable combination of empirical information perceived through mental constructions. However my aim here is not to argue over this abstract philosophy, but rather to alert the reader that there are serious defects in the simplistic empiricist-inductivist dogmas.

Note that there is also the possible fourth combination of:

- *Analytic a posteriori knowledge* tells us nothing substantive about the world, but is gained from empirical observation.

I do not know any examples claimed of this – Kant thought it vacuous.

Note also there are two other categories commonly used by philosophers: *contingency* and *necessity*. A proposition is *necessarily true* if it is true in 'all possible worlds' – if there is no possible situation in which it can be false. It is *contingent* if there are possible worlds where it is false. Logical tautologies *(e.g. If P and Q, then P)* and truths of mathematics (e.g. *1+2=3*) are examples of necessary truths. Empirical facts (e.g. *the Earth is the third planet from the sun; the dinosaurs were made extinct by the effects of an asteroid strike*) are *contingent,* because there are possible (counterfactual) worlds where these are false.

There is a close connection between these and the Kantian categories. Generally *a posteriori propositions* are *contingent propositions,* and often called *empirical propositions.* Alternatively, *a priori propositions* are *necessary propositions* and are called *non-empirical.* Although the classes may be co-extensive, the concepts are distinct however. E.g. *contingency* refers to a concept of *possibility* (a modal concept), while *a posteriori* refers to a concept of *epistemology* (how a proposition is known). (In Kantian terms, the claim that *a posteriori propositions* are *contingent propositions* would itself be a *synthetic a priori* proposition.)

Most important, there are multiple concepts of *necessity and possibility.* For logicians, *logical necessity* is the central concept, but in science, *nomic necessity* is paramount. 'Nomically necessary' just means 'entailed by *the laws of nature'.*

Fundamental physics and some other theoretical sciences are attempts not merely to record facts, or regularities, but to discover principles of *nomic necessity,* to discover laws of nature.[41] Of course it is a metaphysical claim that there is such a thing, and even more contentious that we can determine it.

---

analytic and empirical. Propositions cannot be judged in isolation, but depend on their coherence with larger frameworks.

[41] We may hope to establish certain principles of *nomic necessity* without knowing the full laws of nature. E.g. many scientists think symmetry and conservation principles (such as *time reversal symmetry; mass-energy conservation*) are known to be nomically necessary, even though we do not know the full laws of physics.

It is also generally assumed that *nomic necessity* is itself *logically contingent.* This means the laws of nature can only be established empirically, not deduced by reason alone. Here we begin to get into the twilight zone of metaphysics – in this case, the *metaphysics of empiricism.* For this is now a *meta-claim* about the nature of scientific knowledge itself, not a scientific claim *per se.* Empiricists of course want to deny that there is significant metaphysical knowledge – but what is the status of *that claim?* Most attempts to state such claims are prone to internal logical contradictions – e.g. see Appendix 5.

The main modern technical tool for reasoning about these concepts is called *possible world theory.* Because this is central to modern metaphysics, but practically unknown outside a small group of philosophers, I present some details in Appendix 6. It is the only way presently known to philosophers to make many fundamental metaphysical and semantic disputes clearer.

Having clarified some basic concepts of philosophy that often cause confusion, I now return to the main stream of the discussion.

The ether: an exception in López Corredoira's examples.

Now I have left out one of López Corredoira's examples of "wrong scientific ideas", viz. *the proposal of an 'ether'.* Unlike orthodox physicists, I do not think there is conclusive proof that this is wrong at all. There are a number of reasons. It can be a little vague what an 'ether' is, but I will take it as equivalent to the claim that *there are absolute facts about spatial position.*[42] This is also equivalent to saying there are facts about absolute velocity w.r.t. space, rather than merely relative velocities of objects w.r.t. each other, as claimed in Special Relativity.[43] Now practically every text-book on mechanics and every popular presentation of physics tells us that *motion is relative,* and that this was proved by the Special Theory of Relativity. How could this possibly be wrong if all these scientific experts agree on it? Here are nine possible reasons:

---

[42] This is essentially the Einstein-Lorentz interpretation of what an aether means; viz. that *space is a 'substance' or concrete entity, supporting point-like identity through time* – not that space is filled with yet another substance called the aether. Einstein's *STR* is commonly thought to rule out an aether; but see Appendix 2 for Einstein's *affirmation* of an aether concept in 1920, on the basis of *GTR.* However he affirms the aether in a weaker sense, roughly that space has certain significant irreducible properties, but not point-like identity through time.

[43] And also in the *relational* interpretation of Galilean space, as conceived by Liebniz. Newton believed in absolute space. For him there was no contradiction with the fact that absolute linear motion relative to the 'aether' of space could not be detected. Most modern physicists take for granted that anything empirically undetectable is unreal or meaningless, but this is a Positivist fallacy, based on a superficial understanding of logic. In fact, every substantial theory of physics assumes there are unverifiable facts.

1. The Special Theory of Relativity is empirically wrong and incomplete.
2. The interpretation of STR is conceptually wrong.
3. The CMBR[44] defines a unique universal stationary frame for the universe.
4. A closed curved universe typically requires a unique frame of reference for its consistent global description of space.
5. Quantum wave collapse is super-luminal and thus requires an absolute frame.
6. The phenomenon that would be causally explained by an ether is not causally explained by any other theory – merely postulated as a fact.
7. Time flow (implying an absolute frame of simultaneity, and absolute distinction of past, present and future) is real, and combined with the empirical content of relativity theory, this forces an absolute frame for space.
8. A higher-dimensional theory of space forces a closed finite local boundary, and this is equivalent to an absolute space or "ether".
9. There is something fundamentally wrong with the two partial theories of QM and GTR, and when they are unified, the resulting theory may result in surprising and unforeseen consequences, which may require the reinstatement of an absolute space or "ether".

How plausible are these possibilities really? Well, they are all plausible. They are not just academic possibilities. For instance, the first one is true. STR postulates a *flat space-time metric,* and this is explicitly contradicted by GTR, which entails a *curved space-time metric.* STR is empirically false, and strictly speaking, no inferences from STR are sound (since the premise is false). I will not go through all of these here: rather, the point is that I believe there is ample room for real doubt in this case. Why?

Because, unlike phlogiston, *the explanations in this area are still not stable scientific truths.* The foundational theory of physics (in the opinion of other physicists too) is still prone to *revolutionary* change. The phenomenon the "ether" was meant to explain still have no conclusive causal explanations, and if the whole framework of relativity theory is disrupted, who knows what might eventuate.

Now López Corredoira might say: *"Yes it is academically possible, perhaps, but I still think it is highly unlikely".* Fair enough. That is his intuition. We all have our intuitions. We can only state things as we see them. But with heterodox views in physics, we are talking *exactly* about the fact that certain possibilities remain open despite strong orthodox intuitions to the contrary. López Corredoira is a tolerant person and would likely say: *"Well if you have strong arguments about such a thing, then by all means go ahead and investigate it. It is your time and*

---

[44] Cosmic microwave background radiation.

effort after all." But most orthodox scientists are the opposite. They will say: *"I do not want your silly ideas corrupting my science!".* And if a suspicion of real doubt creeps in, their inner voice will shout: *"... this heterodoxy would be a disaster for me if anything comes of it ... it will undermine my prestige ... people will question my work ... this must not be taken seriously ... at least until after my retirement ... ".* What is the real subject of their science? The personal pronoun?

About phlogiston, I don't think there is any plausible scientific doubt it is wrong. It has indeed become a trivial historical example. But what about when we encounter more *realistic heterodox beliefs* – for example, closely connected to the ether example, the proposition that there may be an absolute frame of simultaneity for time, and hence, an absolute frame of rest for space? Who is going to judge such cases? Remember López Corredoira used the ether as an example that he believes is effectively beyond question.

López Corredoira emphasises that such cases of heterodox theories exist – otherwise successful heterodox science, or revolutionary science, is not possible. Because he is relatively more conservative than I am about fundamental physics, or at least about relativity theory[45], he sees the *plausible limits of heterodoxy* as more limited than I do. Most orthodox physicists are far more dogmatically conservative than either of us. So again: who is going to judge such cases? What methodological principles can we use to make our *pre-scientific judgements of plausibility?* I do not think there are any such principles, certainly none in the practise of science. In practise today, it is left to the raw power-struggle between individuals and ideological cliques: the law of the jungle. If we do not want to live by the law of the jungle, we must instead have some principles of tolerance for ideas and beliefs of others that we do not believe in ourselves. The power brokers of modern science lack such tolerance. They wish to destroy the natural diversity of ideas, just as cultural supremacists wish to destroy the natural diversity of cultures.

Heterodox water science.

I mention another real recent example – an especially vivid one among a number of possible examples from recent science. I agree with the phlogiston example, that conventional chemistry has explained the phenomenon of burning adequately and robustly enough to rule out phlogiston. But how robust is conventional chemistry in general, really? What about, for instance,

---

[45] Actually he may be more heterodox about cosmology, because I do not think there is much real possibility of a steady-state cosmology, whereas he does. But this is more a matter of degree for me: I don't think a steady state universe is very plausible right now, but I certainly accept it as a real possibility, and I am happy to listen to anyone who wants to argue for it. At the same time, I think contemporary cosmology will prove to be wrong in other fundamental respects.

the theory of Brownian motion, osmosis, capillary transport, the structure of liquid water, the formation of clouds, the process of freezing, and other such common phenomenon involving water? Now the modern theories of these subjects are about 100 years old. Water, you would think, has been intensively investigated to death. If science has any claim to produce certain and stable knowledge, it should surely be written in an area such as this.

And yet a revolutionary theory of water appeared in 2013 that contradicts orthodox theories on so many fundamental points it is shocking. Gerald Pollack's *The Fourth Phase of Water* makes astonishing claims in the context of orthodox water science – backed up by careful, replicated experiments, and many references to other heterodox and marginalised scientists. There have been many water scientists who have discovered various of these anomalous phenomenon over the decades, and advanced alternative theories about water, and been professionally destroyed for their trouble.

Pollack is not some marginal heterodox 'crackpot' – heterodox certainly, but he is an eminent, award-winning professor from Washington State University in Seattle, with numerous publications and his own research laboratory. He has multiple empirical demonstrations that liquid water takes on a quasi-crystalline structure, a 'fourth phase', in certain common conditions, and this state is the essential cause of numerous phenomenon, including Brownian motion, capillary action, osmosis, freezing, and many others. These are phenomenon that were assumed to have good scientific explanations already, for decades. See (Pollack, 2013, Preface, Section 1) for some account of the controversial history behind this too.

I pose this as a real, recent, highly surprising example of revolutionary science. If Pollack is right, then many basic *explanations* given in orthodox water science, as taught for decades in high schools and universities, are fundamentally wrong; and moreover, the science has been paralysed by dogmatic intolerance – and quite deliberately so, for personal and political motives.

For a personal connection to this subject, in 2014 I became intrigued by these claims, including a related controversy over effects of radio-frequency electromagnetic radiation treatment of water, which specifically goes back to a controversy over 'magnetic water memory' that started in the 1990's. Such claims are widely dismissed as 'bunk', 'crackpot', 'pseudo-science' by establishment scientists today. It is widely claimed by the 'sceptic' community that any such effects are quite impossible on the basis of known physics. (They are driven by enmity to 'alternative' health movements, fearing such results will legitimise enemies there).

But the 'scientific literature' is so contradictory, polarised and biased between opposing views, that I finally decided the only way to make any realistic judgement would be to test the claim myself. I spent about three months doing a simple experiment that involved reviving about 1500 pairs of wilted dandelions in EM water treatments, to see if there are detectable effects on the speed of revival.

Revival of wilted flowers involves water transport through the vascular system, and osmosis into dehydrated cells. I found strong effects on revival times, and conclude that radio frequency EM treatment of water has very distinct effects on properties of water, lasting for at least several hours. See: (WEBREF Holster (a)) for a report of the experiment[46] and (WEBREF Holster (b)) for an account of some recent controversy. An example of results is shown in the graph insert below, comparing treated and untreated samples. I have little doubt that the EM treatment in suitable conditions has a major effects on water transport properties.

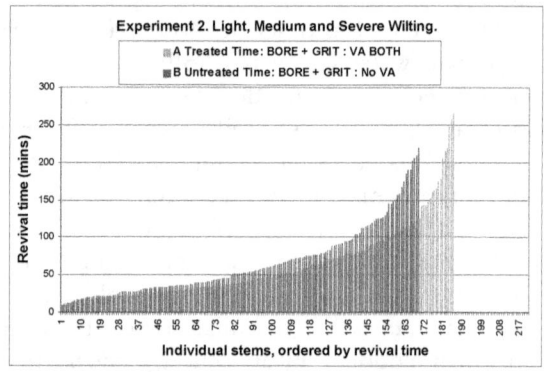

That such a simple amateur experiment can empirically contradict the theoretically based opinion of orthodox scientists on the subject is a big red flashing light: the 'orthodox scientific view' has been formulated from theoretical beliefs, and lacks basic empirical verification. The familiar scientistic propaganda claiming that science is evidence-based – that scientists are constantly testing their claims beyond reasonable doubt - is starkly contradicted. Scientific claims are usually based on *authority*.

In case the reader is sceptical about this, it also recently came to light[47] that this low-energy radio frequency treatment allows salt water to burn – apparently releasing sufficient amounts of hydrogen to create a steady flame. The EM treatment appears to cause dissociation of H_2O into hydrogen and oxygen – only the second method discovered, the first being by applying electric voltage, discovered some 180-odd years earlier.

> "*Faraday in 1831 first established that water could be dissociated in a DC electric field with hydrogen emitted at the cathode and oxygen at the anode. To the best of the authors' knowledge, besides the use of the electric field current in some form, no other vectors have been found to dissociate liquid water into hydrogen and oxygen near room temperature. The use of weak electromagnetic radiation to completely dissociate water into hydrogen and oxygen is, therefore, the key innovation discovered by Kanzius.*" Roy *et alia*, 2008, p.5.

---

[46] The report of this experiment is still in its second review by a peer-reviewed journal.

[47] See Roy *et alia*, 2008.

The fact this effect was not noticed until the C21st should be a scientific scandal! It also reinforces questions raised about effects of EM radiation on health. Our brains and organs are now exposed to low-energy amplitude-modulated radiation from many common electrical devices: how do we know it is safe? The main scientific argument is that the EM energy is too low to have any effects. But this argument now appears to be false.

Heterodox brain science.

I briefly relate a second example that I think is of personal interest to many people, especially those who have had multiple hangovers, watched a lot of daytime tv, or spent a long time in management meetings, and may be concerned that an excessive number of their neurons have perished as a result, never to be replaced. From at least the 1970's it was widely touted as scientific fact that all organs regenerate cells, *except neurons in the brain*. Your quota of neurons, the story went, is fixed once your brain has developed its full quota (apparently by very early childhood), and your neuron count is all downhill after that. As a young student I found this concerning; having also seen alarming estimates of the number of brain cells killed per glass of beer. I also found it odd, but I accepted it implicitly because it was repeated with such scientific authority. More exactly, the claim was that the brain does not produce new neurons; and even when that began to be overthrown from the late 1990's, the claim that the human brain has no neural stem cell pathway, no way to deliver new cells to brain areas needing repair.

But this has recently been dispelled as a scientific myth – after some half a century as a widely popularised *Modern Science Fact*. Recently Professor Richard Faull and his small research team at the University of Auckland, in collaboration with Professor Peter Eriksson and his team at the Arvid Carlsson Institute for Neuroscience, in Gothenburg, Sweden, showed that the human brain does have a pathway to deliver neural stem cells – the Rostral Migratory Pathway. This was a scientific heresy in 2007 when they submitted their findings, based on eight years of careful research, to *Nature*. Their landmark paper was rejected by the peer reviewers, leading authorities in brain research, who maintained this was impossible. Eventually it was accepted by *Science*, making the front cover (Faull *et alia*, 2007).[48]

What seems especially peculiar is that it was already accepted at this stage that other mammals, like rats, have a distinct pathway for neural stem cells: the human brain was considered a unique exception! Faull and his team found that the same pathway also exists in humans; it has simply been dis-

---

[48] See the TED Talk (WEBREF Faull, 2013) for Faull's sense of outrage at his initial peer reviewers.

torted a little as the brain has evolved and enlarged. It seems the long-standing myth never had any real scientific *evidence* in the first place. Yet many scientific experts refused to contemplate that their long-standing myth *could be wrong* when presented with real evidence. *This* kind of dogmatism, I would say, is the typical behaviour of *real* scientists.

The fact our brain can regenerate neurons is a result that I find personally comforting, and I expect some readers will too. But the main point is that this example shows that a small team working at an obscure university even in sleepy NZ can still make radical discoveries that overthrow the orthodoxy of powerful scientific gatekeepers at prestigious international institutions. It may be hard to get such heterodox discoveries past the gate-keepers; but *radical discoveries* still remain to be made.

This example is also interesting to me from another vantage. Brain science has a huge theoretical gap, in that *there is no network model to understand the logic of neuronal wiring.* There is no 'wiring model' of how neurons, which are highly interconnected, *logically represent information.* At least, there is no model that I am aware of. Computational scientists have long assumed that the logic of information representation in the brain will follow a *digital programming logic.* The simplistic assumption is that it will map to some equivalent of the Boolean logic of computer circuits. This assumption is seen repeatedly in academic discussions of artificial intelligence - "once we map the neurons of a human brain into a digital computer simulation, the computer will become intelligent and conscious in the same way as the human".

I sincerely doubt this is any more possible than flying pigs. There are many possible information coding methods beyond the simplistic Boolean logic of a digital computer. 'Quantum logic', using entangled quantum states, is one example of a fundamentally different kind of representational method – and proven to work technologically in the last few years. But there remains a longstanding assumption among materialists that the human brain is functionally equivalent to a *classical digital computer representing a Universal Turning Machine.* Turing and von Neumann were influential in establishing this idea in the early period of digital computer development, in the 1940s and '50s. This paradigm has lasted about three academic generations *purely on the basis of flimsy a priori assumptions, without a shred of scientific evidence.*

I have come to think instead that the *logical representation of information* in natural systems will have a quite distinctive functional network model, based on a specific kind of lattice-like network, built on a specific type of recursive mosaic element. I have found this structure acts as a generalised information model – and one that precisely reflects Tichy's theory of recursive logical constructions. As far as I know, this kind of formal network structure, although ideal for representing Tichy's higher-level logic, and thus natural language capability, has not been considered before, either in information theory or brain science. Indeed the possibility is presently beyond the *imagina-*

tion of experts in these fields, because they have not yet stumbled upon the paradigmatic *network model* required.[49]

The Faull discovery confirms suspicions that despite impressive brain scanning *technology* and such like, brain *science* is still very primitive. I am sure there is a real possibility that such alternative network models are viable. I think progress will be limited until *some* such theoretical model is established.

This also illustrates an important point about drawbacks of scientific specialisation. Such network models can be conceived as abstract structures by *mathematical logicians*. Information theory and brain science are driven in their different ways by applied scientists (computer scientists, neuroscientists), with only a limited vocabulary of such models to bring to the party. Few genuine mathematical logicians will cross the boundaries into these disciplines – and few computer scientists or brain scientists will learn the advanced mathematical logics required. An outsider proposing a novel underlying mathematical model will have great difficulty being taken seriously in such subjects. Professional opposition to outsiders *crossing boundaries of specialisations* is one of the defining features of modern science, and one of the most powerful forces against heterodox thinkers. But combining insights from multiple fields is often the *essential* ingredient for making progress.

In any event, brain science, cognitive science and information science are still in their infancy, and there will be major theoretical revolutions – and philosophical revelations - before good, stable scientific explanations of cognitive abilities can be claimed in these fields. This is a huge area of real importance and interest for the immediate future.

The prospect of revolutionary science.

Pollack of 'exclusion zone water' fame has also started an *Institute for Venture Science* (WEBREF), specifically to fund high-risk, revolutionary research proposals. He has a simple, sensible, and clearly stated *practical philosophy* in place for his own research laboratory. But most important, he has the conviction that *revolutionary scientific ideas remain a realistic prospect*. By contrast, I think López Corredoira's suggestions that most scientific gold in the natural sciences has been mined, and there is not much of real significance left to find, is too conservative. The example of a new age of water science is a striking example, and I think that foundational science in many domains is far more dubious than orthodox scientists generally think is possible.

This is especially true of foundational physics, which prescribes a framework of possibilities for other natural sciences. The situation to my mind is now very similar to that at the end of the C19th, when classical physics ap-

---

[49] I obtained a US patent for a simplified version of this structure, see (Holster Webref 2007/08 (q)).

peared to most scientists as the fundamentally correct *foundational theory* of nature. Lord Kelvin, one of the greatest scientists of the final era of classical physics, is often mistakenly quoted as declaring foundational physics to be settled in the 1880's. This is unsubstantiated,[50] but other physicists, like Michelson, made such declarations, and it was a common article of faith in the final phase of classical mechanics that the classical foundation for physics was final and irrevocable. In the same period, similar sentiments were expressed that mathematics was effectively complete, and others thought that all useful technology had been invented, and suggested closing patent offices. How wrong can you be? These sentiments were expressed by *leading scientists* of the day.

We might think this view was justified at the time, but closer inspection shows that such opinions were completely unfounded on the basis of available evidence. By the 1890's, classical physics was fraught with unresolved anomalies, and classical physics actually left the vast bulk of familiar phenomenon without any *real explanation* – it left *almost every detail of atomic structure and chemistry, optics, electromagnetism, astronomy and cosmology unexplained.* Science in 1890 was profoundly ignorant of practically everything we would come to take for granted over succeeding decades – including electrons, protons and neutrons, sub-atomic structure, Plank's constant h and quantum theory, radioactivity, relativity theory, the strong and weak nuclear forces (two of only four fundamental forces known today), the energy source of stars, the existence of galaxies, the expanding universe and Big Bang cosmology, the vast bulk of modern medical science, the vast bulk of cellular biology and genetics, the vast bulk of chemistry and materials science, tectonic plate theory and continental drift, geological processes and the internal structure and age of the Earth, genetics and detailed mechanisms of evolution; the list goes on indefinitely. We easily forget how *recent* all the major discoveries of science are.[51]

---

[50] The Wikipedia article on William Thompson (Lord Kelvin) observes:

> *"The statement "There is nothing new to be discovered in physics now. All that remains is more and more precise measurement" has been widely misattributed to Kelvin since the 1980s, either without citation or stating that it was made in an address to the British Association for the Advancement of Science (1900). There is no evidence that Kelvin said this, and the quote is instead a paraphrase of Albert A. Michelson, who in 1894 stated: "... it seems probable that most of the grand underlying principles have been firmly established ... An eminent physicist remarked that the future truths of physical science are to be looked for in the sixth place of decimals." Similar statements were given earlier by others, such as Philipp von Jolly."*

[51] See Bill Bryson, *A Short History of Nearly Everything* (2003), for a funny but accurate overview of scientific history to the C20th. Bryson is primarily a travel writer, not a scientist or philosopher: yet this book gives a better perspective on the real development of the present scientific world view than any popular presentations of scientific history by leading scientists or philosophers. The latter invariably project scientific

How could the late-C19th scientists be so wrong – and so blind to the glaring gaps in their account of nature? It is partly because of a mistaken arrogance, found in all ages of modern scientific culture, that *we now represent the most sophisticated culmination of knowledge in history*. It is also because of a particular arrogance among physicists about the finality of *foundational theories of physics*. Scientists had developed a powerful belief in classical physics as the foundational framework, translated into a misguided dogma that all phenomenon *must ultimately be explained in such terms - as classical atoms interacting through classical forces in classical space and time*. Shortly afterwards, physics began on the path to a profound revolution. This was advanced by a tiny group of heterodox physicists, resisted by the orthodox professors of the time.

My view, that science is far more vulnerable to revolutionary change today than orthodox scientists think, is partly because I believe foundational physics of our day is overdue for another fundamental revolution. Its vastly overcomplicated and paradoxical theoretical superstructure will collapse, and be replaced by a radically different conception of nature *as a whole*. It is also partly because I believe that the simplistic *materialist* 'scientific world view', which still dominates the scientific imagination in the same form as it did at the end of the C19th, is radically wrong as a vision of nature - and certainly wrong as a vision of scientific method. It is based on a simplistic idea of the *part-whole structure of objects,* and a simplistic idea of the *causal connectivity* found in nature.

The materialist paradigm explains nothing about mind and consciousness, and yet these are intrinsic to the natural world. An account of nature that leaves them out cannot be taken seriously as a basis for a comprehensive metaphysical belief system, such as ideological materialists prescribe. I believe science must undergo a radical transformation to incorporate mind into its theories.[52] When scientists are forced to look again at the *real world*, instead of at their text-books, they will find that what they thought was a mechanical world is really more like a magic reality.

In fact, a transformation of scientific metaphysics has already begun, through quantum mechanics. QM shows the world is interrelated in ways that are commonly described by leading physicists as 'weird', 'magical', 'spooky', or simply incomprehensible. This 'weirdness' has led to numerous conflicting interpretations of what quantum mechanics means – what it implies about the nature of reality. Some interpretations are metaphysically very radical. The

---

history as a process illustrating their preconceptions about the success of 'scientific method', with banal sermons on the importance of comparing theories against empirical evidence. The scientistic message typical in such accounts is that science is just a simple matter of using 'common sense' to overcome the dogmas of religious and metaphysical beliefs, and philosophers and intellectuals in the past were too thick to understand that theories must be evaluated on evidence.

[52] See the *Manifesto for a Post-Materialist Science* (WEBREF), which I have signed.

best example is Everett's *'many worlds' quantum theory,* which holds that the world continuously divides into an infinity of different realities, and that *everything physically possible happens in some world.* There is no longer a single *actual world* being realised from among the host of possible futures: every possible future relative to our present state will happen. The division of worlds happens every time a microscopic probabilistic quantum event – a 'collapse of the wave function' – occurs.

This is one of the most radical *metaphysical* visions ever seriously contemplated in science or philosophy. Nonetheless it is a brilliant piece of creative scientific speculation, and is taken very seriously by many physicists. But it sparks strong emotions, and most physicists today would still scoff at it (as if they personally know the answers to these controversies that the leading specialists admit to being baffled by). Its initial reception in physics was scorn by the dominant Positivists, and it took some decades to achieve any recognition.[53] But whatever the resolution turns out to be, quantum mechanics shows empirically and theoretically that the physical universe is far weirder than 'scientific common sense' of today can imagine. What practically every specialist in the field agrees on is that quantum mechanics cannot be simply reconciled with our 'common sense' vision of classical materialist reductionism. Yet the latter persists as the archetypal metaphysical vision among the large majority of scientific materialists.

Equally, attempts to reconcile quantum theory with relativity theory have led thousands of leading mathematical physicists to believe in *string theory,* which postulates that physical space has nine or ten dimensions, instead of just the three dimensions we see. It must be emphasised that string theory is only a mathematical speculation, not an empirical theory of physics. It has no empirical evidence, and it has appeared for many years now to be a failed research program. Nonetheless it is a profound idea, and the prospect that space is multi-dimensional must be taken seriously. The failure of string theory may give multi-dimensional space a bad reputation – especially because string theorists themselves dogmatically maintain their theory is only realistic possibility of this kind. But there are many other possibilities for a hyper-dimensional model of the universe, overlooked because of string theory - and I believe many features of modern physics indicate that *the dimensionality of space must be generalised.*

Again physicists have strong personal opinions on this subject. Personally I think it would be very surprising if space turned out to be only three dimen-

---

[53] Everett was a heterodox polymorph scientific genius, and the 'many worlds' interpretation was developed as his PhD thesis (1956), under the famous theoretical physicist John Wheeler. It was met with such scorn by the physics community that Everett abandoned physics permanently, and went to work for US military. Who knows what other ideas Everett might have developed in physics if he had been able to continue in the subject.

sional, as it appears to us phenomenologically - just as it would have been very surprising if the phenomenal material substances (like stone, wood, water, snow, clouds, air, blood, flesh, leaves, gold, copper, iron, etc) that appear as natural categories to our senses, had turned out to be the fundamental types of matter. I do not believe in string theory, but I think that space is multi-dimensional, and this will be the most radical scientific revolution of all time, because we will discover there is far more of the structure of the universe hidden in the hyper-dimensions than revealed to our senses or to reductionist physics. Just as microscopes revealed an astonishing world of structure in the realm of the very small, crawling with microscopic life, eventually fundamental physics will reveal an equally astonishing structure in other dimensions of space, rich with order and information. And I believe this must be central to consciousness too; showing that the present materialistic paradigm is as misguided as classical atomism was.

Because of the number of severe and unexplained anomalies in modern physics and cosmology, foundational physics is open to revolutionary change. Indeed, I believe it would have gone through such a revolution over the last few decades, were it not for the severe repression of ideas in modern physics. And any such revolution will deeply affect other natural sciences, including theories of mind. Yet while I believe a revolution is immanent, I also think it is unlikely to be made through our present scientific institutions. It will be made by outsiders. The new approach required to advance physics will be attacked from within conventional physics, because it *has to revolutionise certain foundational concepts of quantum theory and relativity theory*, but this goes profoundly against the interests of professional physicists.

Having painted themselves into a corner, prominent scientists cannot emotionally afford to allow new ideas they have ridiculed to be recognised. They have too much to lose personally.[54] And this brings us to the second crux of the problem. Scientists typically complain of the bureaucratisation of their subject, and of attacks on science by forces of 'pseudo-science' and anti-scientific culture from outside. But the greatest enemy of science is from within: the bigotry and mediocrity in the professional culture of science.

---

[54] Put yourself in the position of an established string theorist, who has spent all his professional life in this speciality, and still has a decade or two or three left before retirement. Any new foundational theory will immediately wipe string theory off the map. What future career prospects will string theorists have? Their research careers will be over. The same goes for all the other big theoretical research projects. Whether they admit it or not, for practically all theoretical physicists over about 30, by which time they have committed themselves to some program, the prospect of a successful new foundational theory represents a personal disaster. What do you think it is like for a middle-aging academic with only *one* specialist talent to be suddenly cast on the scrap-heap? This is really why scientists fear the suggestion of heterodox theories so much. Since the scientific institutions are so powerful, and have such huge self-interest at stake, they simply *cannot* let this happen.

The scope of science.

One important subject that López Corredoira does not talk about much is the *scope* of science: the different kinds of sciences, the different roles and functions of science. He focuses mainly on pure sciences, using physical or natural sciences (physics, astronomy, biology) as core examples. He is most concerned at how the bastions of pure science have been corrupted, in spirit and practice. But the vastly expanded scope of science in the modern world is central to understanding its death-spiral into mediocrity, and the corruption of its values and quality. I briefly talk about this as part of the broader social context.

Science like art has a number of different functions, and is seen in quite different ways by different people and in different sub-cultures. As broad categories we can identify the following.

(i) The non-scientific public - including corporate executives and government bureaucrats, financiers and politicians, the sub-cultures of power, business and finance – have little knowledge of science, and see it in a cartoon-like image as technology.

(ii) Applied scientists usually see its function as accumulating factual knowledge in their specialist domains, providing an authoritative system of proven beliefs, and enabling technologies.

(iii) 'Pure scientists' in more abstract fields, notably theoretical physicists and cosmology, as well as some chemists, evolutionists and microbiologists, often see its function as discovering knowledge of 'fundamental natural laws', describing the 'essence of physical reality', including the nature of life. But specialist researchers in the pure and applied sciences rarely work on anything outside their own specialisations, and often take their own field as representative of the scientific model. They often deprecate other fields of science.

(iv) Generalist science teachers and educators usually have broader views, presenting science as a general method of enquiry, that can be adapted to multiple fields of knowledge. But they teach from textbooks, rely on scientific authority, and usually have little first-hand experience of scientific research or discovery. Thus while they emphasise the *source* of scientific belief as experiment and observation, they are prone to actually teaching science in a dogmatised fashion, as a set of established scientific facts within a curriculum. Of course students in science classes are rarely actually *doing science*, they are learning what has been done before.

(v) Scientistic propagandists are a special prominent group today, preaching 'science' as *the* final and ultimate form of knowledge, and seeking to replace philosophy, metaphysics, religion, and other traditional systems of meaning, with their own brand of 'scientific metaphysics', combining materialism, reductionism, positivism, nihilism, atheism and an intellectual elitism. They have gained the central power within scientific culture over recent decades, and now have unprecedented influence over the presentation of science, proliferating books, documentaries, magazines, journalism and internet blogs.

(vi) Most other intellectuals, outside science – writers, dramatists, artists, anthropologists, social and political philosophers, political economists, historians, educationalists, journalists, moral philosophers, metaphysical philosophers, theologians, social activists, psychologists, spiritualists - see science as having a limited sphere of validity. Most feel naturalistic science has little to say directly about larger questions of morality, value, political organisation, economics, social justice, art and aesthetics, spirituality and meaning. Many today have become negative about science.

Many ordinary people today also see science negatively, as a set of dogmas preached by elitist academics, and used to attack their own social, moral, religious or spiritual beliefs. Thus there is a wide spectrum of beliefs about what science is and what it means – and not just by the general public, but in different scientific and intellectual sub-cultures.

Science as *technology* is of course the stereotypical image in our culture. Think of all those TV ads for toothpaste, cars, tyres, breakfast cereals, painkillers ... with people in white laboratory coats representing *Scientific Authority*. Science is often proclaimed as the modern cornucopia, giving us all sorts of consumer goodies, medical technologies, weapon technologies, etc. That is the function that makes sense of science to bureaucrats and executives and marketers. After all, if science doesn't give us new technology to increase production, new power to dominate nature, new ways to increase consumption, then how is it going to *make money or enhance power?* And if it doesn't make money or power, what value could it possibly have? "What is it going to do for *me?*" is what people in our modern ego-centred culture instinctively ask. Few people are really interested in knowledge for its own sake: most are satisfied they already understand the nature of the world perfectly well without reference to science. For most, it is not science but social ideologies that define *their meaning and values and importance.* The rich and powerful also resent the potential of science to question their beliefs, which are already fixed to their satisfaction, and see it primarily as a technology factory, enhancing their material wealth or power. But of course that is a purely exploitative materialistic view of science. It ignores the real transformational power of knowledge.

This modern corruption of science to a *technology slave* already sounds alarm-bells. Roman civilisation gives us a powerful image of the dead-end of technology when it is isolated from 'pure science'. The Romans inherited science and natural philosophy from the Greeks (and other cultures). They established a technological civilisation of great power from this, with an Empire based on bureaucracy, law and order, material comfort and military security. But they failed to develop *pure science* – or natural philosophy – any further, and that fundamentally limited their technology. They were on the cusp of developing modern science, with its advanced technology of complex machinery, steam power, metallurgy, electricity, crop science, medicines, etc, but they failed to take the opportunity. Their materialistic, hedonistic, militaristic, technocratic culture killed openness to intellectual discovery, as exemplified by the Classical Greeks. The Romans thought they *knew it all already*. They came to exemplify an extreme and fatal arrogance. Our own materialistic culture, the global syndrome of the modern West, resembles the Roman arrogance.

Technology is enabled by pure science, as the by-product of intellectual discovery: the result of a deeper understanding of causality. The deeper technological implications are rarely envisioned when new science is being discovered. For the most iconic example of our time, atomic weapons of mass destruction, developed in the 1940's by the US, were enabled in theory by the abstract relationship we know as: $E=mc^2$, first proposed by Oliver Heaviside in 1889, and famously established as a general law of physics by Einstein in 1905. But for the next three decades after this equivalence of *mass and energy* was established, no one dreamed that the energy locked up in fundamental particle masses could be harnessed to create the most destructive weapons in history.

The same is true of every major branch of pure science: physics, chemistry, geology, biology, genetics, and so on. They all began as discoveries of abstract knowledge, long before their insights gave rise to practical technologies.

Science starts with the development of *knowledge, understanding, comprehension, insight, explanation* – ephemeral and unquantifiable goods. Such knowledge provides a platform to understand the deeper *possibilities inherent in nature*, leading to a new mastery of cause and effect, which can be exploited for practical goals. But pure science rarely originates as technology goals. For discoveries about the possibilities inherent in nature cannot be *prescribed*, like product developments. You can demand that your technologists (using their scientific training and scientific knowledge) develop a more powerful bomb, a new vaccine, or find a way to reduce cattle flatulence (a top priority scientific project in NZ). But they have to start from a place defined by present science. You can't demand that your scientists develop an anti-gravity compound, or find a new fundamental force, or create a new form of energy - any more than you can demand that your explorers discover a new continent. You can only *discover* what the fundamental forces are, you can't develop new ones. You *might* discover an 'anti-gravity compound', but only if the possibility for such a thing actually exists in nature – which on the basis of present fundamental physics is unlikely – but who knows what will happen in a new unified theory? But in any case, you are not going to develop such a thing directly from a 'technology project' – there is only a potential to discover it indirectly, by first discovering a deeper set of natural laws than we presently know.

When the function of *developing knowledge* is overridden by the function of *developing products for money and power*, the ideal of science as knowledge is lost, and the organisation of scientific research is lobotomised. This is the reality today. Modern bureaucracy, politics and business controls the funding and goals of science, and the arrogance and ignorance of authorities in these domains is partly responsible for the decline of science as a form of knowledge. But that is hardly news: every genuine scientist of every era has probably lamented this corruption; and a significant number of executives and politicians and bureaucrats also understand the importance of pure science as a form of knowledge, too. Pure science today commands huge resources, in name at least. Not everyone is so crass as to think science is purely utilitarian. Science has had many benefactors from business tycoons, aristocrats, and politicians in the past and present. The utilitarian force for the corruption of science has always been with us, but science has survived for three hundred years in the present era.

What is fatal today is the combination of such forces with the corruption of pure science from within its own ranks: its corruption as an open domain of knowledge *by the scientific community itself*. When we consider the most central function of science, as a domain of knowledge, the striking feature is its vast expansion over the last century. From roughly the start of the C20th, core natural sciences (physics, chemistry, biology, medicine, geology) expanded to take in social sciences, political sciences, psychological sciences, anthropological sciences, business sciences, computer sciences, information sciences, sports sciences, media sciences ... By the mid-twentieth century, it seemed that every domain of human activity could be turned into a 'science'. This also reflects the idea that science is not limited to a naturalistic *subject matter*, but is characterised by a *generic method of investigating and organising knowledge*. Natural sciences led the way, but soon any systematic and methodical investigation into any subject became claimed as a 'science'. Thus we now have 'sciences' of sports, war, law, business, farming, marketing, politicking, sales, management, and everything else under the sun.

We must wonder whether this generic extension of science to cover every activity is realistic. Is there really a '*science* of management'? Or a '*science* of business'? Or a '*science* of war'? Certainly there are principles and techniques that can improve our performance in these areas. But how does that count as *science*? Indeed, are the 'social sciences' really science? 'Political science'? 'Economic science'? 'Sociology'? 'History'? Their methods are very different to the natural sciences. They are similar up to a point: they set up systems of technical concepts, and sometimes attempt to make models, and try to measure some things. But does that make them 'sciences'? The big difference with the natural sciences is that their subject matter is itself a result of human artifice: they describe systems of human organisation, cultural constructions and ideologies, deliberately created by conventions of law and social agreement, and varying from culture to culture. Are there really 'scientific generalisations' or 'laws' about such things?

Well, the 'social sciences' so far are very unsuccessful in producing theories of any predictive or explanatory power. They lack *causal explanations,* and they lack clear *ontologies* (models; systems of theoretical entities). Theories of social science or business or economics are intimately tied up with ideologies or 'philosophies', with social *interpretation.* How exactly is 'political *science',* for example, separable from 'political *philosophy'* or 'political *history'?* How do *historical explanations,* which appeal to motives and goals and intentions behind human actions, count as *scientific explanations?* Where are the *counterfactually valid causal laws* of history, sociology, economics? Such laws really don't exist.

Once we go outside the core natural sciences, and into social sciences, what we have are culturally relative frameworks for the description and *interpretation* of complex high-level systems of behaviour. Of course these subjects can benefit by an application of scientific discipline, but that doesn't make them into 'sciences'. Rather, they use some 'scientific' techniques, but their central modes of thought are not 'scientific' in the normal sense: for the subjective realities they deal with cannot be 'scientifically' objectified. This is not to say that subjective realities cannot have objective descriptions – of course they can – cognitive psychology is about such things. Rather, what is *important and meaningful to us* in these domains are interpretations and judgements based intrinsically on subjective points of view – subjective *human* points of view.

By the mid-C20th, it became a fad to call every realm of interest a 'science', but it is misleading. All the 'social sciences' require judgement of values and ideologies and psychology; subjective interpretations through our empathic understanding of human motives and behaviours; teleological explanations in terms of goals and aims. They are closer in many respects to literary interpretation or historical interpretation than to the natural sciences. Indeed, the best work in 'sociology' to my mind is presented through novels and drama. I feel that good social novels give me a better *intellectual understanding* of past culture and society than practically all the millions of academic papers churned out as 'sociological research' in conformity to a fake quasi-scientific tradition.

Over the first half of the C20th, science become a victim of its own popular success – or its own propaganda. It had grown in popular credibility over the previous century to become a positive term for a progressive system of knowledge that was carefully worked out, and systematically verified by experts, and so on. The function of the modern *Oracle.* Science became very much the *authority* against superstition and bigotry. It became a force for enlightened civilisation, for progress.

But it overreached itself, and the *scientific outlook*, where science *informed popular opinion about social, medical and technological issues* soon enough turned into the *scientistic world view*, where scientific opinion is used to *prescribe beliefs about the nature of reality, meaning and purpose*, taking over the role of philosophy, mythology and religion of past eras. Instead of remaining a specialist sphere of human activity, suited primarily to studying objective questions about nature, and informing social debate with reason and evidence, it turned into an ideology of its own, as its proponents began to claim the 'scientific method' as the only legitimate method of obtaining knowledge about *anything*.

We can see two major effects in the second half of the C20th. First, the sheer proliferation of 'sciences' and expansion of the 'scientific professions' led to a mechanical specialisation into trivialities for their own sake, and a degradation of intellectual quality – the kinds of degradation that López Corredoira describes. It is as if we expanded the number of professional athletes from a few thousand to a few million – and paid them all comfortable salaries, regardless of their performance. Or likewise with musical composers, or artists – domains with *critical performance quality*, where a few gifted persons clearly outshine everyone else in *quality*. What would be the result of rampant proliferation of these fields? Severe decline in *quality* of course. This decline of quality is fatal for science. Universities and research institutes are now swamped with incompetent academics who *en masse* destroy quality. Science and academia in general today is dominated by the equivalent of karaoke bar singers and Sunday social football players. If you add such people into professional bands or professional football teams, they do not enhance power by adding numbers: they destroy quality. Likewise the huge amassment of numbers does not add to the creative power of science: it destroys it.

Many physical scientists (typically physicists and chemists) also criticise other sciences – economics, social sciences, and the like – as failures of *quality*. Well, they are failures *as pure sciences*, if taken *on the model of physics* – because their subject matter is far more difficult than that of physics. But the same critics also express scientistic sentiments that *only the scientific method can provide any valid form of knowledge*. So on the one hand, they see social *science* as the only valid form of knowledge about society – and yet they denigrate the achievements of social science in practice. They are often very mocking about subjects outside their own – they are elitists. They will reject social science *as science* because of its failure. But this is surely hypocritical. If science provides the valid method for developing knowledge, then *why hasn't it worked for social sciences too?* If science is meant to be a universal method for successful research, you can't just pick the areas where it has worked and call them 'sciences' and pick the areas where is has failed and call them 'non-sciences'. The scientistic elitists tend to blame researchers in other disciplines, and denigrate their efforts. The problem, they think, is that these other areas attract *poor scientists* – only their own disciplines, of physics and chemistry, produce the 'intellectual geniuses' of science. But this is nonsense. Economics, social sciences and humanities attract just as much 'intellectual genius' as the pure physical sciences. Given 'economic science' has failed for 100 years, surely we must ask: *does the scientific method work?* There is no point making the excuse that *it wasn't done right. If it doesn't work for human scientists in practice, then it doesn't work.* If it doesn't work except in the hands of rare geniuses then how is it *a systematic method of enquiry?* I turn now to the hypocritical attitude of *scientism.*

Scientific bigotry and the scientistic ideology.

A major phenomenon of modern science is the dominance of a *popularist scientistic ideology* within the scientific professions.[55] The project of expanding 'science' to cover every domain of human activity is reflected in the juvenile ideology that science is the *only* valid form of knowledge. This gave rise to the distinctly modern 'scientistic world view', based on a prescription of material-ism, reductionism, nihilism and atheism as a 'scientific philosophy', and belief that these ideas should also dominate theories of value and ethics, economics and politics, metaphysics and religion – thereby making these subjects 'scien-tifically valid'. This *scientistic ideology* has gained increasing popularity within mainstream science today, and combines a number of ideologies – or what may be called *secular theologies.*

One is *empiricist epistemology*, the banal idea that physical observation, measurement and experiment is the sole basis for real knowledge. A second is *materialist reductionism*, the idea that the only real things are *material,* and conse-quently all domains of enquiry and understanding must be reduced to materi-alist terms to be meaningful. A third is *nihilism*, the belief that value, morality, and purpose are subjective illusions, or social conventions, and have no objective reality, as they do not reduce to a material reality. A fourth is *athe-ism,* in recent decades expanded into a militant movement, culminating in the 'New Atheist' attack on religions and spiritual beliefs generally.[56] This attitude is expressed in attacks on moral philosophy, metaphysics, religion and spir-itual belief *in the name of science.* These are attacked because they provide

---

[55] What we now call 'scientism' has a long history, but only became a dominant cultural force around the mid-C20th. "Scientism [is] a term of abuse since Friedrich Hayek first popularized it in the 1940s", according to Robert Bannister, 1991, p.8.

[56] New Atheism is described in Wikipedia *https://en.wikipedia.org/wiki/Secularism* as "a social and political movement that began in the early 2000s in favour of atheism and secularism promoted by a collection of modern atheist writers who have advocated the view that 'religion should not simply be tolerated but should be countered, criticized, and exposed by rational argument wherever its influence arises'." Ref: Simon Hooper, 2006. "The rise of the New Atheists". CNN. *http://edition.cnn.com/2006/WORLD/europe/11/08/atheism.feature/index.html*

competing forms of intellectual authority to materialistic scientism, and encourage realism about mind, value, ethics, morality and purpose in nature.

Here we see science flagrantly overreaching itself, and becoming increasingly arrogant, dogmatic and authoritarian. The foundational movements in C20th philosophy of science – logical positivism, empiricism, instrumentalism, behaviourism, reductionism, nihilism, all various forms of anti-realism - are central props for the explosion of the scientistic ideology we see today. As apologetics for science, the 'scientific philosophies' provide legitimacy for the increasing arrogance of an authoritarian scientific culture. Thus we see science turn from a legitimate domain of enquiry into nature, and become trapped by a metaphysical ideology: the *scientistic world view*. This thrives today within scientific institutions – it has become the *norm* of science educators, journalists, broadcasters and academics.

A typical prominent example of this is Leonard Mlodinow, with statements reeking of arrogant condescension like this:

> "I don't normally quote long-dead physicists, for a couple of reasons. For one, unlike religion, physics doesn't put much weight on authority. Certainly physicists listen carefully to the arguments of brilliant colleagues, but then we check their equations. More important, because science marches forward, every decent physics graduate student today knows far more than Schrodinger, Heisenberg, Bohr, Planck, Einstein, or any other pioneer of quantum theory ever knew about quantum theory, or any other fundamental theory in physics." Leonard Mlodinow in Mlodinow and Chopra, 2011, P. 98.

Mlodinow is an archetype of the extreme 'scientistic evangelist', and his statements here are quite absurd if you stop to think about them for a moment. Schrodinger, Heisenberg, Bohr, Planck, Einstein were among the most brilliant scientific minds of any age; they discovered and developed the concepts of quantum mechanics for themselves, and they spent their whole lives thinking very deeply about the conceptual problems involved. Einstein *et alia* are a couple of *standard deviations* ahead in creative intellect than Mlodinow and his 'decent graduate students': you don't make up that kind of intellectual gulf by any amount of text-book learning!

Mlodinow's statement that *"every decent physics graduate student today knows far more than Schrodinger, Heisenberg, Bohr, Planck, Einstein, or any other pioneer of quantum theory ever knew about quantum theory, or any other fundamental theory in physics"* seems so ignorant that it is hard to credit. He makes this ridiculous statement because he is imagining physics as a simple accumulation of facts – of *scientific truths*. He wants to make the point that science progresses, but he lacks the subtlety to articulate it in any reasonable manner, and he lacks the insight and honesty to see that physics has deep conceptual problems. So he ends up making a strident statement denigrating the greatest scientists in the field in trying to claim supremacy for his own knowledge.

A recent physics graduate should indeed know more facts about certain things today than scientists 50 years ago – e.g. about the families of sub-atomic particles, discovered by modern particle accelerators – but this does not constitute knowledge and understanding of the fundamental concepts and conceptual problems that beset theoretical physics today. Most physics graduates have *specialised* into narrow domains, and remain ignorant of areas outside that; they have spent years memorising 'facts' and rehearsing known solutions to text-book problems in their sub-domains, but they have very little opportunity to contemplate the deep conceptual problems of physics – and especially not if they have teachers like Mlodinow, intolerant of concep-tual questions. By the time they become graduates themselves, they have generally had all the intellectual curiosity beaten out them.

In reality, Einstein, Schrodinger, *et alia* – and we can add Paul Dirac and David Bohm as two other pioneers of modern physics who were profound critical thinkers - uncovered fundamental conceptual problems in the heart of physics that deeply disturbed them, and remain unsolved. These conceptual problems challenge the positivistic ideology espoused by technocrats like Mlodinow. He wants to discredit their critical and philosophical views as being unworthy of attention because he cannot understand them himself, and these thinkers all became highly critical of the shallow scientific positivism in quantum mechanics that he supports.

There are many prominent scientists with the same extreme scientistic at-titudes as Mlodinow – Richard Dawkins, Peter Atkins, Daniel Dennett, Lawrence Krauss, Stephen Hawking are prominent examples. They are high priests of 'New Atheism'. These people should be regarded as a disgrace to atheism itself, because they have advanced a false view of science in an attempt to force their own pseudo-religious view on society. It is important to separate religion from science and politics – not to conflate them.

This scientistic movement is comparable to the rise of fundamentalist re-ligious ideologies. Scientific ideology has been deliberately politicised and emotionalised, and reaches its most reactionary form through militant groups like the 'guerrilla sceptics', an intellectual terrorist gang specialising in internet attacks on dissenting thinkers; or in the silliness of the 'Brights', a cringingly geekish atheist club.[57] The scientistic ideology now has huge influence over scientific media, with its language of scientistic faith becoming the editorial stock-in-trade of publications, from popular magazines like *Scientific American* to serious journals like *Nature* and *Science*. You have to speak in the scientistic rap to become the editor of an orthodox science journal today. The leaders of this ideology are notably ignorant of philosophy and history, but they do

---

[57] Even Christopher Hitchens, a prominent atheist himself, called it a "cringe-making proposal that atheists should conceitedly nominate themselves to be called 'brights'."

understand one principle very well: the power of personal ridicule and abuse of opponents as a propaganda weapon.

Thus the 'scientific philosophy' that started way back in the Enlightenment as a positive force for realism and sanity, has ended up becoming a self-destructive ideology, a virulent nihilistic metaphysics, led by atheist propagandists. The community of 'scientistic atheists' is philosophically and socially naïve in the extreme – the equivalent of extreme religious fundamentalists on the opposite side of the spectrum. The 'science wars' they have created, of 'science versus religion', does huge damage to the image of science in the general community. This radicalised scientistic movement is now widely popularised through 'sceptic' and 'debunker' communities that spawn across the internet on 'science community sites'. They attract immature science students and immature academics. They worship the *authority of science,* and their stock insult is to brand opponents as *pseudo-scientists,* the equivalent of religious blasphemers.

This movement is a betrayal of the very spirit of science itself, which originally sought to *separate objective and dispassionate evaluation of factual questions about the natural world from emotive and political and metaphysical judgements.* The scientistic movements of our day, spawned from within mainstream scientific institutions – by science academics, teachers and philosophers radicalising their students – seeks to emotionalise the debates about philosophical and scientific issues, to demonise their opponents, using a stock vocabulary of contempt and ridicule; the common tricks of political ideology.

The rise of modern 'sceptic' groups is a prime example of the power of propaganda. The term 'scepticism' has reversed its traditional meaning, of *questioning theories and claims made on the basis of authority, and demanding an evaluation of evidence,* and come to mean the opposite: *insisting on positive belief in materialism and atheism as ultimate 'scientific truths'.* This extends into attacks on any genuinely sceptical heterodox challenges to conventional scientific authority. We now find virulent scientistic dogmatism popularised as 'scepticism' in numerous subjects. Bloggers in this sphere eagerly denounce independent intellectuals as 'pseudo-scientists'. This is their signature phrase: every heterodox idea is attacked as *pseudo-science.* You only have to sound suggestive of having an open mind about subjects like electromagnetic effects on water[58], alternative health and medicine, organic farming, alternative

---

[58] I note Pollack's EZ water theory has debunker attacks too, e.g. this from a popular chemistry site:

> *"Recently the quack medicine folks online have been promoting the research of a certain Dr. Gerald Pollack who claims to have discovered a "forth phase of water", and who has recently published a book on the said topic. Although it's clear that his research is being deliberately misinterpreted, it's not clear to me whether or not his own claims are valid in the first place. Although some papers on the topic were published in peer reviewed journals, it doesn't seem like anyone else in the scientific community has acknowledged or replicated his results.*

archaeology, theories of mind and consciousness, the completeness of evolutionary theory, non-materialist metaphysics, psi phenomenon, non-terrestrial origins of life, ufos and aliens, mysterious creatures like Big Foot, homeopathy, not to mention spirituality, God, religion or mysticism, to be branded as a 'pseudo scientist' and become an object of sceptical contempt.

And yet these subjects of attack either lie outside scientific evaluation at all and depend on personal experience and faith – like beliefs in God and spirituality – or they provide challenging examples of perplexing empirical phenomenon that science has been unable to resolve – areas where science has so far *failed to establish conclusive explanations or consensus* - including the explanation of consciousness, esp, psychic phenomenon, NDEs (near death experiences), the origin of life, the existence of aliens. The pathetic quality of sceptic and debunker material is well illustrated by the following extract.

"So, is "EZ-water" a real breakthrough or just the brainchild of a deluded scientist? EZ-water is not a breakthrough.

"Answer: It is not new, nor is it valid. This appears to be one of the many claims about the healthful benefits of drinking "ionized water".

http://chemistry.stackexchange.com/questions/5925/ez-water-fraud-or-breakthrough

But most scientistic commentators have ignored it, recognising that it has too much credibility to be dismissed just by polemical ranting. See Holster (WEB REF (a), (b)) for some detail on the controversy over EM water treatment and sceptic attacks.

New Age Atheism: The New Frontier of Scientific Ideology
http://www.ted.com/conversations/16141/new_age_atheism_the_new_front.html

There is criteria which one should follow in order to be a neoatheist:

*Understanding science, rel igion, supernatural and atheism
~Science is trying to figure out what is true through methods of logic and rationality, while religion does the same thing but through dogmas and old scriptures.
~Supernatural is the silly notion things cannot be explained   by science but religion, and sometimes pseudoscience (which is fake science).
~Atheism is the lack of belief in deities
~Religions are the enemy. Buddhism gets a pass because they are hardly a religion   - more of a philosophy.

*Make sure to know prope r arguments to distinguish atheism from religion:
~"So by the lack of belief in God, I have a belief? So my lack of belief in Santa Claus is "Aclausian?"
~"Is being bold, a hairstyle?"
~"The television being off, a channel?"

*Check out Dawkins, Rosenberg, Dennet and Harris:
~These guys pave the way for what it is be a rational, logical and non -dogmatic person.
~They demonstrate how belief in a God is just nonsensical through science!
~They prove logic is EVERYTHING to how to think properly.

*Being militant does not mean physical actions
~Only extremist harm others for their beliefs, and since we have none there is no need for violence.
~Never allow 'faith' to be an acceptable reason for the other person to avoid an argu ment.
~Don't be afraid to debate, you are right! Religion is a destructive practice!
~The burden of proof is on those who claim truth!

Always keep in mind something a leader of our movement had to say, which proves powerful:

"That which can be asserte d without evidence, can be dismissed without evidence."
-Christopher Hitchens

Let's how a strong discussion here on what it REALLY means to be the neoatheist everyone should be!

As an active blogger and forum user to discuss new age atheism, there are   a couple of websites I can share. Once a week we have a podcast for lectures with live com menting! Join in the movement!

Related Talks: Richard Dawkins: Militant atheism
Alain de Botton: Atheism
Nicholas Lukowiak   Belleville, NJ   United States

You might think this is just some purile rant by a crackpot science student. But it is a very popular *leading conversation piece* from *TED Conversations* – it came up as the third Google search result on 'New Atheism and Science'. This excerpt is representative of the dismal quality of thousands of such 'conversations' carrying on in 'science blogger' forums. These forums have become very popular with science students and academics seeking validation for a belief system. This is also a very mild piece compared to levels of abuse often found on sceptic and debunker sites.

The scientistic sentiment dotes on the authority of orthodox materialistic science, and denigrates entire *subjects and domains* it is prejudiced against. The hit-list of 'pseudo-scientific subjects' is steadily extended, to rule out more and more traditional questions of philosophy, and rule out sceptical doubts about present scientific certainties.

Of special interest to me is the encroachment of this kind of ideology into the philosophy of time, one of most poorly understood subjects of modern philosophy, where the central question is about the nature of *time flow and temporal directionality*. In the philosophy of physics, there are now powerful attempts to denigrate questions about *time flow* to the status of 'meaningless metaphysics'. Most of the leading authorities from the philosophy of science now ridicule it as a pseudo-scientific concept. The concepts of *past, present and future,* so deeply embedded in our language, and universal to our experience of existence as temporal, are ridiculed as *pseudo-scientific* and *nonsensical* by leading physicists and philosophers of science.[59]

Beyond any arguments about specific ideas, this is a deep mistake about the nature of science itself. It is not *subject matter per se* that defines science: it is *method and principles*. You cannot possibly define science by prescribing subject matter *unless you believe science is finished*, because new science introduces new subjects and concepts. Science can study false beliefs as well as true ones – being discovered to be *false* does not turn a theory into 'pseudo-science'. You can do a 'scientific study' of Santa Claus if you want. "Oh that is surely meaningless pseudo-science, everyone knows Santa Claus doesn't exist!" Well, I agree that Santa doesn't exist: *and that is the factual conclusion you should reach when you objectively evaluate the evidence.* But this conclusion is not 'pseudo-science' – it is a factual claim, just too trivial to count as significant knowledge for adults. And in fact, everyone *doesn't know it* – there are millions of pre-school children who don't know it. For them, it becomes a *serious problem of empirical belief* when they come to question it. Indeed, it is an iconic intellectual puzzle for young children in the West. They must engage their rationality to decide what to believe about Santa.

More significantly and controversially, you can do a scientific study of esp or psychic phenomenon. A number of serious scientists have. "Oh that is pseudo-science, done by crackpots, everyone knows *esp* doesn't exist!" the sceptics will angrily retort. Well, first of all, everyone *doesn't* know that, and the empirical evidence is actually very difficult to judge. It is not a pseudo-scientific question: it is a perfectly real question, and one that *prima facie* should be able to be studied scientifically. "But it is pseudo-science because esp would contradict fundamental physics! It appeals to mystical powers!" cries the sceptic. But that is wrong. As a widely reported *empirical phenomenon,* esp does not necessarily imply 'mystical powers' at all, any more than consciousness itself does. If the *phenomenon of information transfer by esp was strongly empirically verified,* many scientists would expect it have a purely naturalistic explanation – they would look for causal connections beyond the normal senses. Of course it would raise questions about the nature of consciousness,

---

[59] These are closely related to some examples in Part 2 of this essay.

and it would encourage some to propose 'mystical' or non-materialist explanations. But consciousness simply *is mysterious anyway;* and if esp was validated, finding an explanation would be a genuine challenge to materialist science. A non-materialist explanation may still be *causal.*

The sceptics' claim that esp is impossible because it would 'contradict fundamental physics' is a characteristic fallacy. *When* exactly did scientists establish that it contradicts fundamental physics? The year 1600? But no scientists had any real idea of what fundamental physics allows as possible at all then. By 1700? But no scientists had any scientific idea of the nature of *life* then, let alone any idea of the functioning of the brain or consciousness. By 1800? But as well as the previous lack, there was not even a theory of electromagnetism then – Coulomb's description of the static electric force was only a decade old, there was no theory of magnetism, and there was no idea that electromagnetic radiation could be used for communication-at-a-distance. By 1900? But no scientists had any idea of quantum theory then, with its strange and 'spooky' effects that are central to *every detailed phenomenon of chemistry.* By 1950? But there was still no understanding by then of two of the four present fundamental forces, or of quarks, or of the 'Standard Model' of particle physics and forces – how could anyone know if there might be some other fundamental interaction that could be involved in consciousness, and potentially give rise to esp?

From our present point of view, no scientist could have realistic arguments to rule out esp on the grounds of *fundamental physics* until at very least the 1960s-70s. And in fact, fundamental physics is still so incomplete today that it rules out nothing conclusive about the possibility of esp. Indeed, the vast majority of 'sceptics' who would claim esp is ruled out by 'fundamental physics' have no understanding of fundamental physics themselves. This opinion is based purely on *science gossip.* It is parroting authority. The only meaningful *scientific evidence* for or against esp has to be *direct evidence of phenomenon,* such as evidence from replicated scientific attempts to try demonstrate esp effects under controlled conditions.[60]

Sceptics typically appeal to 'fundamental facts of physics' to attack threatening concepts, ignoring the scientific principle that *empirical evidence of phenomenon* has to be more primary than theoretical arguments. Whenever you read claims by sceptics that some phenomenon or other is 'ruled out as impossible by basic physics', you are almost always having the wool pulled over your eyes.

---

[60] This does not necessarily settle matters either, but it is the obvious initial scientific approach. If some people sporadically experience esp phenomenon, and can think of no other explanation to rationalise their experiences, they are perfectly justified in believing in it personally, without having scientific evidence. Anyway it may lie beyond conventional experimental proof for a variety of reasons.

The typical contemptuous attitudes towards esp, alternative health, and a host of other scientific border-line subjects, illustrates how a dogmatic bigotry in the name of science, and a hypocritical attitude to scientific principles, has become dominant within mainstream science culture. It also illustrates a culture of extreme *judgementalness*. 'Sceptics' present it as *stupid and despicable* to believe in certain concepts – or even to rationally question them or discuss them. By making such claims they back themselves into a corner: when evidence does come to light for various heterodox ideas, as it inevitably does in some cases, those who have ridiculed these ideas cannot allow the evidence to be recognised. Their emotional egos override scientific objectivity. This investment in emotional egocentricity is the primary mechanism behind the abuse of heterodox ideas within conventional science.

I think this reflects a general flood of *emotionalised judgementality* in our society. It reflects the deeply *ideological* pattern of thinking that has become the norm of our fake modern 'enlightenment'. The C20th was the century of ideology. Political thinking is dominated by ideologies: Fascism, Nazism, Communism, Marxism, Capitalism, Socialism, Individualism, Democratism, Bureaucratism, Conservatism, Liberalism, Globalism, Nationalism, Internationalism, Patriotism, Anarchism, Racism, Feminism, Environmentalism, Egalitarianism, Utilitarianism, Secularism, as well as Atheism and other militant religious fundamentalisms. The list goes on through a host of philosophical ideologies; and we find science deeply infected with Scientism, based on Positivism, Materialism, Physicalism, Empiricism, Inductivism, Nominalism, Verificationalism, Falsificationism, Instrumentalism, Operationalism, Pragmatism, Behaviourism, Logicism, Formalism, Nihilism, Atheism, on one hand; and in various degrees of opposition to these, Realism, Platonism, Rationalism, Kantianism, Intuitionalism, Structuralism, Cultural Relativism, Subjectivism, Idealism, Phenomenalism, Scientific Anarchism, Post-Modernism, Irrationalism, etc.

Most ideologies start with some core positive values, and typically as reactions against extremes of other philosophies; they have to provide something positive to get off the ground. But all ideologies are prone to *Extremism*, to having their positive values corrupted by *Fundamentalism*, to excessive *Literalism*, where their core principles are taken as universal laws.

It is certainly necessary to appeal to more fundamental principles of one sort or another to counter destructive ideologies. But it is dangerous to use these to create new anti-ideologies, and impose these in turn as foundational moral principles. For all ideologies are prone to abuse. The error lies in thinking that we can rationalise all our judgements and beliefs from fundamental principles – like a logical axiom system.[61] Chess gives a good illustra-

---

[61] And note that axiomatics, in the form of Euclidean geometry, was one of the earliest scientific developments – the idea of simplifying knowledge into a few unquestionable principles that give rise logically to all other facts appears to be a

tion of the futility of this. The principles or rules of chess are explicit and simple: there are a small number of moves, and an explicit goal. But people are incapable of *rationally calculating* their best moves in chess, even though they do have such simple principles to start from. We are too limited intellectually. We can rationally calculate one or two, perhaps three or four moves ahead – but it is impossible for us to calculate 10, 20, 30 ... moves ahead, which is what we would need to do to give a rational solution to chess. It has proved exceedingly difficult even to program computers to play chess adequately, despite their vast calculating power. Chess-playing by human beings relies on *judgement and intuition*. It is not just rational calculation.

Some people have a special gift for this. They will always beat people without that gift. Ideological thinking is similarly a drive to reduce individual judgement to calculation from a set of prescribed rules. In public life it is seen in the cult of *bureaucracy* – the attempt to create totalitarian systems of rules to take over all our decision-making. In scientism it is seen in the cult of a mythological 'scientific method', meant to take over the individual judgement and creativity of gifted intellectuals.

Ideological thinking is so deeply embedded in our culture that its unnaturalness has become largely invisible; we now are trained to *conform and rationalise* our behaviour and feelings to social ideologies throughout our lives. We learn as teenagers that our taste in music or fashion will see us admitted or ostracised from 'in-groups'. Education is insidiously designed to engineer ideological conformity. Television advertising, news and current affairs programs are saturated with ideological messages, overt and subliminal. As adults, political or religious conformities come to replace music and fashion as central to our social identities. The power of this identity can hardly be over-estimated. The result is that people today are extraordinarily *judgemental* about each other's beliefs.

These are not just causal judgements, or abstract disagreements. In key social interactions you will be *scrutinised* for whether you might be a Communist or a Capitalist or a Socialist or a Nationalist or a Globalist or a Democrat or a Republican or a Feminist or an Environmentalist or a Christian or a Muslim or an Atheist, and so on and on. And on the basis of prejudices for or against these *isms*, you will be *morally judged* – with powerful prejudice if you are suspected to fail in some ideological conformity. It is a recipe for petty-minded neuroticism. Academics are the most intensely neurotic about such things – with bitter wars waged for years between academic philosophers and scientists, over variations of ideological belief that seem utterly meaningless to outsiders. It is pathetic to see grown-up adults squabbling like this.

Demanding such conformity is part of forming the social solidarity of groups. It is not only that your judges hold personal feelings about such

---

strong intuitive drive. It is even more obvious in the development of religious dogmas.

beliefs themselves – it is also that they cannot afford to let others in their in-group think they will associate with those holding alternative beliefs. People are emotionalised into public displays of approving and disapproving beliefs for social acceptance. Thus it is within the professions of science. The claim of science to *objectivity* is really a myth. Objectivity is an *ideal* of science, but it is overridden by egocentric psychology – sustained more than ever by the scientistic ideology.

What is most striking is how insidious this phenomenon is in modern society, and how quickly and easily people are sparked to strong emotions of judgement, especially negative emotions, contempt, anger and hatred, about very abstract beliefs. In the academic world such judgements are virulent. There are surely psychological or cognitive reasons for this that should be central to understanding social behaviour. I think it is partly because we take *emotions as perceptions of value*, and conversely, we *perceive values as emotions*. (If we *like* something it is *good*. If we *hate* something it is *bad*. If something is *good* we *like* it. If something is *bad* we *hate* it.[62]) So we cannot separate our value judgements from our emotions about them, or at least not without great effort.

In any case, these overpowering and irrational judgements are hardly mitigated by modern education.[63] On the contrary, they are prominent within intellectualised professions – academics, scientists, executives, teachers, lawyers, bureaucrats - those who have taken the social engineering messages of our education system most to heart. This mentality has now taken over in the domain that was meant to be separated from emotionalism, and science has become saturated with emotional ideology.

What is also evident is that the orthodox academic philosophy of science has dismally failed to be of any help. It is more responsible for creating ideological conflicts than resolving them. I agree with López Corredoira that it has ceased to have any positive value within scientific practice. It has abdicated interest in the real problems of science in relation to society. It has failed to give any meaningful guidance on the problem of incorporating scientific knowledge into a wider philosophical understanding. Exiling philosophy outside science has left a terrible vacuum within science. It is nowhere more evident than in the modern attitudes to Materialism and naturalistic metaphysics, as noted next in the concluding section to this Part.

---

[62] How many people that you *dislike – or hate* - do you also regard as a *good person?* It is almost impossible psychologically to combine these feelings. Thus people rationalise their hatreds (which are uncontrollable emotions) by finding reasons to classify the objects of hatred (people, or ideas, or races, or organisations, etc) as *bad*.

[63] They are mitigated by certain kinds of religious principles however – Buddhist principles of compassion and non-judgementalness, Christian principles of forgiveness and humility for example.

Materialism and naturalistic metaphysics.

Science is a system of factual knowledge, and primarily suited to developing empirical knowledge of the physical world. But its most important and fascinating implications are in *metaphysics*. Science informs key aspects of our larger world view: questions about *where we came from, what we are, what our fate is*. Particle physics, astronomy, cosmology, thermodynamics, chaos theory, evolutionary theory, biology, genetics, micro-biology, cognitive science, information theory, logic and linguistics, all have a deep impact on our understanding of the nature of ourselves and the world we live in.

These sciences have transformed our views about *fundamental laws of nature and causation, the nature of matter, the nature of time and space, the origin of the physical universe, the nature of living processes, the origins of life, the basis of intelligence, the nature of cognition, thought and language.* These are areas of pure science that have the deepest interest for our larger metaphysical conception of the world and our place in it. Philosophy and metaphysics cannot be done in a vacuum, without recognising knowledge we have gained from science. Making sense of scientific knowledge is a crucial part of philosophy.

But it is very difficult to make sense of science. The intrinsic difficulty is because science is in process: it is fundamentally *unfinished and unstable.* Thus far in history, scientific *metaphysics* has proved to be the most unstable aspect of science. For this relates to its formulation of broad comprehensive theories, proposed as *theoretical ontologies.* We have made steady progress in giving scientific explanations of many phenomenon – and these can certainly become stable. But science has only partial *theories,* and is far from having a complete view of the nature of physical reality at a deep level.

The instability of scientific theories is obvious when we look back over the last 350 years of Western science – from the time of Newton's theory of mechanics and gravity. It is obvious when we look back over the last 80 years, or the span of one person's life-time. Science has changed *dramatically* in this short period. Practically all the 'modern views' about the subjects listed above have formed within this short period – the expanding universe, Big Bang cosmology, Standard Model of particle physics, DNA and genetics, neo-Darwinian evolution theory, information theory, chaos theory, modern logic and linguistics, cognitive psychology and neuroscience. We should remember that the *proton* was not identified until 1919! (Rutherford). The *neutron* was not identified until 1932! (Chadwick). (But then within 13 years we had started using nuclear bombs). The fact that the Milky Way is only one *galaxy* among numerous others was not established until 1920s. The concept of an expanding universe was not established until 1929. It seems amazing to me that the CMBR[64] was not identified until the 1960's!

---

[64] Cosmic microwave background radiation.

Cosmology today is full of anomalies and paradoxes and speculative 'dark substances'. It should be stressed that there is no understanding of the ultimate origin or fate of the physical universe *at all*. It is completely unknown whether our universe is in a cyclic expansion-and-collapse process, or due to expand forever, or to collapse permanently in the future. Physicists will pontificate on their 'best guess' about this, and today usually tell you that it will keep expanding; but in fact they have *no more realistic idea than you do*. No scientist has a clue about where the universe came from. Even the dimensionality of space is completely unknown. Fundamental particle physics today is similarly incomplete, with multiple anomalies, conflicting theories, and a basic failure to tackle foundational issues.

Progress in the life sciences has historically been even slower. The double-helix structure of DNA, and encoding of genes, were not discovered until the 1950s![65] The origin of life on Earth remains unknown, and the existence of extraterrestrial life is unknown. It is clear that species evolve, but detailed mechanisms, including interaction with intelligence, are unknown.

Neuroscience is also in its early infancy, with no good theory yet of the logical structure of the neuronal network. The example noted in Part 2 below of Richard Faull shows that neurologists were unaware until very recently (2007!) even of the basic fact that the human brain reproduces and replenishes neurons – with leading scientific authorities refusing to admit it when scientifically demonstrated. The basis of consciousness remains completely opaque to physical science – particularly the interaction of *consciousness* with the physical machinery supporting *intelligence* – two quite distinct concepts that are regularly conflated by scientists.

Mainstream research into linguistics and natural language – theories of the 'deep structure' of natural language – has stalled since the 1970s, and represents a failed paradigm. This translates into a profound failure in AI (artificial intelligence) research paradigms, because attempts to generate recursive intelligence in computer programs must be able to solve the problem of natural language semantics. What we have so far in AI are just elaborate pattern-recognition machines based on *ad hoc* programming techniques in an application of brute force and ignorance.

So what metaphysical conclusions *can* we draw from science? Despite its radical incompleteness, which realistic scientists will admit to, there is nonetheless one metaphysical view widely drawn from science: *Materialism*. This holds that *everything real reduces to a material reality*. There is surely a problem claiming this as a scientific conclusion given the incompleteness of so many fundamental sciences claiming to support it. But the first problem to my mind is that it is itself is a very vague idea, encoding a number of different conceptions under one umbrella.

---

[65] Although DNA itself was identified in late 1860s by Friedrich Miescher.

I think the claim that many scientists really want to express is a principle of '*causal unification*': that everything real is *causally connected in a single framework of natural laws and substances.* The point is that science has continued to expand the boundaries of what counts as '*material things*', or '*substances*', to incorporate expanded evidence of *causality*. E.g. in the C19th, Faraday and Maxwell and others introduced the idea of *electromagnetic fields*. These appeared be ghostly *non-material* entities, diffused through space. They literally represent *potential-energy fields*. As such they met resistance from Materialists of the time, who did not believe in things that were not 'material substances'. They might allow 'fields' as mathematical constructions perhaps, but not as referring to real physical things.

Eventually these objections were overcome (and it was recognised as a metaphysical dogma to identify: *material = corpuscular),* notably when light was shown to be a free electromagnetic field. Individual particles of light can be transmitted across space, and it seems the 'free electromagnetic fields' representing them must be real *material things.* Today of course physics is full of 'fields' – EM fields, quantum fields – indeed the quantum vacuum is full of *virtual particles, representing vacuum fields,* so there is not even such a thing as 'empty space' any more. Indeed, some philosophers think that *there are no particles, only fields* (e.g. there is a paradox in quantum field theory, that your *count of particles* in a volume of space depends on your state of motion.) These fields seem suspiciously ethereal, *non-material* – but they have a role in unifying the causal connections, providing the mechanism for quantum particles to exchange energy and momentum. So materialists simply expand their definition and call them 'material'.

Thus science will expand its conception of what is 'material' *according to whatever is required to model causality.* How could we hope to *prescribe* what kinds of 'materiality' physical things have, before *empirically* investigating them? Nowadays, cosmology proposes 'dark matter', 'dark energy', 'quintessence', mysteriously undetectable 'substances'. Quantum mechanics is defined by *quantum waves* which have disturbingly non-materialistic properties – they are spread over space, they can 'jump' instantaneously from one place to another, they can be entangled with each other over large distances. And relativity theory postulates *curved space-time,* defined by what is called a *metric field.* Is *space-time* a 'material substance'? Most relativists would say that it is *physically real:* it is quite distinct in kind from the material particles embedded in it, but it affects their motions, and is affected by them.

So what does 'materialism' mean any more, when such a host of ephemeral 'substances' are inferred to be real? The classical concept of *material substances* has long broken down. The classical vision of materialism was a simplistic one of 'atoms-in-a-box', a simple common-sense dualism of *matter (atoms)* in *space (container).* This is based on a simplistic concept that the *part-whole composition of physical objects is simple spatial decomposition* – defined by breaking things down into smaller and smaller bits, that are mutually impenetrable, and cannot overlap. Causality is seen in this vision as being strictly local – things only affect each other when they 'touch'.[66] It is a simplistic conception derived from ordinary perception of macroscopic objects. Nowadays, physical objects are seen as intrinsically *interpenetrating, overlapping fields*. And causality is distinctly *non-local*: in principle, every physical system in the universe can instantaneously affect any other system. In this sense, the simplistic materialist vision that most Materialists still visualise has actually been long overthrown in physics. Modern Materialists cannot define their belief in terms of prescribing the *material* of the universe, because no one knows what it is any more!

It is easier to ask what kinds of things Materialists *reject*. The first thing they reject are 'transcendental causes' or 'supernatural agencies', such as 'life forces', anima, interactions with spirits, souls, God or gods, etc. They reject these as having any place in a unified causal account of physics. In traditional thought, such agencies lie outside 'laws of nature', and are said to cause things by a mysterious exertion of will or intelligence. It is this kind of *ad hoc* explanation that scientists primarily reject as being scientific.

But it is a big jump from rejecting *ad hoc* causal agencies in physics, to the vast metaphysical implications claimed by Materialism. For rejecting appeal to 'supernatural agencies' in *causal laws of physics* does not rule out that souls, spirits or God may yet still exist; that some entities that *appear supernatural* to us might end up having real causal physical connections to us after all. Esp, as mentioned earlier, is an example: the demand that physics is causally unified by general laws does not rule out the possibility of esp, or a non-material source of personal identity, or many similar things rejected by Materialists.

What Materialism certainly does *not* do is to *scientifically explain* the phenomenon of subjectivity: *consciousness, mind, personal identity*. This is what ideological Materialists claim: that science has explained how these reduce to 'brain activity'. This claim is just not true. What Materialism claims is that *there must be a purely physical explanation* of subjective phenomenon, given in the same terms as other physics, referring only to fundamental particles and forces. It does not follow that science *has given* such an explanation. It most certainly has not. And while the nature of subjectivity lacks any real scientific explanation or model, so does *value, morality, purpose*, and so on, i.e. the concepts of moral philosophy that Materialists so much want to destroy in their eagerness for a nihilistic doctrine.

Another claim implicit in Materialism is *monism*, the idea that there is ultimately only one kind of substance, which is physical. Today this is often

---

[66] Of course Newtonian gravity is already non-local in this sense. This aspect of his theory disturbed Newton. He consequently felt it described gravity, but did not explain it. Of course he was not a Materialist in a general sense, he believed in God and other metaphysical things; but he thought that *physics* must be governed by strict causal laws that did not require reference beyond material things.

described as *energy* (or *mass-energy*), rather than just 'matter'. We will be told that everything that exists is really a form of physical energy. Again, this contradicts the classical Materialist dualism that the universe consists of *matter* in *space and time*. For space (or now space-time) is not itself reduced to 'energy' in conventional physics, it is a container for energy.

I would also object to the idea that *monism* implies any of the traditional implications drawn from Materialism. In this respect, I support a heterodox theory of physics, which is a strong form of monism, where everything is actually composed of *'space'* – shapes and wave-motions of an underlying 'aether', which appears to us as *both space and matter*. It is similar in a way to the C19th theory of 'vortex atoms', which goes back to Descartes' vortex theory of matter. Similar suggestions that a form of monism will be the ultimate end of physics have been suggested regularly, e.g. by Wheeler in the 1970's[67]. The aether-based theory I have proposed is a genuine type of monism, it is causally unified, and it has simple precise mathematical laws. So if this was established, one might think it would justify Materialism.

And yet this theory contradicts all the usual prescriptions of Materialism! It is a causally unified theory but it entails far more physical connectivity than recognised in conventional physics, identifying real structures lying *outside ordinary three-dimensional space* – for it requires a multi-dimensional theory of space to model particle physics and gravity. See (Holster WEBREF (c), (e).) I think this is an essential kind of theory to consider – not just as physics, but because it shows that radical *metaphysical possibilities* remain entirely consistent with our present knowledge of fundamental physics. For it allows realistic correlates for entities such as 'souls' or 'spirits' and even 'God'.

The ideological Scientistic Materialists dislike this because they have obnoxious views about *metaphysics*. They believe they have conclusively solved all the problems of philosophy and metaphysics through their materialistic 'scientific' doctrine, and they want to *close down any debate*, declaring 'metaphysics' to be nonsense. But there are many areas of real metaphysics, traditional and modern, that are central to understanding.

Metaphysics may be thought irrelevant within narrowly defined scientific specialities: but it is central to natural philosophy, to the larger goal of drawing out a philosophical understanding from science. I have illustrated the major form that modern metaphysics takes, in the form of *possible world theory*, in Appendix 6. This is a point where modern *logic, analysis, and naturalistic science* all meet. It is not some kind of abstract speculation about meaningless pseudo-scientific generalities: it is a way of making the analysis of central concepts of physics, as well as theories of mind and existence and logical semantics, precise and explicit.

---

[67] Misner, Thorne, Wheeler, 1973.

In summary, there are different shades of 'materialism', and different shades of 'metaphysics'. Many ordinary people and many scientists have a general conviction that *reality is all based in material reality, i.e. in the physical world.* Thus they say consciousness is based in the brain, and not in some transcendental realm of spirits or souls. And methodologically: *when we come to investigate consciousness, we should start with what we know and can measure, the brain.* We do not start with assumptions about something transcendental. I think this is a reasonable belief, even though I do not believe the reality is ultimately so simple myself. But as far as I am concerned, people are perfectly entitled to adopt such beliefs – and brain science *is* a primary starting point to investigate *consciousness* (only not the only one). This is 'materialism' with a small m. I would say López Corredoira is a materialist in this sense.

But the philosophy is dominated by much more radical ideological Materialists, with a big M. These are ideological extremists (typified by the New Atheists or NeoAtheists), who believe fervently in a bundle of scientistic Positivist-Materialist doctrines, and want to force their world-view on everyone. They want to rule the domain of Science itself under the iron hand of their own orthodoxy. They claim Science as proof of Materialism. They openly despise metaphysics, philosophy, morals, religion. Like political ideologues, they are usually impassioned, arrogant, intolerant and loud. Turning science into a domain of personal ideology, they destroy the fundamental principles of *scientific objectivity.*

These people are a powerful destructive force *within* orthodox science, and within orthodox philosophy of science. They are the real enemies of creative heterodox scientists, and the true enemies of philosophy. Over decades, they have taken increasingly control of scientific culture, dominating propaganda, media, gate-keeping roles and power structures. They have destroyed the philosophy of science. They represent one kind of the death of science. The death of the spirit that comes from driving true scientists from Einstein's Temple.

PART 2. DEATH BY PEER REVIEW

The first part of this essay has focussed closely on López Corredoira's critical views, and then on some key themes in the philosophy of science underpinning his account. In the second part, I go on to give real examples of what he is talking about from my personal experience. These illustrate the problems and typical behaviours of academics first hand. There is no substitute for real material, as opposed to second-hand descriptions. This is inspired in part by López Corredoira's attitude. In his introduction he explains his background and motivation. After observing that he is actually relatively successful as a scientist himself, he says:

"My major frustration is not about my own creations but perhaps about the lamentable show I have to contemplate, in which intelligence and creativity are disdained whereas technology and money occupy the privileged position, in which poor science is applauded whereas extraordinary ideas are not even commented on. It is indeed a general frustration about the whole culture in most parts of our world: Capitalism gives all the force to people with money, and ideas are only important insofar as they can generate great amounts of money. It is frustrating to see how unfair and how harmful is a world dominated by these market criteria and in which we cannot do anything to stop it. That is the fatal circumstance of our present, and certainly there is not much we can do to save the world from it, but at least we can complain, and this is what I intend to do in this book. P.17-18.

He is not afraid to put himself on the line, not just with criticism of the system in general terms, but by giving personal examples. Of course everyone has such examples, in every area of professional life, but people rarely speak out. There is a fear that this is 'bad form', a social taboo against complaining, fear that you will be ostracised by your academic colleagues, and so on.

But I am not an academic, and in the following sections I will break this taboo, and publicise peer reviews of some of my own work, as real examples. There are multiple peer reviews of each paper, all highly contradictory of each other. I have made the papers available on preprint sites, so the reader can check if they agree with the judgements of the peer reviewers. These examples illustrate the problems with the peer review system, and I hope they serve as useful examples for other researchers, and especially younger intellectuals who are still naïve and idealistic enough to believe that the system will treat their own work fairly and objectively.

The examples I give are also from the *philosophy of physics* – these paper address real questions in physics, but they are conceptual and theoretical, not experimental. López Corredoira is scathing about modern philosophy, and rightly identifies its diminished quality. Talking of philosophy of science he says:

> "In the few fields where some important aspects of unsolved questions have arisen, powerful groups control and manipulate the flow of information and push toward a particular ideology. Obstruction of the freedom to initiate a line of research or develop an ideology is more prevalent in the faculty of philosophy than in science. Philosophy congresses are simply imitations of scientific congresses. Censorship of publications is more evident (most being confined to local dissemination rather than international); they have practically no objective criteria and there are no empirical data, so a paper can be rejected whimsically, without even producing an explanation for the rejection. Work positions are nearly always handpicked. P. 154.

This criticism is precisely borne out in the following examples. I have a slightly different attitude about *philosophy in physics* or *philosophy in science* than López Corredoira, because I think this is a critical mode of thinking essential to advance creative physics or creative science itself – indeed, it is a component of *natural philosophy* that was excluded in turning it into the specialised sciences of our day. López Corredoira would likely agree that this is a mode of thinking that has been lost from within science, but rejects its separation into a new discipline. I have come to agree with him: pursuing this mode of thought as a separate enterprise, 'philosophy of science' or 'philosophy of physics', outside the sciences themselves, is a failure and a waste of time. The following examples confirm exactly what he states above.

López Corredoira on peer review.

López Corredoira's Chapter 3 "The Institutionalisation of Science" and especially his criticism of the peer review system, "3.3 Publications, referees", is one of the most interesting parts of his book, especially for researchers themselves. Peer reviewed publications in powerful journals is the arena where ideas are directly admitted or excluded from circulation. This is the front line where the agendas of science are fought out.

> As is well known, control of communications and practice of power are closely related. I do not think that I have revealed anything new with such a statement. Thus the system, far from allowing free publication of results among professionals, works hand in hand with censorship. Theoretically, this control is presented as a quality filter but its functions are frequently extended to the control of power. Those researchers who want to publish in these journals are subject to the dictates of the chosen referee and the

journal editors, who will say whether the paper is accepted or not: this is the peer review system. P.68.

I was fascinated to discover this anecdote about Einstein.

The system of peer refereeing is relatively recent in science. In continental Europe it was mostly unknown before World War II. Rather, decisions about publishing were taken by editors, based on trusting the unquestionable authority of the author of the paper, or, in the case of the author being some young scientist without reputation, on the recommendations of some senior scientist whom the author directly had contacted. The United States was one of the first countries to adopt the policy of "peer reviewing" and, since science was "Americanized" after World War II, this, together with the use of English as the scientific language, extended to all scientific journals in Western Europe. There is a story that when Einstein first submitted a paper to an American journal, *Physical Review*, in 1936, the paper was rejected after a peer review. Einstein replied to the editor, "Dear Sir, We had sent you our manuscript for publication and had not authorized you to show it to specialists before it is printed" (Kennefick). Certainly, Einstein was not familiar with the peer review system. P.68.

It appears from this that Einstein's *first submission to a peer reviewed journal* – in his mid-50's, and long after he had won a Nobel prize – was rejected for publication! No wonder perhaps: at that time he supported an unpopular and controversial critique of quantum mechanics, arguing that it is incomplete. In any case, the peer review system is fraught with problems, widely observed over recent years.

Apart from minor details—changing the structure in order to better present the argument, rewriting a paragraph to be clearer, citing some other paper (in many cases the referee advises the author to cite some paper of their own, or by collaborators, etc.)— objections are very often to do with the referee's own opinion or how convinced they may be about the content of what is going to be published. Generally, according to my experience and that of people I have spoken to, the more controversial the topic, and the more of a challenge it is to established ideas, and the newer the approach, then the more difficult will be the problems in publishing it, and the higher the probability of its being rejected. Gillies argues that when a researcher makes an advance which is later seen as a key innovation and a major breakthrough, a peer review may very well judge it to be absurd and of no value. P.69.

López Corredoira observes an impulse for this lies in the vanity of experts.

The peer review process is flawed in many aspects; it presents many problems, and there is no evidence that it works efficiently to select the best science (Thurner & Hanel; Bauer). The background problem is as follows: referees are people who have dedicated their whole life to doing research on the few problems of a particular field. They are usually people widely recognized in their field and their social status is due to their contributions in the field. ... As people with experience and prestige, sometimes accompanied by an excess of vanity, they usually think along the lines of "I am a great specialist in this field. I know the interesting and crucial ideas about it. If a new idea were presented, either it is wrong, it is of little interest, or I would have thought of it before. Therefore, if somebody presents a new work that tries to tap into crucial questions, either it is a continuation of my own work and ideas and those in which I was involved, or it is wrong". Moreover, it might be misconstrued as competitive, that is, argumentative or contrary, for somebody to publish a theory or interpretation different to that held by the referee. P.71.

The system protects itself first by making the peer reviewer anonymous. This may seem necessary in a world of academic spitefulness – to protect peer reviewers from disgruntled authors. However I have come to agree with López Corredoira, that peer reviewers should be identified. They are making judgements of the work of others, after all: they should be held responsible. And the reality is that the peer reviewers, being in the position of power, are the most likely to be spiteful against authors and ideas they do not like. Authors presently have no recourse whatsoever against bad reviewers.[68]

More insidiously, as a power structure, it protects itself by giving no redress against poor reviews, and by a convention that peer reviews are confidential. Thus no matter how badly misconceived or deliberately manipulative peer review are, they are usually never made public. This is the taboo I will break here, making a set of academic peer reviews pubic. The reviewers I criticise are anonymous. The material is well over ten years old. In any case, these are not from physics journals but from a small group of powerful *philosophy of science* journals, controlled by powerful cliques of academics, generally from prestigious US and British institutions.

Background of peer review examples.

These papers are on the boundary between *theoretical physics* and *philosophy of physics,* being about time symmetry and the interpretation of quantum me-

---

[68] See the TedTalk in (Faull, WEBREF 2013) for an illustrative recent example.

chanics and relativity theory. Why not just send this material to *physics journals* you may ask? In one case, after a long series of rejections by philosophy journals, I did, and it was published by the first journal I submitted to - the *New Journal of Physics*, a high-ranking journal (Holster 2003 (a))[69]. Probably today some other papers might be publishable in physics journals too, as the questions they address have become more open to scientific enquiry. However, coming out of the late 1980's and early 1990's when I did the work behind these studies, physics was even more bigoted than it is now about allowing anything in the way of *critical conceptual studies*, and there appeared little chance of publishing such material within mainstream physics. Although it still is deeply conservative, a new wave of theoretical desperation over the last ten years has started opening the doors to *some* heterodox questions in physics.[70]

Also, after training originally in philosophy of science and logic, and only subsequently returning to study mathematical physics more seriously, it was natural to think first of publishing this kind of material in philosophy of science journals. The subjects in question are a regular topic in these journals. In fact, this is a rare area where philosophy of science publishes material that is actually relevant to a real science. (Their standard fare is grandiose meta-philosophical foundational systems, and endless commentaries on scholarly trivia in each other's work).

And finally, at that time, I thought that the tradition of *philosophical and logical analysis* was of real value in areas of theoretical physics (and other sciences), and that it was a good idea to emphasise that the origin of the ideas came through *philosophy*. I still think that philosophical and logical analysis has this real value within science, but I no longer think the institutions of philosophy of science are viable to do such work. What is needed is for this kind of analysis to be taken back into science itself – for the sciences to return to the tradition of *natural philosophy,* equipped with modern analytic tools. 'Philosophy of science' as an independent academic enterprise is brain-dead, and just waiting for someone to switch off the life support system (funding).

I observe in this respect that the philosophy of science clique that publishes widely on the topic of time and reversibility suppresses acknowledgement of fundamental errors that are *widely claimed as proven scientific results in the standard literature,* such as the error of adopting a false criterion for time reversal debunked in (Holster, 2003 (a)). Science has always made errors in the past: making errors is an intrinsic part of its process. Its salvation in the past has always pivoted on its willingness to correct its errors. When the

---

[69] After being hauled over the coals by four referees, and forced to reply to multiple objections. But most of these referees also had valuable comments and suggestions that improved the paper.

[70] At least as long as you do not question the most fundamental dogma, the *space-time manifold.*

research community in scientific philosophy begins to deliberately conceal scientific errors then its role in science is over.

Peer review example 1.

The quantum mechanical time reversal operator, 1997 – 2003.

I start with four distinct reviews of a paper called *The Quantum Mechanical Time Reversal Operator.* (WEBREF Holster 2003 (m)). I wrote a version in 1991, following my PhD (WEBREF Holster 1990). This first version was rejected for publication, and I lost it (along with electronic copies of all material I had written up to 1993), and rewrote it in 1997. As explained below, I added a little further material in the 2003 version, after publication of a flawed paper on the same subject by (Callendar, 2000). I first submitted it to *Synthese* in early March 1997, a leading philosophy of science journal. The paper was rejected on 30 July 1998, after almost one and a half years in review. Two reviews were obtained by *Synthese*. They are reproduced in full below.

The paper argues that the *orthodox time reversal operator adopted in quantum mechanics (which is different to the time reversal operator adopted in all other theories of physics) has never been justified properly, that justifications that are given are based on flawed principles of reasoning and semantics, and that the correct time reversal operator in QM is the same as in all the other theories of physics, viz. the transformation mapping time to its inverse: T: t* $\rightarrow$ *-t.* What is really unclear is not what the *time reversal transformation* is, but what the *interpretation of QM is.*

Ironically, physicists hide from this problem by playing with semantics – something they routinely accuse philosophers of. The debate is explained a little further in Appendix 3 for anyone with an interest in this topic.

Reviewer 1. Synthese.

Kluwer academic publishers
Journals Editorial Office

Achterom 119
P.O. Box 990
3300 AZ Dordrecht, The Netherlands
Tel.: +31 (0)78 6392392. Fax.: +31 (0)78 6392555

AUTHOR'S COPY; REF.1

```
Review Sheet SYNT: General Judgement
=====================================================================
Manuscript no.  SYNT114-97
The Quantum Mechanical Time Reversal Operator

Type of article:
0 review       Ø general paper     0 short communication
0 ...........................
-------------------------------------------------------------
                                          yes no see comments
1  Is the subject of the article within    α   0   0
   scope of the journal?

2  Are the interpretations/conclusions     ✗   0   0
   sound and justified by the data?

3  Is this a new and original              ✗   0   0
   contribution?

4  Does the title clearly and              ✗   0   0
   sufficiently reflect its content?

5  Are the presentation, organization      ✗   0   0
   and length satisfactory?

6  Can you suggest brief additions or      0   0   ✗
   amendments (words, phrases) or an
   introductory statement that will
   increase the value of this paper for
   an international audience?

7  Can you suggest any reductions in       0   ✗   0
   the paper, or deletions of parts?

8  Is the quality of the English           ✗   0   0
   satisfactory?

9  Are the illustrations and tables        ✗   0   0
   necessary and acceptable?

10 Are the references adequate and are     ✗   0   0
   they all necessary?

11 Are the keywords and abstract or        ✗   0   0
   summary informative?

12 Is the paper acceptable for publication?
         (a)    in its present form?  ✗       0
         (b)    with minor revisions?  0      0
   Should the paper be reconsidered after
   major revision?                         0      0
   Is it unacceptable for publication?     ✗      0

13 Please list any other general comments or specific
   suggestions on the enclosed comments sheet in a form
   in which they can be forwarded to the authors.

14 Have you made any annotations on the    ✗   0
   manuscript? AND I WANT TO KEEP IT
```

Summary page of Reviewer 1.

Reviewer 1 p.2.

Kluwer academic publishers
Journals Editorial Office

Achterom 119
P.O. Box 990
3300 AZ Dordrecht, The Netherlands
Tel.: +31 (0)78 6392392. Fax.: +31 (0)78 6392555

Review Sheet SYNT: Comments
===
Scientific quality:
Ms no:SYNT114-97   X high 0 average

Point-Ms Remarks
no.   page
        - - - - - - - -

COMMENTS ON «THE QUANTUM MECHANICAL TIME REVERSAL OPERATOR»

This is a very lucid paper clarifying an important issue -or rather the point of departure of an investigation needing to be pursued.

No additional explanation is recommended because it could only be too short or too long. But pursuit of the investigation on the following points is recommended :

1° PT invariance of Maupertuis' action and of de Broglie's phase. Stealthy step-in of complex numbers, to invade later the probability realm.

2° CPT invariance of the Dirac electron equation.

3° CPT invariance of the Dirac-Feynman transition amplitude:
PT , $<A|B> \Leftrightarrow <B|A>$; C, $<A|B> \Leftrightarrow <A|B>^*$; CPT, $<A|B> = <B|A>^*$.

4° Prediction-retrodiction symmetry in Feynman graphs.

5° Reciprocity of cybernetical coding and decoding ; reversibility of the negentropy-information conversion.

6° Aristotle's twin symmetries between information as knowledge and information as organization , and between efficient and final cause ;

7° Mehlberg's « factlike irreversibility » and smallness of Boltzmann's k .

 The authors very efficient « down to earth » method should help clarifying deep issues in the philosophy of science.

Comment page, Reviewer 1.

Reviewer 2. Synthese.

Kluwer academic publishers
Manufacturing Department

Achterom 119
P.O. Box 990
3300 AZ Dordrecht, The Netherlands
Tel.: +31 (0)78 6392392. Fax.: +31(0)78 6392555

```
Review Sheet SYNT: General Judgement
=====================================================================
Manuscript no.  SYNT114-97
The Quantum Mechanical Time Reversal Operator              Ref. 2
                                                           authors copy
Type of article:
0 review        ✗ general paper      0 short communication
0 ..................................
----------------------------------------------------------------
                                           yes no see comments
1  Is the subject of the article within     ✗   0   0
   scope of the journal?

2  Are the interpretations/conclusions      0   ✗   0
   sound and justified by the data?

3  Is this a new and original              ✗   0   0    new &
   contribution?                                         wrong

4  Does the title clearly and              ✗   0   0
   sufficiently reflect its content?

5  Are the presentation, organization       0   ✗   0
   and length satisfactory?

6  Can you suggest brief additions or       0   ✗   0
   amendments (words, phrases) or an
   introductory statement that will
   increase the value of this paper for
   an international audience?

7  Can you suggest any reductions in        0   ✗   0
   the paper, or deletions of parts?

8  Is the quality of the English           ✗   0   0
   satisfactory?

9  Are the illustrations and tables        ✗   0   0
   necessary and acceptable?

10 Are the references adequate and are     ✗   0   0
   they all necessary?

11 Are the keywords and abstract or        ✗   0   0
   summary informative?

12 Is the paper acceptable for publication?
             (a)     in its present form?  0       ✗
             (b)     with minor revisions? 0       ✗
   Should the paper be reconsidered after
   major revision?                         0       ✗
   Is it unacceptable for publication?     0       ✗

13 Please list any other general comments or specific
   suggestions on the enclosed comments sheet in a form
   in which they can be forwarded to the authors.

14 Have you made any annotations on the    ✗   0
   manuscript?
```

Summary page, Reviewer 2.

Reviewer 2. P. 2.

Kluwer academic publishers
Manufacturing Department

Achterom 119
P.O. Box 990
3300 AZ Dordrecht, The Netherlands
Tel.: +31 (0)78 6392392. Fax.: +31(0)78 6392555

```
Review Sheet SYNT: Comments
============================================================
                              Scientific quality:
                              0 high  0 average  X low
Manuscript no: SYNT114-97
------------------------------------------------------------
Point  Ms.   Remarks
no.    page
------------------------------------------------------------
```

Comments enclosed

*The printing is poor, unfortunately.
No lines are crossed out.*

Comments on "The Quantum Mechanical Time Reversal Operator,"   E3REF2   619462

This paper suffers from several conceptual errors.

1. It does not take into account that the time variable is not an observable either in elementary quantum

mechanics (in which spatial position is an observable but time is a parameter in the evolution of the state) nor in relativistic quantum field theory (in which four-dimensional position is not an observable but a parameter) For this reason, H defined in terms of a time derivative, as on p.7, is not an operator on the Hilbert space

associated with the system, nor is T, defined in terms of a sign change of t, an operator on this Hilbert space. Of course, when one has a time parametrized family of vectors in a Hilbert space one can differentiate

with respect to time, but that does not make time an observable that is represented by a self-adjoint operator.

2. The kinematics of time reversal can be treated in quantum mechanics without reference to the time reversibility of the dynamics of a particular system, the dynamics depending upon the Hamiltonian. The kinematical treatment is correctly done by Messiah, Baym, and other authors, and the author's rejection of this approach on p.5 eigenkets 1x> constitute a basis. If T were a linear operator it would suffice to have T defined on this basis to give a definition for all vectors in the Hilbert space. In accordance with classical kinematics, which must be preserved by the correspondence principle, T!x>=!x>, and then linearity would guarantee that TIntegralexp(ipx)dx = Integralexp(ipx)dx, which has the immediate consequence that T!p> = !p>, contrary to the classical kinematical relation that time reversal changes the signa of the momentum. We need T!p> = !-p>. If we require that transition probabilities are invariant under time reversal, then Wigner's theorem requires that the time reversal operator be either a unitary or an anti-unitary operator -- hence providing a link between operations on the projective Hilbert space to the Hilbert space, in answer to the query in the last sentence on p. 27. Since unitarity has been eliminated, the time reversal operator must be anti-unitary. For a spinless system, the complex conjugation operation in the position representation

suffices, since obviously the complex conjugate of Integralexp(ipx)dx is Integralexp(-ipx)dx, wexcept for normalization. For systems with spin the situation is somewhat more complex, and one may choose among several possibilities. Messiah is good on this point.

3. The argument on 16 that a physical process can be equally well represented by a wave function and its complex conjugate in no way shows that quantum mechanics is generically a time reversible theory. It only shows that there are two different ways to represent the same physical reality mathematically, and indeed there are infinitely many ways. Time reversal invariance is established if the Hamiltonian is invariant under the time reversal operator, i.e. H = THT' (where for reasons of typing I use T' to mean the inverse of T). In that case, if u(t)

is a time-dependent family of vectors satisfying the time-dependent Schroedinger equation, then v(t) = Tu(-t)

does so also. Thus both u(t) and v(t) are dynamically possible trajectories in the Hilbert space. This is particularly interesting, of course, when u(t) and v(t) are linearly independent of each. The paper neglects the difference between quantum mechanics generically and the quantum mechanics of a specific system with a specific Hamiltonian. However, the kinematical reasoning in (2) is independent of the Hamiltonian.

4. The new interpretations of quantum mechanics on pp. 17ff cannot be taken seriously unless full details are presented, with arguments showing agreeing with experimental data.

For these reasons I recommend that the paper not be published.

First comment page, Reviewer 2.

Reviewer 2. P.3.

$$(T\psi, PT\psi) = \cancel{} (\psi, T^\dagger P T \psi)$$

$$\| $$

$$-(\psi, P\psi) \qquad S. \quad P = T^\dagger P T$$

(But this assumes linearity of T

$$T\,|\vec{x}\rangle = |x\rangle$$

$$T\,|\vec{p}\rangle = |-\vec{p}\rangle \qquad \text{not only} \atop \text{form } |x\rangle \therefore \atop a \text{ basis}$$

$$\vec{p} = \frac{1}{(2\pi)^{3/2}} \int e^{i\vec{x}\cdot\vec{p}} |\vec{x}\rangle\langle$$

$$\text{if} \qquad T\vec{p} = \frac{1}{(2\pi)^{3/2}} \int e^{i\vec{x}\cdot\vec{p}} \underline{T\,|x\rangle}\langle x$$

$$= \vec{p}$$

$$\text{If antilinear} \qquad T\vec{p} = \frac{1}{(2\pi)^{3/2}} \int e^{-i\vec{x}\cdot\vec{p}} T|x\rangle \atop \langle x$$

(This def.
is bounded, $$= \cancel{} |-\vec{p}\rangle$$
inst of mann
of H $\qquad e.g. \quad \mathcal{H} = \frac{\vec{p}^2}{2m} + \frac{1}{8}(\vec{x}\cdot\vec{p} \atop + \vec{p}\cdot$

$$T H T^{-1} \neq T$$

Second comment page, Reviewer 2. (The second reviewer's explanation of the 'correct' solution to the problem.)

Comparison of the Synthese reviews.

Both reviewers judge the article to be within scope, well written, a new and original contribution, properly referenced.

The first reviewer is enthusiastic about the article, judges it of "high scientific value", thinks the interpretation and conclusions are "justified by the data", and recommends publication "in its present form". He comments: *"This is a very lucid paper clarifying an important point – or rather the point of departure of an investigation needing to be pursued. No additional explanation is recommended because it could only be too short or too long. But the pursuit of the investigation on the following points is recommended...."* He gives a list of seven points for future investigation, all very relevant points. He concludes: *"The author's very efficient down to earth method should help clarify deep issues in the philosophy of science."*

This first reviewer was the late Olivier Costa de Beauregard (1911 –2007), the very eminent French physicist and philosopher of physics, author of the two-volume work *Time: The Physical Magnitude (1987)* a masterpiece of C20th philosophy of physics. (I sent back comments through the editor, and he subsequently wrote back to me.)

The second reviewer declares the paper 'NEW & WRONG', "unacceptable for publication", and of "low scientific value". He provides a set of technical objections. At first glance they look sophisticated, but closer inspection reveals they *do not relate to the argument in paper,* and simply *reiterate the orthodox analysis that is analysed and criticised in the paper, as if it is true.* He provides a page of scrawled equations representing his own idea of the 'correct' solution: this is simply repeating the conventional analysis. It is clear to me that he has not bothered to read my paper, thinking it is sufficient to *reiterate the conventional analysis.* I do not know who the second reviewer was. The paper was rejected after one and a half years in the review process.

The reviewer's error: censorship.

The glaring question is of course: *how do two peer reviewers come to such diametrically opposite conclusions, over a paper that revolves on clearly defined technical arguments?* Both are apparently specialists, but in complete disagreement. The second reviewer is affronted that anyone could question the orthodox analysis of his subject. Well of course: he has been teaching this analysis for years, he has probably written text-books about it, his reputation *depends on it being true.* He looks very silly if his analysis turns out to be wrong *now.* He probably cannot admit the *possibility* that it is wrong. Perhaps he has that anger that commonly arises when we are questioned about something that we cannot *afford* to be wrong about?

We may recall López Corredoira's description: *"As people with experience and prestige, sometimes accompanied by an excess of vanity, they usually think along the lines of 'I am a great specialist in this field. I know the interesting and crucial ideas about it.'"* The second reviewer makes a fatal mistake *as a peer reviewer*. The paper raises a serious question – a *very serious question* if the argument is correct. (And one that a number of researchers have now taken up as research projects - but remember this is in 1997-98, before it has been accepted as a real question). He is intent on *forcing his own opinion about the stupidity of the question*. He makes the most fundamental error of *putting forward his own resolution of a complex controversy in a peer review, as being the conclusive resolution*. He is suppressing what is an open issue, censoring the question itself. This question remains a deep controversy in the philosophy of physics today.

In point of fact, his arguments are very poor: all he does is to repeat the orthodox arguments, the very arguments that are critically analysed in the paper itself, without responding to the text or material in front of him. This is the classic response to heterodox ideas by 'experts': *restating the orthodox position as if it is accepted fact*. His attitude shows he should be disqualified as a peer reviewer. I sent the following rebuttal of his points back, but received no response.

My rebuttal of Reviewer 2.

Synthese, Journals Editorial Office
Kluwer Academic Publishers
Achterom 119
P.O. Box 990
3300 AZ Dordrecht, The Netherlands

31(a) Lewis St.
Hamilton
NEW ZEALAND

Ph. 0-7-843 7521

18/9/98

Dear Editor,

RE: SYNT114-97

Thank you for informing me of your decision not to publish my paper, entitled "The Quantum Mechanical Time Reversal Operator". I have some serious concerns about the review procedure for this paper, particularly the conflicting views of the reviewers.

My paper was reviewed by two independent reviewers, whom I will call Reviewer1 and Reviewer2, Reviewer1 being the first, *favourable* reviewer, Reviewer2 being the second, hostile reviewer.

Reviewer1 provided an initial favourable review of the paper, scoring it positively in each category, including that the paper is "acceptable for publication in its present form". Reviewer1 describes the paper as "a very lucid paper, clarifying an important issue", and recommends publication without any changes to the paper or reduction in size. This reviewer includes helpful and positive comments concerning further investigation of the issue raised in the paper.

A second review is also included, from Reviewer2. This reviewer gives a completely different evaluation, holding that the paper is "original, but wrong", that it "suffers several conceptual errors", and "for these reasons I recommend that the paper not be published."

The conflict in judgement between these two reviewers seems to require some explanation. It must surely be wondered which of them is right? I will give my view below, which is that Reviewer2's review demonstrate serious lapses of judgement, and should be disregarded.

Reviewer2 provides five points of criticism, including the scrawled 'proof' of the conventional result on an attached sheet, but unfortunately, *fails to provides any specific*

Rebuttal

p1.

(Rebuttal cont.)

criticism of any specific hypothesis, or principle of reasoning, or derivation, from the
paper under review.
Reviewer2 neither quotes any statement from the paper, nor summarises any view or
conclusion reached in the paper.

Reviewer2 provides 4 typed comments. Unfortunately, *none of these is a viable*
criticism of the paper under review. Eg. the fourth, and most intelligible point is, in full:

4. The new interpretations of quantum mechanics on pp. 17ff cannot be taken seriously unless full
details are presented, with arguments showing agreeing with experimental data.
For these reasons I recommend that the paper not be published.
(Reviewer2).

But this comment completely misconstrues the point of the 'new interpretation', which is
stated very clearly on p. 4, and on p.22 of the paper - note Reviewer2's reference to
pp.17ff is incorrect. It is intended to be merely a mathematical model illustrating a logical
point, and is quite explicitly *not* required to be experimentally verified, merely to be
consistent. Similar flaws of logic and accuracy are found in *all* Reviewer2's comments, and
they make it clear to the author of the paper that Reviewer2 has little grasp of either the
content or the conclusions of the paper. Reviewer2's comments appear, superficially, to be
well-informed, but in fact, to the extent any of them is accurate in itself, their *application*
to the paper is obscure or non-existent.

Perhaps the most important point is that in his main argument against the paper, (see
his scrawled additional page of proof), Reviewer2 commits a common fallacy of
reasoning, called 'repeating the argument'; he simply repeats the most common
conventional argument, which is the argument analysed in depth on pp. 6-12 of the paper.
In this analysis, I have *quoted the argument from a number of textbooks,* then analysed,
and eventually rejected it. However, Reviewer2 *neither summarises nor quotes any*
passage from my paper. He does nothing to evaluate the arguments given in my paper,
which are precisely the objections required to dismiss the argument which he gives in his
comments. Instead, he simply presents yet another version of this well-known argument, as
being known fact.

I do not think it is possible to properly establish that my paper is wrong *without at the*
very least quoting or summarising something from my paper.

I also note that the problem analysed in my paper has been pursued at some length already
by the eminent physicist, Costa de Beauregard, among others. If the problem is so trivially
disposed of as Reviewer2 thinks, then why has de Beauregard spent so much time and
energy on it, and also suggested reasons for adopting a different solution? I suggest that
de Beauregard has a real point to make, while Reviewer2 has failed to see either his point,
or my point, which is different again.

Reviewer2 shows serious lapses of judgement throughout, and his review should be
disregarded.

Rebuttal p2.

My objections to the second review were returned to *Synthese.* No response
was received.

Submission to BJPS.

A couple of years later, in 2000, Craig Callender managed to publish an article on this issue, and David Albert managed to publish a book arguing that even time reversal of *classical EM theory* is mistaken, raising the broader issue that *physics lacks principles for deciding time reversal transformations*, and it flared into a controversy. While I agree with both up to a point, my views are somewhat different to theirs. In 2002, I added some introductory paragraphs to the original paper, referring to Callender's paper (which I criticise in parts) as a point of reference to what was now an ongoing controversy in the field, and submitted it to BJPS.

The British Journal for the Philosophy of Science

Editor
Peter Clark

Deputy Editor
Katherine Hawley

Associate Editors
Alexander Bird
Peter Milne
Adam Rieger
Denis Walsh
Michael Wheeler

Editorial Assistant
Patricia Barrie

Department of Logic and Metaphysics
School of Philosophical and
Anthropological Studies
University of St Andrews
St Andrews, Fife, Scotland
KY16 9AL, UK

Tel: 01334 462486
Fax: 01334 462485
email: bjps@st-andrews.ac.uk

Dr Andrew Holster
1 Haunui Road
Pukerua Bay
Wellington
New Zealand

17 June 03

Dear Dr Holster,

Thank you for submitting your paper, The Quantum Mechanical Time Reversal Operator, to the British Journal for the Philosophy of Science. We regret to inform you that we have decided not to accept it for publication. A selection of comments gleaned from our referees' reports is enclosed. We hope that you find this helpful.

Yours sincerely

Peter Clark

Rejection from BJPS.

THE DEATH OF SCIENCE

The premise is this: in implementing a desired transformation (in this case
t -> -t) one is not at liberty to simultaneously transform other quantities (in this case i ->
-i) appearing in the equations so as to obtain a symmetry.

If this philosophy were correct, then neither are Galilean boosts
symmetries of non-relativistic quantum mechanics, nor is time reversal a
symmetry of Maxwell's equations (to give just two examples). The latter
question was in fact recently raised by David Albert in his book "Time and
Chance". David Malement gave a rebuttal of this claim at the recent
Maryland conference, winning (I gather) wide agreement on the error of
Albert's ways.

The author needs to set his philosophy in this wider context,
or else to make clear why it applies to time
reversal in qm, but not to other (alleged) symmetries.
If the author is inclined to pursue the matter I would also urge him to
make a clear statement of what it is for a theory to have a given kind of
symmetry. He makes a stab at this, on p.1-2, but in my view the crucial
point is not clearly stated; when he says that L is symmetric under T if
whenever world W satisfies L, so does world TW, he needs to make clear
that (to interpret T as time reversal) TW is the time reversal of W with
respect to all the physically real quantities (or, in more operational
terms, the measurable quantities). It is up to the theory to say what are
the real quantities (respectively, the measurable quantities) - subject to
experimental checks - and quantities that are not real (respectively
measurable) can be transformed any way you like, so long as the real
quantities defining the real events have their time order reversed.

This gets to the heart of the debate. It is very simple and clear in the
electromagnetic case. There Albert assumes that the direction of the
magnetic field vector is real, and, because the time reversal of a
solution to Maxwell's equation (as the reversal of the sequence of field
configurations one starts off from) are not a solution, that

Reviewer 3 Comments, P1.

(Reviewer 3 cont.)

electromagnetism breaks time reversal symmetry; he is not prepared to reverse the direction of the magnetic field vectors as well (in which case the reversal of the sequence of field configurations is a solution). But who says the direction of the magnetic field is real? Unlike the electric vector, its sign is settled by a convention - and the convention can be changed for the time-reversed sequence. What is completely unambiguous in this context are particle configurations: the time reversal of any sequence of particle configurations that satisfies Maxwell's equations is a sequence that satisfies Maxwell's equations.

This same analysis - determining first what are to count as the real physical quantities - was carried out by Wigner when he first defined a time reversal operator.
The quantities there were expectation values. It was on the assumption that these were unchanged that he deduced symmetries must be implemented by unitary or anti-unitary transformations.
By all means let the author challenge this assumption! That seems to be what he has in mind when he speaks of the need to have a "complete and definitive interpretation" of quantum mechanics. The author makes some comment on this in Sec.12, but I find the discussion in these later sections so confusing - in Sec. 11 he speaks of an "interpretation" of qm in which "the operators T and T* become indistinguishable" (which strikes me as just nonsense), and in Sec. 12 that psi and psi* become physically identical in QM2(what is QM2? again this seems nonsense) - that I cannot really follow the argument. Anyway, it is only the last paragraph on p.21 that is germane to the point I am making, and it hardly offers any reasons for rejecting Wigner's assumptions as to the measurable quantities in qm.

Reviewer 3 comments, p.2.

Comparison of BJPS with Synthese Reviews.

This third review has different arguments again to the hostile *Synthese* review. This reviewer knows that there is now a recognised public controversy about this issue. He does not know the orthodox analysis in QM that the *Synthese* reviewer recites – he is much less technically sophisticated. But he pulls very similar stunts.

- First, he does not directly address any of the arguments in the paper, he *recites his own opinion about the solution.*

- Second, he takes a bunch of *positivist assumptions* for granted, illustrated by gossip that Malament won an argument at a conference against Albert (on an issue with classical electro-magnetic theory) *by giving similar positivist arguments.*

- Third, he thinks he can adequately solve this problem himself in the space of a brief review, and *shut down the debate.*

While the details are different, his response is similar to the first reviewer in terms of these three main stunts. I sent a response back as follows.

My rebuttal of the Reviewer 3.

Comments on BJPS Review of "The QM Time Reversal Operator".

I found this review to be very poor indeed.

The first and major problem is that the reviewer does address any of the arguments or analysis given in the paper: instead, he rejects the paper in a very general way, essentially by giving *his own (one-sentence) 'proof' that my conclusion is wrong.*

In Paragraph 2: *"If this philosophy is correct, then neither are Galilean boosts symmetries of non-relativistic quantum mechanics, nor is time reversal a symmetry of Maxwell's equations".*

The first point is a blunder: time reversal is not related to Galilean boosts. Adopting T for time reversal does not affect the Galilean symmetries.

The reviewer then goes on to spend much of the review summarizing his opinions about the case of electromagnetic theory. But my paper is about quantum mechanics.

The issue about EM theory is in fact also contentious; the reviewer justifies his own conclusions by reporting hearsay: "David Malement gave a rebuttal of this claim [by David Albert; not by me] at the recent Maryland conference winning (I gather) wide agreement on the error of Albert's ways." This is just a second-hand opinion about a complex issue. And it appears irrelevant.

The reviewer's summary of the EM case is poor anyway; he traces the solution to the idea that *"the sign [of the magnetic field] is settled by a convention – and the convention can be changed."* This is not the solution to the problem at all; it shows a superficial confusion about the logic of the problem. My paper aims to analyse the logic in detail for the QM case.

Paragraph 3, as well as his final sentence, show the reviewer bases his views on the background assumption of a 'positivistic' or 'operationalist' or 'instrumentalist' philosophy: he repeats his assumption that the 'real quantities' are the 'measurable quantities', etc; and in his final summary says that "it hardly offers any reason for rejecting Wigner's assumptions as to the measurable quantities in QM". Positivism is of course still overwhelmingly common among quantum physicists, even though widely rejected in philosophy of science and semantics for decades. A lot of my paper is taken up with disentangling exactly this part of the debate. One error that positivism leads to in this case is the failure to distinguish the concept of a formal (e.g. model-theoretic) interpretation of a formal theory, from the 'empirical' interpretation of the empirical predictions of the theory. My paper goes into detail about the formal interpretation.

His criticism of my definition of TW in Paragraph 3 is based on his own philosophical demand that I explicate T by giving an instrumentalist or operationalist interpretation. In fact, the kind of general definition of T I give is well known, as well as being very

My rebuttal p.1.

(My rebuttal cont.)

obvious; and my paper contains references to detailed studies by Earman and others who provide similar general definitions.

In the final Paragraph, he complains that a key point of my paper is "just nonsense". In fact the point in question is developed and discussed in considerable detail; it is the fact that, although the operators T and T^* are of course mathematically different; they *have the same effect as each other* on some suitably defined theories, and in this sense become indistinguishable. In the same way, the effect of a clockwise rotation of an object through 360 degrees is indistinguishable from an anti-clockwise rotation through 360 degrees. There is nothing very mysterious about this.

The reviewer fails to give any detailed discussion of the actual *content* of my paper. Even his first paragraph stating "The premise is this:..." is inaccurate (for we are perfectly "at liberty" to define the operator T^*; my main argument (not premise) is that $T: t \rightarrow -t$ is *the time reversal transformation,* while T^* is a different transformation.)

In summary, this review is completely misleading; it contains serious errors of fact; it takes a number of mistaken and opinionated assumptions as facts; it fails to address the content of my paper in any serious manner; and it is philosophically naïve.

My rebuttal P2.

My objections to the review was returned to *BJPS*. No response.

Submission to AJP

I then submitted the paper to the Australasian Journal of Philosophy. While not a specialist journal in philosophy of science like *Synthese* and *BJPS*, the *AJP* regularly publishes papers on similar subjects, including philosophy of quantum mechanics and time. The paper was duly rejected on the basis of the following review.

Reviewer 4. AJP.

Referee Report: "The quantum mechanical time reversal operator," *Australasian Journal of Philosophy*

It has often been claimed (in the physics literature) that quantum mechanics — at least without the addition of any sort of "collapse of the wavefunction" — is time-reversal invariant (TRI). That is, for any sequence of events, or states, that is permitted by the laws of QM, the time reversed sequence of events, or states, is also permitted by the laws of QM. This article contests that claim.

The author follows Callender in claiming that two main arguments have been given for the time reversal invariance of QM. According to the first argument, the time-reversed theory is really the same theory as QM. (The author claims that this argument slides from the empirical equivalence of the two theories to their physical equivalence; but I think that this is an uncharitable reading of the argument.) The second argument claims that the naive time-reversal operation simply doesn't make sense in the quantum-mechanical context, and that the only meaningful transformation that reverses time is the "Wigner reversal." But since QM is obviously Wigner-reversal invariant, this second argument claims that QM should be regarded as time-reversal invariant.

This paper discusses an interesting issue in an interesting way. Although the conclusion of this paper is not novel — the same conclusion was argued for by Callender (2000) — the reader is led to believe that the author has a novel argument for this conclusion. However, as a result of some pervasive unclarities in the argument, it is difficult to assess this claim for novelty. (However, the discussion in Section 5 of definitions and empirical laws in QM is of independent interest. To my knowledge, this issue has not been addressed elsewhere in the philosophical literature.)

The main flaw of this paper is that it does not rise to the level of clarity and precision that is expected of papers published in *AJP*. First, nowhere in the paper does the author give anything resembling a *definition* of time-reversal invariance; nor does the author clarify what sorts of things can have or lack the property of time-reversal invariance; nor does the author emphasize the crucial distinction between the time-invariance of laws versus the time-invariance of laws plus particular initial conditions. This latter distinction is crucial to motivating the project of investigating the time invariance of particular physical theories (as beautifully illustrated by Callender's discussion).

I will now add a few more scattered comments.

1. The author needs to clarify his use of the word "state." In particular, the author speaks as if some states are disallowed by the dynamical laws of QM (or by the time-reversed dynamical laws). However, it is more typical — at least in the philosophical literature on these topics — to think of states as "initial conditions," so that any state is compatible with the dynamical laws. It's *sequences* of states, or histories

Reviewer 4 p1.

(Reviewer 4 cont.)

of states, for which it is meaningful to inquire whether they are compatible with dynamical laws. Of course, if we're working in a theory with deterministic dynamical laws, the two notions of state are interchangeable; but using the same terminology for the two notions obscures the distinction between instantaneous states and dynamical laws.

Once the distinction between instantaneous states and sequences of nomologically possible states is made, the issue at stake can be reformulated as follows: the time reversal operation takes the history $\{s(t) : t \in \mathbb{R}\}$ of states to a new history $\{s^T(-t) : t \in \mathbb{R}\}$ of states, where for fixed $t \in \mathbb{R}$, $s(t) \mapsto s^T(t)$ is the time-reversal of the instantaneous state. The question, then, is whether time reversal in QM should be taken as acting trivially on instantaneous states (as Callender claims, and as the author seems to claim):

$$\psi(x, t_0) \xrightarrow{T} \psi(x, t_0),$$

or whether time reversal in QM should be taken as "conjugating" the instantaneous state (as claimed by Wigner et al.):

$$\psi(x, t_0) \xrightarrow{T} \psi^*(x, t_0).$$

2. The author needs to be more clear about what he takes to be the domain and range of the time reversal operation. For example, in some places the author applies the time reversal operation to quantum states, and in other place the author applies the time reversal operation to the Schrödinger equation (which is supposed to be a law of QM). But if we apply the time reversal operation to the laws themselves, then it is no longer clear what we mean when we ask if a theory is time reversal invariant. In particular, the typical definition says that a theory is TRI just in case for any state of affairs that is compatible with the laws, the time-reversed state of affairs is also compatible with the very same laws. So, it seems that we should never alter the laws themselves when discussing time reversal invariance. (The author might wish to define a theory as TRI just in case its laws are invariant under the time reversal operation. However, if this definition is chosen, it should be consistently used throughout the paper, and the author should also include a discussion of the relation of this definition to the more standard definition in terms of transformations of states.)

. Callender argues that the quantum mechanical time-reversal operation should not change the sign of momentum. (His reasoning is that the quantum mechanical momentum observable is not a derivative w.r.t. t, and so is not logically supervenient on t.) On page 2, it

Reviewer 4 p2.

(Reviewer 4 cont.)

sounds like the author agrees with Callender on this point. He notes that the "standard argument" for Wigner-reversal invariance is based on the purported fact that time reversal should switch the sign of momentum; and then he claims that, "this conventional argument... is unsound."

However, on page 17, the author says that, "the reason Callender gives in the quotation above is also not correct: time reversal does reverse the classical property of momentum." But now it seems as if the author is granting the main premise of the conventional argument. How is it that the conventional argument is then blocked?

4. I suggest that as an aid to clarity, the author set $\hbar = 1$ and $m = \frac{1}{2}$ throughout the paper. This will make all the equations (e.g., on the bottom of p. 3) much more transparent, and it seems to me that no essential content will be lost.

5. (page 6, bottom) It appears that there is a conflict between equations T(2) and T(4). In particular, since $H^* = -H$, and since H is a positive operator, it follows that the spectrum of H^* consists of only non-positive numbers. However, T(4) states that $H^* = (P^*)^2$, and the square of a self-adjoint operator is a positive operator. So, it follows that the only possible number in the spectrum of H^* is 0, and so H^* is identically zero.

Perhaps I have missed something here; but I think the trouble in these equations may be a consequence of a more fundamental problem. In particular, the author claims that $T(H) = -H$, where T is the time-reversal operator applied to observables. But now measured energy values will always be negative. The author needs to fix this problem, or explain why it isn't a problem. (Here's another case where a bit more clarity would be useful. The time reversal operator is, in the first place, a mapping of states. Is it linear, anti-linear? Is it unitary? The author never tells us these things. Furthermore, it doesn't really make sense to apply T to observables like H; we would would need something like the "dual" of T. The author should explain how the the mapping of observables is related to the mapping of states.)

6. (p. 8, bottom) The author says that the problem with the argument can be found in premise (iii). But the author indicates later (on p. 15) that premise (i) is also false (since he claims that $T(H) = -H$).

7. (Section 6, p. 17) Regarding the first argument for the time-reversal invariance of QM: the author claims that this argument is based on faulty positivistic assumptions. In particular, the argument is supposed to claim that the time-reversed theory is empirically equivalent to quantum mechanics, and since empirical equivalence entails physical equivalence, QM is time-reversal invariant.

Reviewer 4 p.3.

I agree (along with most sane philosophers of science) that empirical equivalence doesn't entail physical equivalence. *But isn't there a similar argument for TRI that doesn't use this premise?* In particular, suppose we adopt Glymour's definition of theoretical equivalence ('Theoretical Realism and Theoretical Equivalence', in R. Buck and R. Cohen (eds.), *Boston Studies in Philosophy of Science, VIII*, pp. 275–288). According to Glymour, two theories are equivalent *if they are intertranslatable* (and this translation must include both the empirical and the theoretical vocabulary of the two theories). But now couldn't a proponent of TRI argue that QM and T(QM) are theoretically equivalent — i.e., that they are notational variants of each other? So, while the laws of the two theories might not be *identical*, these laws might say exactly the same thing about the world (i.e., they might have the same truth-conditions). [In fact, the author says on page 5 that *(QM) is isomorphic to QM; and I assume that this comment would also apply to $T(QM)$. Isn't isomorphism a reasonable notion of physical equivalence between two theories?]

Note also that, to refute this argument, it's not sufficient to point out (as the author does on p. 4) that the space of solutions to the Schrödinger equation is disjoint from the space of solutions of the time-reversed Schrödinger equation. This still leaves open the possibility that these two theories are just different ways of describing exactly the same thing. (Compare with the Stone-von Neumann uniqueness theorem, which shows that matrix mechanics and wave mechanics are theoretically equivalent.)

Reviewer 4 p.4.

End of AJP Review.

Comparison of Reviewer 4 with others.

This review contradicts the two previous hostile reviews, asserting that: *"This paper discusses an interesting issue in an interesting way."* The reviewer himself gives a long exegesis of his own ideas about its solution – essentially following Callender's treatment, which he seems to take as definitive. In any case, his views are completely at odds with the dismissive response from the previous hostile reviewers.

But he is equally at odds with the first reviewer (de Beauregard). His reason for rejection is that: *"The main flaw of this paper is that it does not rise to the level of clarity and precision expected of papers published in AJP".* Yet de Beauregard thought the opposite: *"This is a very lucid paper clarifying an important point – or rather the point of departure of an investigation needing to be pursued. No additional explanation is recommended because it could only be too short or too long. But the pursuit of the investigation on the following points is recommended...."*

So now we have a fourth point of view entirely. But although this reviewer recognises the problem as significant, in terms of imposing his own agenda in his review, he is similar to the other two hostile reviewers.

The reviewer is an admirer of Callender's (2000) view. I support a similar criticism of the orthodox view - but criticise Callender's own conclusions, arguments and analysis. This reviewer gives a lengthy exegesis of aspects of the problem from Callender's point of view, and is hostile to my counter arguments.

> *"Although the conclusion of this paper is not novel – the same conclusion was argued for by Callender (2000)[71] – the reader is led to believe that the author has a novel argument for this conclusion. However as a result of some pervasive unclarities in the argument it is difficult to asses this claim for novelty.[72] (However the discussion in Section 5 of empirical laws and definitions in QM is of independent interest. To my knowledge this issue has not been addressed elsewhere in the philosophical literature.)"*

By 'unclarities' the reviewer really means *errors in my view from his perspective.* He demands that I explicate certain concepts *the way that he thinks they should be explicated.* He is already captive to a certain view and a certain agenda, which

---

[71] Actually in my article I emphasise that the problem has a longer history, and that others, including de Beauregard and Racah, and Watanabe in other contexts, have focussed on the issues decades earlier. Callender has a habit of failing to acknowledge earlier writers. On the other hand, he has succeeded in raising attention to these issues, while the earlier and more sophisticated writers have been ignored.

[72] But I don't make claims about novelty, I explicitly refer to de Beauregard, Racah, Callender. *"The reader is led to believe that the author has a novel argument ... "* is just a sneaky way of insinuating that somehow I am being dishonest. In reality however my argument is quite different to these others.

he thinks now defines the subject. But he spends almost no time examining my argument – he quotes practically nothing from *the text in front of him* - preferring instead to press his own version of how he thinks this problem should be solved. On all these points, his assumed explanations are *not* how I think the concepts should be explained at all. In any case, it is *not* the point of the article to explain all the things he talks about: the point is to present a certain analysis and argue for certain conclusions. The problem is that my paper does not fit the reviewer's broader *agenda* in this field.

This reviewer spends pages raising complex points and explaining his own ideas. His review is almost long enough to be a paper itself! Although more sophisticated than the other hostile reviews, it shows the same bias *of assuming the reviewer's own preferred solutions are correct, and censoring the discussion because it does not conform to them.* Like the two previous hostile reviewers, he does not seem to realise that *it is not the reviewer's job to settle such a complex level of dispute in a review.*

If these points are interesting and challenging enough to write so many pages about, perhaps he should allow publication of my argument, and *submit his objections in an article of his own.* Specialists typically make this mistake: thinking that solutions that seem obviously right to them are thereby conclusive. But the primary point of publishing material on controversial topics to allow different points of view to be put forward.[73]

It should be said that this reviewer does raise real issues, and they need to be dealt with in detail – his views are by no means to be simply dismissed. However they can be replied to in detail. I agree with de Beauregard's judgement that while there is a lot left to say on the topic, "No additional explanation is recommended because it could only be too short or too long." I specifically address some of these problems in (WEBREF Holster 2003 (l)), and more systematically in (WEBREF Holster 2003 (n,o)). But this is not the point of my paper: I am giving a positive argument of another sort. Reviewer 4's insistence that discussion must revolve around the framework of his own ideas reflects his preoccupation with steering the discussion to his own agenda, not with evaluating the paper for publication in its own right.

---

[73] The same censorship is well known in discussions of the interpretation of quantum mechanics. How many papers have been suppressed because an author argues for some solution to real problems of incompleteness, e.g. that conditions for wave function collapse are not specified in the theory, but the reviewer is convinced that the positivist-Copenhagen interpretation solves all these problems, and rejects all other arguments as worthy of publication? How many of the brightest students have left physics over the last decades because of such attitudes?

The power of precedent.

What have seen in these reviews is the typical behaviour of the modern peer reviewer: *censoring ideas that do not support their own agenda.* Academic *philosophy of science* is overwhelmingly about *agendas,* and once a paper on a certain idea is published, it represents a *precedent.* Once a view or a theory is published, others are able to treat it as authentic, refer to it for support, etc. It doesn't particularly matter what its quality is, it gives a quasi-objective point of reference. In philosophy especially, everything can be twisted with jargon, and academics know that publication of a *plausible* alternative view can harm their own agenda. *This* is the primary filter that peer review represents.

Proposing a *new idea,* getting it published the first time, is very hard: but once published, it is much easier for others to write about the same idea, and it can quickly snowball out of control. If my paper was published, it could help undermine the AJP reviewer's agenda. By suppressing publication, it may take years or decades for anyone else to realise there is an alternative solution, and by then his own agenda will be well advanced. Modern philosophy is primarily a competition between personal agendas, wrapped up in ideological programs. It is like party politics. This is really why heterodox views are so severely repressed, both in science and philosophy.

I now give a second example of the *power of precedent,* that I used to my advantage to get another paper in this field published. The previous paper is by no means my central criticism of the modern treatment of time symmetry in physics. In fact, it is on a rather abstract issue, and does not directly explain the phenomenon of irreversibility. (It is the fact that physicists and philosophers have *failed to come to terms with the logic of the analysis of time symmetry* that is so hard for them to admit). But there is a far more striking and important failure in the orthodox treatment of time symmetry of *probabilistic theories,* of which QM is the prime example of a *foundational theory,* and thermodynamics is the prime example of a 'secondary theory'.

The story in brief is that physicists have long adopted (since Lewis, 1927) a certain criterion for *judging the time reversibility of probabilistic theories,* and applied this to quantum mechanics. On this criterion, QM appears 'time reversible'. But the criterion is wrong – it does not represent time reversal symmetry at all – in fact it does not even represent a *symmetry transformation.* The physicist Satosi Watanabe emphasised this in a number of papers in the 1950's and '60s (Watanabe, 1955, 1965)[74]. He showed that on the correct time reversal of probabilistic laws, QM is *irreversible,* and argued that this is the reason the physical world around us is filled with irreversible processes (as thermodynamics, the second irreversible theory, also tells us). Watanabe's work is known to specialists, but has been long ignored: overridden by a

---

[74] Also the only physicist I think I have ever seen who refers to Frege.

brute force of opinion in physics in the mid-1960's, and hidden under the blankets by most philosophers who are aware of it. I also note (Healy, 1981) is a classic forgotten paper on this topic by a philosopher of physics. This earlier material is much better than more recent work in the 2000's that tries to reanalyse the problems all over again.

In my PhD thesis (WEBREF Holster 1990) I gave logical proofs that *the orthodox criterion is neither a necessary nor a sufficient condition for time reversibility*. I demonstrated the correct criterion for time symmetry, and that QM fails this symmetry, confirming Watanabe's views. I should add that the orthodox criterion is blatantly wrong: it is like claiming that being an *odd number* is the criterion for being a *prime number*. (It is neither a necessary nor sufficient condition: *2* is prime and not odd (so oddness is not *necessary* for primeness), while *9* is odd and not prime (so oddness is not *sufficient* for primeness).) This blunder is contained in a vast number of papers and books on physical time symmetry, up to the early 2000's, when for some reason appeal to the orthodox analysis large dried up – with no explanation or acknowledgement that anything was wrong with the earlier 'authoritative treatments' by writers like Grunbaum, Davies, Sachs, Sklarr, and many others.

This result is published in the New Journal of Physics (Holster, 2003). It took 15 years of attempts and multiple submissions to get it published - even though it is an easily verified mathematical-logical proof. A key mistake I made was submitting it to *philosophy of science journals*. Reviewers gave an astonishing mixture of responses.

- One referee (for AJP) recommended publication – others for the same journal rejected it.
- About 30% of peer reviewers rejected the question out of hand as absurd.
- About 30% rejected it by providing counter arguments of various kinds: repeating the conventional analysis; arguing the result was inconsequential; appealing to fallacious positivist principles.
- About 30% thought the result uninteresting, the paper too difficult to appeal to their readers, that it lacked clarity, etc.

In these versions, submitted to philosophy of science journals in the 1980's and 90's (as well as the AJP), I referred to Watanabe's papers from the 1950s, so there was a definitive point of reference. But because this work was so old, it was not taken seriously, probably being thought to have only historical interest. Anyway, the physics community had cobbled up its own dismissive response to Watanabe's arguments in the 1960s – which was partly why I set out to give a more rigorous logical analysis.

However Callender managed to publish his (2000) paper, questioning the orthodox criterion, and proposing a new criterion - but another fundamentally wrong one. His paper has no formal proofs for his claims, and lacks rigor. But it flows nicely in the casual style of argumentation that philosopher like. This gave me a *recent precedent for discussing the problem.* I could now introduce my analysis up-front as *correcting an error in Callender's treatment,* and go on to give the rigorous treatment of time reversal that I wanted to in the first place. In the end, the *NJP* obtained four reviews, of varying quality, and raising a host of different issues. I was probably lucky that the first couple of reviewers were positive. In the end, all recommended publication. I still had to answer a bundle of objections and questions, shorten the article, etc.[75]

This example illustrates why reviewers with agendas resist publication of alternative points of view so strongly. Were it not for Callender's flawed paper in 2000, I may not have had any chance to publish this paper. If I had only Watanabe's (1955) and (1965) references to fall back on, referee's would typically say something like: *"Watanabe's arguments were criticised in the 1960's and there is no further scientific dispute possible about this question."*

The advice for the newcomer trying to publish heterodox ideas is: *Avoid claiming originality or acknowledging heterodoxy of ideas – reviewers hate that! Remember: the peer reviewers are the <u>authority</u>! At least in their own minds. Try to find precedents that you can present as needing improvement, discussion, rejection, etc. Claim originality only for minor points of argumentation.*

---

[75] One disturbing event was that the sub-editor who reformatted and micro-edited the paper replaced all instances of the technical logical term term: *if and only if* with the single term: *if,* throughout a series of logical arguments! This rendered the arguments invalid. The sub-editor thought these were logically equivalent, and s/he was being helpful by simplifying the style!

Peer review example 2.

Formal incompleteness of object languages of physics, and intensional semantic for physics applied to the time reversal transformations. Part 1 and 2.

I now move on to another paper, split into two (WEBREF Holster 2003 (n), (o)), which illuminates the previous problems from a deeper perspective. This study is an example of another characteristic of heterodox thinking: *the synthesis of ideas from different specializations.* In this case, I apply a specialised logical tool, *intensional logic,* to analyse theories of physics, using the troublesome treatment of time reversal as an example. The result illustrates both areas: it shows the deeper logical origin of problems of analysing time reversal, and it shows how intensional logic applies to theories of physics, an area it has not been applied to before as far as I am aware.

The result, unfortunately, illustrates another primary force against heterodoxy: the intense dislike orthodox specialists have for anyone crossing the boundaries of their specialisations – and their amateurism when they try to review such work. This is reflected in the extreme difficulty of getting anyone to review such work – when logicians know nothing of physics, and physicists know nothing of logic or semantics, who on earth is going to review such a paper that brings the two together? We see below.

To explain the context a little more, the foregoing problems with time reversal in physics show there is a big hole in the conceptual foundations of physics. *There are very poor proofs in the subject,* resulting in ongoing controversies that no one is able to settle with ordinary 'proofs', as normally done. You would expect that the application of time reversal to states, processes, laws and theories would be formally defined well enough to allow decisive proofs of results. But this is patently not so. People end up in endless *semantic* arguments about the basis of such proofs.

My paper shows that there is a gaping hole in the *formal language and formal logic of physics itself.* How can this be? Isn't physics done in *mathematical equations?* Isn't its logic well established?

Well, this is where modern logic enters the picture. The watershed in modern logic, in 1971-72, was the formal development of *intensional logics,* independently by Montague and Tichy. This reflects the realisation that *for* _empirical_ languages – *as opposed to purely mathematical languages - we need an* _intensional_ *logic – as opposed to the extensional logic that suffices for mathematical languages.* Few physicists will have a clue what this statement means, and quite understandably. But philosophical logicians of any competence should know what it means. Most intensional logicians use 'Montague logic', which is one way to construct an intensional logic. I prefer Tichy's *TIL (transparent intensional logic),* which has explicit objectual quantification over worlds (or world-times), treated as a domain of entities, rather than Montague's symbolic quantification, as well as other serious advantages. But despite these technical differences, the main lesson of both is the same in this respect at least. We must quantify explicitly over worlds (or world-times) when we come to analyse or formalise the logic of *empirical* languages. Physics is an empirical language.

I show in this (pair of) papers that a fully adequate *formal treatment* of time reversal symmetry (or space reversal symmetry, or any similar meta-properties of theories) requires the extension of ordinary object languages of physics to an intensional language. Part 1 shows that the usual extensional language is incomplete. Part 2 shows that an intensional language is adequate – at least to represent symmetry transformations, like time symmetry.

This is intimately connected to the controversies discussed in the previous papers, which relate to the analysis of time reversal. These problems are now put in a larger context. The view supported here is that:

(i)   Time reversal has an objective and universal definition across all theories: it means the transformation that reflects time on itself: $T: t \rightarrow -t$, and induces transformations in all entities that have time as part of their constructions.

(ii)   The problem of analysing the time reversal of a theory is not to decide what *time reversal* is but to *interpret the theory*, in the sense of making explicit what role *time plays* in the constructions its terms refer to.

A common view now held among specialists is precisely the opposite: that *the concept of time reversal is specific to each theory, and time reversal for QM may be arbitrarily chosen in a quite different way to time reversal for classical theories or relativistic theories, etc.* This is what happens in practice: physicists make arbitrary choices of 'time reversal' to suit the conclusions they want to draw. But that means these conclusions are not objective facts about the theories, they are subjective choices. They are made by physicists and philosophers who want the theories to be time symmetric.

The fact that the language of equations in which theories of physics are expressed is *formally incomplete* in this respect does not appear to have occurred to anyone in the philosophy of science, even though it seems quite obvious from the point of view of intentional logic. This is because physics and modern logic are completely distinct *specialisations*. There seem to be no physicists or philosophers of physics who know what intensional logic is – and no intensional logicians seem to have an interest in applying it to physics.[76]

Is it so unlikely that there could be such a logical hole in the formal languages of physics? Well, there was a big hole in the ordinary treatment of

---

[76] The closest most philosophers of physics get to intensional semantics is *model theory*, as used by Earman and the Pittsburg school; but this is quite inadequate.

differential geometry, in the context of General Relativity, that no one noticed for *decades* – reflected in the fact (as López Corredoira notes P.70) that Chandrasekhar's pioneering work on black holes was rejected for decades. This was reflected in controversy over the physical possibility of black holes – essentially whether the event horizon is a 'physical singularity' or a 'removable singularity'. It took until the 1970s for this to be resolved, and the "good ol' boy", rough and ready applied mathematics of ordinary physics was found to be sorely wanting. A new formal treatment was established, as popularised by Spivak (1979). This 'modern treatment' is far more explicitly *semantic* than the rough and ready approach to applied maths we see throughout physics generally. But it is still inadequate as a semantic langauge, and a formalism supporting *intensional logic* is required.

I also have considerable confidence in the arguments in this paper (especially Part 1) because it was meticulously reviewed before submission by *two outstanding logicians in the field of intensional logic,* viz. Dr. Marie Duzi and Professor Pavel Materna of the Czech Republic. They corrected a number of errors of expression I had made in Part 1, and greatly improved some details of the paper. They confirmed the main proofs and conclusions. They were more reticent about my approach in Part 2, not because they disagreed with the main aim or conclusion – viz. that a proper logical treatment allows an objective logic for the time reversal operator to be given in physics - but primarily because of certain subtle philosophical issues, such as the treatment of the *actual world* and *present time* operators. They may well be right: but this is part of the open development of a new subject, and these subtle issues have nothing to do with the childish misunderstandings demonstrated by the JPL reviewers below.

Submission to JPL

I submitted this (pair of) paper to the *Journal of Philosophical Logic* and *Philosophy of Science* in 2003. It was duly rejected, with three reviews.

Official Journal of the Philosophy of Science Association

Philosophy of Science
Noretta Koertge, editor

Dr. Andrew Holster
1 Haunui Road
Pukerua Bay
Wellington
New Zealand

August 8, 2003

Dear Dr. Holster

We have completed the evaluation of your revised manuscript, "Incompleteness of Extensional Object Languages of Physics and Time Reversal Invariance. Part 1." (file number 03.07.08.1). I regret to inform you that it has not been accepted for publication; please see the enclosed reports for considerations offered by the referees.

However the issues you are thinking about are important and worthy of further study. Best wishes in your future philosophical inquiries.

Sincerely yours,

Noretta Koertge
Editor

123 Goodbody Hall, Indiana University, Bloomington, IN 47405-7005
(812) 855-3539   pos@indiana.edu   http://www.indiana.edu/~philsci/

It is difficult to understand why the editor would make the positive comment that 'the issues you are thinking about are important and worthy of further study' on the basis of the following reviews. I begin with the most serious review, which is not completely negative.

Reviewer 5. JPL

Review of "Formal Incompleteness of Extensional Object Languages of

to Time Reversal Transformations"

This paper is made up of two parts. The first argues that physics uses extensional object languages and that this is the source of the problems physicists have had making sense of time-reversal symmetry. The proposal is to provide an intensional semantics sufficiently rich to provide the needed notion of contingency. This is done in the second part of the paper for a toy physical theory.

This paper is relatively clear and interesting and on an important topic, but I cannot recommend it for publication. In the first part of the paper, the case is not well made, I think, that the language of our best physics is purely extensional. Neither do I believe that a good case is made that the extensional features of our physical language lead to bad physics. And finally, the bulk of the second part is concerned with a toy example of the problem of time-reversal for physical theories that may or may not have anything to do with time-reversal symmetry in real physical theories. But it is the second part of the paper that is the more interesting.

The main argument of the second part of the paper seems to be that the intensional semantics cooked up in the context of the toy example will help us with the difficult question of how time-reversal should work in real physical theories. We do not get to the real physics until part two section 13; but then the answer is, as one might have suspected from the start, that the way one understands time-reversal in a specific physical theory must be sensitive to one's interpretation (or understanding of the content) of the theory. Indeed, when one considers theories like classical quantum mechanics, it seems to me at least that one can salvage virtually nothing from the toy example. And the relevance of the toy example is all the more puzzling in the context of any relativistic theory.

When the nature time itself is at stake in our best physical theories, it is implausible to suppose that there can be any single, canonical, trans-theoretical notion of time-reversal. The project then should presumably be to do the detailed work necessary to understand time-reversal in the context of each important physical theory and interpretational variation.

The paper is quite long: about 70 single-spaced ms. pages in all. I would under no circumstances advise that all of this be published in a serious academic journal.

Reviewer 5

My rebuttal of Reviewer 5.

Here is a brief response to the review: please forward to the reviewer as you deem appropriate. I would not normally reply to a review, but in this case it is so poor that it completely misrepresents the paper.

Thanks,

Andrew Holster.
>
> ================================
>
> Review of "Formal Incompleteness of Extensional Object Languages of
> Fundamental Physics, and Intentional Semantics for Physics Applied
> to Time Reversal Transformations"
>
> This paper is made up of two parts. The first argues that physics
> uses extensional object languages and that this is the source of the
> problems physicists have had making sense of time-reversal symmetry.

Not exactly: the first part gives a proof that it is impossible to formulate a deductive system for obtaining time reversals of propositions in an extensional object language for physics; hence the physicist's attempts to establish deductive proofs of their results in this area using an extensional object language cannot succeed.

> The proposal is to provide an intensional semantics sufficiently rich to
> provide the needed notion of contingency. This is done in the second
> part of the paper for a toy physical theory.

Not exactly: I specifically do not claim to make a complete intensional semantics for physics: rather, I demonstrate that an intensional semantics leads to the natural solution of the problem, and more importantly, I offer a proof that in an adequate compositional language there must be a general time reversal operator with a simple distributive syntactic property that allows deductions of time reversals.

>
> This paper is relatively clear and interesting and on an important topic,
> but I cannot recommend it for publication. In the first part of the
paper,
> the case is not well made, I think, that the language of our best physics
> is purely extensional.

Can the reviewer give any examples of non-extensional object language for ph ysics?
The analysis of time reversal is routinely done in an extensional language.

17/04/2003

Rebuttal p.1.

(Rebuttal cont.)

>Neither do I believe that a good case is made that
> the extensional features of our physical language lead to bad physics.

I give some examples of real problems in Part 2, which the reviewer ignores.

> And finally, the bulk of the second part is concerned with a toy example
> of the problem of time-reversal for physical theories that may or may
> not have anything to do with time-reversal symmetry in real physical
> theories. But it is the second part of the paper that is the more
> interesting.

What the reviewer derisively refers to as a 'toy theory' is in fact a simple
example of a theory (a very simple variant of Newtonian gravity): the point
is to prove that the physicist's syntactic method of deducing time reversals
cannot work for this simple theory, since it leads to contradictions. The
choice of a simple theory to prove this is deliberate, and perfectly
adequate to prove this generally. The reviewer appears to misunderstand the
concept of a proof: it is not necessary to reprove this result for every
theory of physics: the result is a general one. It is preferable to choose
the simplest theory adequate for the proof.

> The main argument of the second part of the paper seems to be that
> the intensional semantics cooked up in the context of the toy example
> will help us with the difficult question of how time-reversal should work
in
> real physical theories.

No. First, I offer a general schema for an intensional semantics: it is not
claimed to be complete, but it provides an intensional translation for the
fundamental terms of basic physics, including trajectory functions, time and
space differentials, and real and complex fields. This is clearly quite
general: it is not limited to the simple example that is analysed. Second, I
return to the example examined in the first part of the paper to illustrate
how providing an intensional semantics works to solve the problem. Third, I
go on to consider whether this can be generalised to a range of other
theories, (and I point out that it leads to real differences with the
conventional analysis of time reversal of quantum theory and probabilistic
theories). Fourth, I provide a general proof that in a compositional
language it must be possible to define a general syntactic time reversal
operator. The reviewer ignores the claims actually made in the paper.

>We do not get to the real physics until part two
> section 13;

The logic of physics is a part of real physics.

>but then the answer is, as one might have suspected
> from the start, that the way one understands time-reversal in
> a specific physical theory must be sensitive to one's interpretation
> (or understanding of the content) of the theory.

Absolutely not: this is a blatant misrepresentation; and the real point is
explained in great detail in the paper. The conclusion is that , first of
all, we have to interpret the formalism of a theory before we have a

Rebuttal p.2.

(Rebuttal cont.)

physical theory, i.e. a theory that states anything about the physical world. But once we have provided an interpretation, the time reversal transformation is objectively defined: we do NOT - and we CANNOT - subsequently choose a 'time reversal' operator to suit ourselves.
This is the major point of the paper!

>Indeed, when one
> considers theories like classical quantum mechanics, it seems to
> me at least that one can salvage virtually nothing from the toy
> example.

Misinterpretation.

>And the relevance of the toy example is all the more
> puzzling in the context of any relativistic theory.

Again: the reviewer misunderstands the role of the simple theory.
>
> When the nature time itself is at stake in our best physical theories,
> it is implausible to suppose that there can be any single, canonical,
> trans-theoretical notion of time-reversal.

The main proof in the second part of the paper claims to demonstrate precisely the opposite: that in a compositional language, it must be possible to define a syntactic operator to represent the time reversal transformation.

>The project then should
> presumably be to do the detailed work necessary to understand
> time-reversal in the context of each important physical theory and
> interpretational variation.

This is the (conventional or orthodox) view that the paper disproves: the reviewer takes it as a fact, or a foregone conclusion, without ever touching on the proofs offered against it.
>
> The paper is quite long: about 70 single-spaced ms. pages in all.
> I would under no circumstances advise that all of this be published
> in a serious academic journal.
>

This review is completely inaccurate. It shows a complete misunderstanding of the paper. It is simply a reflection of the reviewer's own preconcieved ideas. It fails to give any analysis whatsoever of the proofs that the claims in the paper are based on. It shows very clearly that the 'experts' on the physics of time reversal are indeed badly in need of some help from logicians and philosophers if they wish to obtain any clarity about these problems.

Rebuttal p.3.

Reply from Reviewer 5 to my rebuttal.

Below is the reviewer's subsequent response to my objections to his review.

> > >The project then should
> > > presumably be to do the detailed work necessary to understand
> > > time-reversal in the context of each important physical theory and
> > > interpretational variation.
> >
> > This is the (conventional or orthodox) view that the paper disproves: the
> > reviewer takes it as a fact, or a foregone conclusion, without ever
> > touching on the proofs offered against it.

I have no comments for the author.

As the author notes (above and elsewhere), I do
indeed believe that there can be no single, canonical,
a priori notion of time-reversal symmetry when the nature
of time is itself a subject of empirical investigation. This
is not a position against which "proofs" can be offered.
Rather, it represents a commitment about the proper
source and nature of physical knowledge. While the
author refers to this as the "(conventional or orthodox)
view," (unfortunately) not everyone smart agrees with
my view of how one goes about learning about the
nature of physical concepts like time. This said, I
don't think the people on the other side of this issue
would recommend publishing these papers either.

My Response to Reviewer 5's reply.

> I have no comments for the author.

In my original comments to the reviewer, I observed that the reviewer blatantly misinterprets the content of my paper; makes highly opinionated claims without offering any justification; and fails to consider the proofs and arguments that are actually offered in the paper. If the points I made were wrong, I have no doubt the reviewer would have been able to reject them, and would have taken the trouble to do so.

> As the author notes (above and elsewhere), I do
> indeed believe that there can be no single, canonical,
> a priori notion of time-reversal symmetry when the nature
> of time is itself a subject of empirical investigation. This
> is not a position against which "proofs" can be offered.

This is deeply ironic as a comment sent to a journal of logic! In reply to my observation that the reviewer has failed to consider the proofs offered in the paper, and merely reiterated their own opinions against them, they have claimed that their own opinions on this matter are simply not open to disproof!

> Rather, it represents a commitment about the proper
> source and nature of physical knowledge.

The reviewer feels that, rather than being required to offer reasoned arguments about controversial views, they are justified simply by making a 'committment' about the 'proper source of knowledge'. God help us if this is what passes for the philosophy of science.

>While the
> author refers to this as the "(conventional or orthodox)
> view," (unfortunately) not everyone smart agrees with
> my view of how one goes about learning about the
> nature of physical concepts like time.

What amazing arrogance. It is against scientific and philosophical principles to reject reasoned debate about controversial questions simply by referring back to your own assumption that you are right. The reviewer is

My response p.1.

(Response cont.)

full of opinions: "I think that...", "my view is that..."; they clearly
think that their own opinions are so authoritative that they require no
further justification.

>This said, I
> don't think the people on the other side of this issue
> would recommend publishing these papers either.
>
This is a remarkable claim: the reviewer thinks that even their opponents on
this issue, who would support the main conclusion argued for in the paper,
would not want the paper published! What a fantastic reason to reject a
paper that opposes your own opinions!

In summary: Given: (i) their blatant lack of objectivity, (ii) their
arrogance about their own infallibility on controversial questions, (iii)
their unashamed misrepresentation of arguments they disagree with, and (iv)
their complete failure to deal with the real content of the paper, in favor
of merely taking the opportunity to express their own opinions, this
reviewer shows themselves completely unsuitable to referee papers on logic,
philosophy, or science.

Such unashamedly subjective and self-serving reviewers exert a dangerous
influence over the publication process, and jeopardise the validity of the
peer review process.

Andrew Holster.

My response p.2.

The JPL provided two further reviews, below.

Reviewer 6. JPS.

Referee #1 Report on 03.07.08.1

"Incompleteness of Extensional Object Languages of Physics and Time Reversal Invariance. Part 1."

The author makes the claim that "ordinary extensional object languages of physics are too weak to construct adequate syntactically-based system of rules for making formal deductions of the effects of general transformations (such as time reversal) on propositions." I don't find the claim, or the attempted proof of it, interesting since an adequate treatment of time reversal invariance in physics doesn't require the time reversal transformation of propositions.

The author generates conundra by assuming that the time reversal invariance of a theory T means that TT is logically equivalent to T, where T is the time reversal transformation. This requires, of course, that T be an operator on propositions. But the usual approach in physics is to take T as a transformation of states and, thence, of histories or sequences of states. Time reversal invariance of a theory T then means that whenever a history satisfies T, so does its time reversed image. This formulation bypasses the various difficulties the author tries to generate. Curiously, he/she refers to this approach on p. 28 of the ms!

The author tries to make a mystery of the standard time reversal operation in QM. He/she claims that "there are no valid deductive proofs that [this operation] is the time reversal operator in QM" (p. 10). I disagree. In the case of a spinless particle, the time reversal operation in QM is picked out by the correspondence with classical mechanics: in classical mechanics T reverses velocities; to get a reversal of the velocity of the quantum wave packet the phase relations have to be changed—this is where complex conjugation comes in.

The author tries to make a mystery of the time reversal operation for probabilistic theories. One problem here is that no fundamental theory of physics is a probabilistic theory in his/her sense. He/she assumes that such a theory of formulated in terms of conditional probabilities. But QM is not such a theory—it uses transition probabilities rather than conditional probabilities which are generally not well defined. When one tries to translate the author's mutant definition of time reversal invariance of probabilistic theories (TRI, p. 9) into a statement about transition probabilities, it becomes nonsensical. The usual definition of time reversal invariance in QM is provably equivalent to the equality of transition probabilities and their time reversed counterparts.

In sum, this paper is an exercise in mystery mongering. There are some genuine mysteries about time reversal, but this paper does not manage to locate them.

Reviewer 6.

Reviewer 7. JPS.

Referee #2 Report on 03.07.08.1

"Incompleteness of Extensional Object Languages of Physics and Time Reversal Invariance. Part 1."

A major weakness of the paper is that it contains surprisingly few references to the earlier literature. For example, on pp. 7-8 the author considers and discards what she calls the "Orthodox Criterion for Reversal Symmetry". If I have understood the contents of this criterion correctly, it is absurd for obvious reasons, and the paper does not contain references to any authors who had ever supported this criterion! Hence, the point of this discussion seems quite unclear.

Reviewer 7.

A response to Reviewers 6 and 7.

I did not respond to these two reviews, because they are so extraordinarily poor, but I will give a response here.

Reviewer 7 (JPL referee #2) does nothing to evaluate the paper whatsoever! Interestingly, they raise no issues about the claims made in the paper about the analysis of time reversal itself (the primary concern of the earlier Reviewer 5). Contrary to their statement, references to the 'Orthodox Criterion' for reversibility *are* given, and anyway it is common knowledge throughout the subject. It is especially ironic the reviewer thinks the Orthodox Criterion is "absurd for obvious reasons" – whereas it is adopted and repeated in the scientific and philosophical literature for decades, and proved difficult to publish a disproof of it for this very reason! And on this point, Reviewer 6 takes the opposite view, and still believes the Orthodox Criterion is correct!

It is obvious Reviewer 7 has no familiarity with the subject of time reversal. But in any case, as made very clear in the text, the point is not to prove the failure of this criterion here, as I refer to other published proofs of this result elsewhere. The point is to use time reversal as a real concrete example, to show something much more general: the logical inadequacy of the extensional object languages for physics.

Reviewer 6 starts with the extraordinary claim that "I don't find this claim … interesting since an adequate treatment of time reversal invariance in physics does not require the time reversal transformation of propositions"…

"[The author's analysis] requires, of course, that T be an operator on propositions. But the usual approach is to take T as a transformation of states…".

These statements are nonsense. First, my treatment generalises T as a transformation on *all kinds of entities* – including *propositions, states and processes*. The point is to be able to define a general *syntactic* transformation on *linguistic terms* that correctly induces the time reversal transformation *semantically*, i.e. in the *entities they refer to*. This is why it is about the *logic of the T operator*, not the *physics of time reversal*.

But in any case, *physical laws are stated as propositions, and the key interest in the subject is <u>precisely</u> in whether various laws of physics are time reversal invariant or not*. I think every serious researcher in the subject recognises that *we need to be able to analyse the transformation of propositions*. The reviewer's claim that we can analyse the reversibility of a theory by analysing whether reversed processes satisfy the theory is precisely a method for analysing the reversal of the proposition of the theory! It is just a simplistic, informal and generally inadequate.

My paper proposes a general time reversal transformation that applies universally to all constructions – propositions, states, processes, etc. The reviewer may as well object to predicate (quantificational) logic by saying "I do not find this treatment of logic interesting because we do not have to use predicate logic in physics". But of course *you do have to use predicate logic in physics – it is represented explicitly in the mathematics, and used implicitly all the time in physics*. Physicists may not think they use it, because *quantificational symbols are not usually written explicitly in physics equations – which are only semi-formal*. But in reality they use quantificational logic all the time.

The paragraph about 'transition probabilities' not being 'conditional probabilities' is also nonsense. In fact practically every sentence in this review is nonsense. The reviewer makes numerous claims without offering justification, and says practically nothing about the content of the paper. It is clearly stated that the point is not to prove the claims about time reversal *per se*: it is to use it as an example to show something about *the logical power of the formal languages of physics, and how this can be enhanced*.

The question that might be asked of both reviewers is: *do you know what intensional logic is?* The honest answer in both cases will be: "*No, I've never heard of it before. I was kinda winging it on that one….*" But this is the most basic requirement to review the paper, since *this is explicitly what the paper is about*. Reviewer 6 appears to be a physicist with no idea of what *logic* is. He is affronted by the fact that the paper calls the orthodox analysis of time reversal in physics into question. Reviewer 7 appears to have no idea about either physics *or* intensional logic. Reviewer 6 concludes: "In sum, this paper is an exercise in mystery mongering. There are some genuine mysteries about time reversal but this paper does not manage to locate them." This is the response orthodox academics give to heterodox thinkers who ask questions that go beyond the dogmas of their specialization.

Summary of peer reviews.

Apart from Reviewer 1 for *Synthese* and Reviewer 4 for *AJP*, all the other reviews show alarming incompetence. This is especially obvious with the last three – where the topic explicitly crosses two subjects, *physics and logic,* and it is clear reviewers have no competence in both (or even one for that matter).

- *There is <u>no value</u> in peer reviews of such low quality.*
- *There is <u>no point</u> submitting to journals that provide such peer reviews.*

Why on earth did these academics agree to review the paper in the first place? They could have said: *I don't know anything about intensional logic,* or: *I don't know anything about physics.* It has occurred to me that perhaps it took *Synthese* so long to process the first paper because they found it difficult to find reviewers. Perhaps some quite sensibly said: *I don't know enough about this issue to give a judgement.* It is then left to the very arrogant, who think they know everything, to be the peer reviewers of the most challenging material. The reason these incompetent reviewers accept the task is the same reason their reviews are so poor: *the arrogance and vanity of many academic experts.* Well, that's not exactly news to anyone.

What strikes me more generally is that most peer reviewers treat articles *as if they, personally, are the intended audience.* Their egos are at the center of the world. *"If it is not appealing to me, it is not interesting to anyone else".* It is obvious in the examples here that the hostile reviewers have no interest, but there are others who *are interested in these topics.* Such articles are intended for *these* people. Reviewer 1, Costa de Beauregard, is obviously very interested – and Reviewer 4 is also very interested, he just does not want to see my view published. Such people, open to entertaining heterodox ideas, are rare enough that it is unlikely you will get one to review your article by chance – and if you do, the Editor is most likely to engage a second opinion, like Reviewer 2, who will still reject it.

There are distinct audiences for the journals: the strictly orthodox (perhaps 95%) and those open to heterodoxy (perhaps 5%?). Journals pander to orthodoxy at the expense of heterodoxy, as the music industry panders to pop music at the expense of 'alternative' music. That is their core business in a world controlled by money and popularity. But if everything is dominated by the tastes of the masses, what outlets are there left for alternative views? Are there any peer-reviewed heterodox journals in physics or philosophy? None with any significant ranking. You do not get job credentials by publishing in heterodox journals – it is detrimental to your prospects. You get credentials by publishing in high-ranking *pop-science journals.*

I can also guess what many academics will say about these reviews:

"This is nothing special, it happens to me all the time, it is normal to be rejected with poor reviews. We all know that academic reviewers accept or reject articles according to their own tastes. Instead of complaining about it, you simply send it to another journal. I myself usually have to send papers to about 5 journals in a row before being accepted... you should see some of the ludicrous things reviewers have said about some of my papers that have then been accepted by other journals!"

Which confirms my point. These reviews are incompetent but *this is nothing special*. What is the conclusion of this? One conclusion is that if papers are sent to 5 journals on average before either being accepted or abandoned, we have a massive duplication – a *5-fold duplication* - of peer review activity for *every paper submitted to a journal*. Only a small faction of submissions are finally published, so despite the flood of published papers, it is still only a trickle compared with the deluge of peer reviewing activity. It is hardly a wonder that quality is lost in this frantic activity – every academic and researchers gets roped into the peer review process. It is a statistical inevitability that most reviewers are incompetent when confronted with something new.

The general failure of peer review.

There is overwhelming evidence for the failure of the peer review system. López Corredoira emphasises that it is especially a problem for original work – for the *best work*, the most *critical work for the future*, work that challenges old dogmas and opens new doors.

"As noted by Van Flandern (1993), peer review in journals interferes with the objective examination of extraordinary ideas on their merits. Maddox (1993), who was editor of the journal *Nature*, has said that if Newton submitted his theory of gravity to a journal today, it would almost certainly be rejected as being too preposterous to believe. On the one hand, there is a failure to select novel ideas (Brezis, 2007; Horrobin, 1990). P.69.

"There are many examples of papers unfairly rejected by journals. Nobel Prize-winner Subrahmanyan Chandrasekhar produced a theory of stellar development which was not published because of strong opposition by the establishment. Twenty years later the theory was developed again and published by others (Wali, 1991). The biologist Lynn Margulis is today recognized for her theory of the origin of eukaryotic cells as a consequence of the symbiotic union among different prokaryotic cells, which was published only after being rejected by about fifteen journals (Brockman,, 1995). P.70-71.

Examples can be multiplied almost endlessly: take any significant novel discovery in the last 50 years, and you will usually find a peer review scandal lying behind it. If so much of the work now recognised as the most significant is suppressed before eventually coming to light years later – how much similar work *remains permanently suppressed, and has never seen the light of day?* How many researchers with genuine originality and insight have their careers curtailed by this?

Yet the peer review process is constantly appealed to as the guarantee of scientific quality. To take the extremist scientistic community, look at any sceptic or debunker site and you will find endless statements that: "these crack-pot claims have not been published in *peer reviewed journals, the gold-standard for scientific authenticity."* To the orthodox, peer review makes perfect logical sense as the process to guarantee credibility: *"Check if the ideas are validated by the professional experts in the author's field."* But this of course assumes that the 'professional experts' are competent to make judgements about novel work. This is scientistic propaganda. (And whenever alternative research *is* published in peer reviewed journals, it is dismissed anyway: *the peer reviewers are then said to be incompetent fools.*)

In preceding sections, I have effectively given *peer reviews of the peer reviews.* Why should this not be required in the peer review process too? For peer reviews are *novel works too: they are novel judgements about ideas.* How do we know that *they* are valid? Shouldn't they be peer reviewed too? To mitigate this problem, journals often get multiple peer reviews. In the examples above, we see two journals with double reviews, and altogether, there are four reviews of one paper and three of another – and they all have quite different judgements! (This is how we can tell that most of them must be unreliable.)

In the *Synthese* example, we have two *completely contradictory judgements* received by the journal editor. Yet this simply results in rejection – the competence of the peer reviewers is not questioned. The reason comes back to the boundary of authority invested in titles and offices: if you define someone in the *office of peer reviewer, then by definition of the office their opinion carries authority.* This is part of the wider bureaucratic phenomenon of *investing authority in titles and offices.*

I think others should follow my example and start *publishing lousy peer reviews they receive* – especially for papers that are subsequently shown to have real quality. I think there should be internet sites to publicise bad peer reviews. I expect the result would be a deluge of material. Of course the system protects itself by normally placing this out of bounds – by the convention that *peer reviews remain confidential.* It is time this convention was broken in the name of truth and honesty.[77]

Even if you have 'clicked' a statement of confidentiality, breaking this is justifiable under *whistle blower* legislation. To back up your points, you can put

---

[77] And some open access journals are now starting to do this: see Appendix 10.

your original papers up on *preprint servers,* as I have done. I turn to this mode of publication next, because I think it offers a potential solution to circumvent the censorship of journals. But to conclude this section, I observe that:

- There is overwhelming evidence that the peer review system in science and philosophy of science *is largely a sham.*
- The ratio of incompetent, biased, and wildly inaccurate peer reviews by major journals is *demonstrably much higher* than the ratio of fair and competent peer reviews.
- This applies in both directions: good articles demonstrating orthodox fallacies and proposing novel ideas – *the most important new papers in the field* – are *typically* rejected by peer reviewers; poor articles reinforcing orthodox views and fallacies are routinely accepted.

What on earth is the use of the peer review system at all in the light of this? The answer to that is obvious too: peer review is a tool to prop up power and authority. It is like the system of staged political trials in totalitarian regimes. Many scientists are similarly critical of peer reviews, but typically for allowing the publication of 'nonsense' papers in science journals. Peter Woit in *Not Even Wrong* (2006), which is a powerful critique of the 'string theory' industry, makes highly critical observations of the peer review and journal system. Recounting the 'Bogdanov affair' of 2002 he says:

"… the nonsensical papers of the Bogdanovs … made it into five [physics] journals, not one. This brings into question the entire recent peer-reviewed literature in this part of physics, since the refereeing process is evidently badly broken.

One unusual thing about the Bogdanov papers was that they were never submitted to the on-line preprint database used by virtually all particle theorists and most mathematicians. Fewer and fewer physicists ever look at print journals these days, since essentially all recent papers of interest are available conveniently on the web from the database. The continuing survival of the journals is somewhat mysterious, especially since many of them are very expensive. A typical large university spends over $100,000 a year buying physics journals, the content of which is almost all more easily available on-line for free. The one thing the journals do provide which the preprint database does not is the peer-review process. The main thing the journals are selling is the fact that what they publish has supposedly been carefully vetted by experts. The Bogdanov story shows that, at very least for papers in quantum gravity in some journals, this vetting is no longer worth much. Another reason for the survival of the journals is that they fulfil an important role in academia, where often the main standard used to evaluate people's work is the number of their publications in peer-reviewed journals … The breakdown in refereeing is thus a serious threat to the whole academic research enterprise." P.222-3.

This is rather typical in that physicists are much more worried about their science looking silly for publishing 'nonsense' papers, than for the most critical failure of censoring important heterodox ideas. But Woit recognises the problem goes both ways to some extent. He vividly portrays the arrogance of string theorists in Chapter 16, "The only game in town". This conveys how arrogant physicists can be, and how real the phenomenon of censorship of ideas in physics is. It is interesting too that much-scandalised 'nonsense papers' usually get past journal referees by being so obscured in jargon and 'orthodox' technicality that they are impenetrable. All the heterodox papers I have had rejected are *lucid*. The arguments and positions are clearly stated. It is this very lucidity that is punished.

Preprint sites: *arXiv.org*.

Preprint sites allow authors to publish articles rapidly, without a peer review process, usually while papers are in submission to a journal. But they must still filter materials somehow, and this process is equally open to abuse. Actually there is only a tiny number of successful preprint sites – just a handful – and their controllers have now become powerful figures in the academic industry. López Corredoira has some interesting comments on preprint sites. He focuses on the major physics preprint site, *arXiv.org*:

> "The most important tool for communicating scientific results in physics is the preprint server *arXiv.org*. It is a monopoly within physics and has no competitors. Even most of the papers published [in] journals are posted on this preprint server, and people read them here. The situation is that papers not posted on *arXiv.org*, will receive scant dissemination within the community, particularly when the papers are not published in a reputed refereed journal, which is often the case for non-mainstream positions.
>
> The development of arXiv.org, first at Los Alamos National Laboratory and later at Cornell University, was a wonderful example of freedom of expression between 1992 to 2004 that provided everybody with an open forum in which to post their ideas. There was a small fraction of papers with exotic ideas, but they were very few (5% or less), so they did not disturb the flow of information. P.73.

This freedom did not last however:

"However, after 2004 there was a change in policy and those responsible for the site decided to block the posting of certain contributions. In 2004, a system was introduced in which in order to post something on the site support was requested from a colleague with experience in the field. The system would become more perverse in the following years, forbidding some scientists from giving support when arXiv moderators noted that they had allowed the publication of very challenging heterodox ideas, and creating committees to reject papers without having read them and with the absence of a referee's report: the committees just read the title and the abstract and, if they did not like the content (and normally they do not like anything that smells of heterodox ideas, they channel the paper from the specialized section to their section of general physics 'physics.gen-ph', which is hardly read by anybody. In some cases, they remove the contribution totally, without further explanation (e.g., Castro Perelman, 2008). When asked for an explanation for a rejection, they usually reply with set phrases: "arXiv reserves the right to reclassify or reject any submission. We are not obligated to provide substantive reasons for every rejection, and usually the moderators do not provide more than a sentence or two, often in a form not appropriate for author viewing". This method of censorship of the promotion of new ideas is on a par with censorship in the Middle Age or in certain totalitarian regimes.

Censorship by arXiv has become a real problem since this policy. Ironically, this was exactly the moment when I first decided to post a paper on arXiv, on the application of an alternative theory of gravity (a variation of the Schwarzschild solution of GTR) to the anomalies in the Pioneer spacecraft trajectories. I was blocked by this very first wave of censorship, by a hostile moderator from NASA.[78]

However López Corredoira does not think censorship is the main problem, and continues:

---

[78] The reason given was that the moderator *thought the theory would fail when applied to the precession of the perihelion of Mercury*. On the first page of the paper I had noted that the theory appears consistent with the Pioneer anomalies and with planetary orbits, but emphasised I had not yet applied it to the precession of the perihelion of Mercury. I am sure the moderator did not read the paper: he simply spotted this as an excuse to suppress it. NASA had repeatedly tried and failed to explain the Pioneer anomalies, starting from the mid-1990s. (They have proposed a new explanation in 2014, but I doubt it is robust: it is merely a face-saving gesture.) The moderator himself had proposed another (failed) explanation earlier in 2004. It is against his and NASA's interests to allow an outsider to provide an explanation. My explanation may or may not be correct – I have no reason yet to think it is wrong – and it would be easy to test it conclusively. The reason it was blocked was not because it is shown be wrong, but purely because it threatens the self-interest of a powerful establishment. In any case, *it is a preprint site, not a peer reviewed journal.*

"Actually, the main problem is not direct censorship itself, but the screening action of the massive overproduction of papers, with millions of scientists producing millions of papers every year, ... the multiple creation of subfields within a field, and subsubfields within a subfield, etc. Even within a microspeciality, the number of papers may be around a thousand per year, still a huge amount even to take a quick look at. This number continues to grow in an uncontrolled way... This means that, once the obstacle of direct censorship in the journals is removed, the researcher who tries out new ideas will have to fight with indirect censorship: the super-production of papers that conceal what is not of interest to the system... p.74.

"Propaganda is the key element in a paper becoming known. For this, the leading specialists again have the advantage, because they control most of the strings which move the publicity machinery; they have the appropriate contacts, they write reviews (summaries of scientific discoveries within a field), they organize congresses and give talks as invited speakers. Moreover, the reproduction of standard ideas is more acceptable because many people are interested in them, while the diffusion of new ideas is of interest only to their creators. This is not something new, it has happened all throughout history. The new thing is the institutionalization and bureaucratization of this process. P.75.

López Corredoira supports preprint sites, but observes limitations.

"There are other ways to publish results [than journals], mainly on the web or preprint servers, but there are also filters and the way to achieve recognition with these unofficial publications is also very limited. Somebody might also steal a researcher's ideas, but that also happens with papers accepted in reputed journals that went unnoticed at the time of publication until years later a prestigious author rediscovers them, and picks up on the ideas. How many authors of the old Soviet Union discovered interesting things, which the world did not know about until a clever North American researcher, with plenty of dollars, rediscovered their work. P.73.

Incidentally, this suggestion that *American scientists* would steal ideas from others may have sounded quite scandalous a few decades ago – but today it does not raise an eyebrow. The public now sees science as intimately merged with big business and politics, and expects all three to be corrupt as a matter of course. There is now an expectation of corruption *especially* in cultures that are most strongly driven by Capitalist values. *If there is money or power involved there will be corruption:* that is probably the one point of politics that everyone agrees on today! Cultural expectations, reflecting the 'spirit of the age', have changed dramatically over the last decades. The spirit of our age is cynicism.

In any case, preprint sites exercise varying degrees of censorship, but not only of material. They increasingly censor *individuals*, on the basis of *organisation or institution*. To register in most sites now you must declare your organisation and role, with sites explicitly rejecting people *without academic or government jobs*. This is a far more insidious form of control than censoring ideas. See Appendix 11 for a new movement in this direction: *de facto* systems to control *membership of the research community allowed to submit publications to journals*!

Preprint sites: *philsci-archive.*

I agree with López Corredoira that preprint sites have a major role to play, but have also become prone to corruption in the power struggle, and I now give the example of *philsci-archive*. The reader may have noticed that I have referenced my earlier articles to two philosophy preprint sites: *philsci-archive* and *philpapers.org*. *Philsci-archive* is the most powerful preprint site in philosophy of science. *Philpapers.org* is a much larger general archive for all kinds of philosophy preprints (with almost two million papers). I subsequently contrast the policies of these two sites.

Philsci-archive is controlled by members of the University of Pittsburg *Department of History and Philosophy of Science*, with intimate connections to *Philosophy of Science*, the leading North American journal in the subject. It is controlled by a small clique of individuals at the center of the North American academic power hierarchy in the philosophy of science, with Earman, Norton and Roberts the major power players directly controlling this site. The site opened in January 2001. I posted five articles on it in 2003 – including the rejected articles seen above - and two more in 2004.

1. Holster, Andrew (2004) Time Flow Physics: Introduction to a unified theory based on time flow. [Preprint]

2. Holster, Andrew (2004) A paradox in quantum measurement theory? [Preprint]

3. Holster, Andrew (2003) An Introduction to Pavel Tichy and Transparent Intensional Logic. [Preprint]

4. Holster, Andrew (2003) The time reversal invariance of classical electromagnetic theory: Albert versus Malament. [Preprint]

5. Holster, Andrew (2003) The Quantum Mechanical Time Reversal Operator. [Preprint]

6. Holster, Andrew (2003) The incompleteness of extensional object languages of physics and time reversal. Part 1. [Preprint]

7. Holster, Andrew (2003) The incompleteness of extensional object languages of physics and time reversal. Part 2. [Preprint]

However the next article I tried to post was blocked, and I have not been allowed to post any further articles since then. I appear to be banned from posting any material on the site. Moreover, I have been able to update papers on the site[79], or update my contact details, which changed in 2004. As a result, despite thousands of downloads of some papers, I have almost never had any enquiry from anyone about my work posted there. It is difficult to tell what is going on as the journal editor will not respond to my enquiries.

The site policy is stated as follows:

"The archive does not referee postings and does not edit them. The archive merely filters minimally to assure relevance to philosophy of science."

Perhaps my papers blocked from the site do not fit the criterion of *relevance to philosophy of science?* Well, the academics who edit and control the site are specialists in – guess what? – *exactly the same areas of philosophy of physics that are central to my concerns.* They do not like my views on the irreversibility of QM[80], but they accept this topic as an open issue, and have not rejected my articles on that subject. But they hold very strong opinions about a closely related subject: *the metaphysics of time flow and space-time.* They support the conventional interpretation of time as merely a dimension of *space-time* (the *'bloc universe'* metaphysics), and they reject any possibility of *time flow*, which is to say, the view that there is an objective distinction between *past, present and future,* and that time is the representation of *change*, rather than a dimension in the bloc

---

[79] The *time flow physics* paper proposes a new approach to constructing a foundational model unifying physics, but this version is now long out of date, contains flaws, and I have found a much better solution. Leaving this version in place without being able to reference subsequent progress and acknowledge flaws is misleading and detrimental to the project. Why put an imperfect treatment up in the first place? Because the proposal is a good idea with implications that need to be explored, and I would like to interest others in pursuing the approach, but such a theory cannot be solved all at once. It is a work in progress. The *Paradox in quantum measurement theory?* paper is likewise incomplete – in fact just a fragment of a much longer and very technical treatment, but this technicality belongs on a physics rather than philosophy site. Academics today are overwhelmingly preoccupied with the *flawless scholarly-looking presentation of trivial ideas,* and unable to distinguish the *value of novel ideas.* Presentation values vastly override any concern with content. But a preprint site is surely for *communicating ideas,* a different function to publishing manicured results for posterity.
[80] Their main projects in this area have now degenerated into *attempts to develop new definitions of time reversal to render QMs time reversible.* Despite being explicitly aware of blatant errors of analysis in earlier work, published in a good peer reviewed physics journal (Holster 2003 (a)), the leaders in this field, and particularly at Pittsburg, have never acknowledged errors, and never made reference to this explicit debunking of the widely published orthodox theory.

universe. They hold that 'time flow' is a subjective illusion, and that this is *conclusively proved by physics,* and *no longer a scientific question.*

They are so certain of their opinions on this matter that they reject further questions or challenges about it as being viable scientific questions – in the same way López Corredoira rejects phlogiston as being a viable scientific concept any longer. They have pointedly expressed strong opinions on this. Roberts, present editor, and previous PhD student of Norton and Earman, states that the possibility of *time flow,* i.e. realism about the *past, present and future,* is irrevocably rejected in modern physics. He is adamant that the concept has no possible role in any future science. Pursuit of this topic, according to the power elite of *philsci-archive,* is the domain of cranks and crackpots, and *no longer falls within the scope of philosophy of science.*

The last article of mine the site accepted, and the next article of mine that was blocked, both maintain that time flow has been rejected too hastily, and argue for its reinstatement. I argue that time flow remains a legitimate concept in physics; and moreover, I propose that *adding it back into modern physics* is not contradictory, as most assume, but leads to a new and very realistic unified foundational model, a type of model that physicists have overlooked. In *Appendix 4,* I briefly explain one starting point for my view. I have developed this view at length in other places.[81]

In fact I had already gone some way to this view in my PhD thesis in 1990, which Earman himself examined.[82] Now this is a heterodox view *par excellence* in the culture of modern physics! Physicists are taught to ridicule the notion of time flow – i.e. the common-sense view that *existence is temporal, based on a changing present* – from their first lessons on relativity theory as first-year undergraduates. Physicists reading this may therefore likely agree with the Pittsburg nepotism, that questions about this should be censored from the literature – for this is surely the *single most fundamental* metaphysical dogma of modern physics.[83] Physicists simply never question it. López Corredoira himself, for instance, takes the *'ether',* like phlogiston, to be a concept that has been irrevocably ruled out of science. Reintroducing time flow means postulating an absolute frame of simultaneity, and in the context of relativity theory, this leads back to treating space as having absolute properties of *position* – which is what the postulate of an 'ether' means in its stronger form.

So certainly this is a heterodox view. But is it still open to question? In fact there are a number of very serious thinkers in philosophy and logic who

---

[81] (Holster WEBREFS (c), (i), (p)).

[82] He passed it as a meritorious thesis – but I had unwittingly made a powerful enemy through my criticisms of the orthodox views he supports.

[83] This is why such a theory cannot be submitted to physics journals. Their response is "We do not review this kind of work". But I am submitting this to a *philosophy of science preprint site,* a domain that is supposed to allow critical questioning of fundamental concepts and dogmas.

have repeatedly questioned this point over recent decades – (Storrs McCall, 1979) being a classic example. There are at least a few dozen recent philosophers with strong academic credentials who think that it is still an *open question*, and that at very least, the conventional reasons given for rejecting time flow are scientifically unsound, and confused. (Which is certainly true: if time flow is rejected, it is still important that it is rejected for proper reasons, not on the basis of fallacies, which is the situation at present).

So this is a question of whether a heterodox view should even be allowed to be *heard* in the philosophy of science. We should remember that as far as heterodoxy goes, the suggestion that fundamental physics might be *time asymmetric (irreversible)* was also heterodox from 1960's through the 1990's, and the suggestion of this would be greeted with ridicule and anger just a few years ago.[84] Yet in the early 2000's, this became a real question again – and specialists in the field now recognise that *the analysis* that the previous 'scientific certainty' was based on is inadequate - in fact riddled with errors. But it still angers most experts if you point these errors out - for recognition of them has been suppressed.

We should also remember that a similar thing happened on a grander scale in the general philosophy of quantum mechanics. In the 1940s-60s, concerns of realists like Einstein, Schrodinger, Dirac and Bohm, about the completeness of QM became a subject of ridicule within physics, while the 'Copenhagen Interpretation', the positivist interpretation, flooded the market as the final and decisive answer. Questions about the *interpretation of QMs* were placed off-limits to enquiry. Only after Everett's 'many worlds interpretation', and Bohm's 'hidden variable interpretation', and Bell's work, etc, in the 1960s, did this slowly became an authentic question again – until by the 1980s-2000 it had become *the* leading question in the philosophy of physics, with many different interpretations being proposed (and still no closer to any resolution today).

And in support of the *time flow* heterodoxy, it is worth observing that 'time flow' is not some abstract or theoretical substance like phlogiston or celestial spheres, proposed as a theoretical device to explain certain phenomenon. It is absolutely central to our experience that *existence is temporal,* that what exists is the present, and it undergoes change. Everybody knows what time flow *means: we live with it.* Everybody thinks and acts as if time flow is real, in their normal lives. And the temporal nature of the world, including the irreversibility of natural processes, remains a mystery in physics. So this is *not* an area where conclusive scientific explanations have been established and

---

[84] Adolf Grunbaum, a positivist, and long-serving Professor of Philosophy at Pittsburg, a generation ahead of Earman, was a prominent early propagandist in the philosophy of science for the view that physics is time symmetric and time flow is unreal. He is followed by numerous others.

widely accepted. It is an area fraught with theoretical difficulties and meta-physical controversies – and most of all, with conceptual confusion.[85]

So with *time flow* we meet López Corredoira's problem of heterodoxy in a real example, and with *philsci-archive* we again meet the serious problem of censorship on a preprint site. *Philsci-archive* is owned and controlled by a tiny nepotistic[86] clique of powerful academics for whom *this specific* heterodoxy is central. Their careers and reputations are staked on the orthodox position. If they are wrong about this, the relevance of most of their own work disappears in a puff of smoke: they have a huge vested interest in the outcome.

Philsci-archive has powerful larger biases too. Its editors and owners belong to a larger philosophical ideology. They are typically materialists, reduction-ists, atheists, with scientistic attitudes, as well as a highly US-centric outlook. Their preprint site contains a large volume of material on the failed philoso-phy of *positivism* for example – Pittsburg being an original hub of positivistic philosophy in the US, through the positivist extremist, Adolf Grunbaum – but little or no material on a variety of alternative traditions. Other heterodox topics unwelcome in their world include non-materialist theories of mind, spirituality in science, questions about intelligent design, evaluations of controversial phenomenon like psychic phenomenon or UFOs, or 'mythical creatures' like Big Foot.[87]

"But that is <u>not</u> science or philosophy of science!" they will immediately say. *"That's just cranks talking nonsense! That is exactly what we <u>do</u> want to exclude!"*

Well, to me, questions about these things *are still open and unresolved scientific questions*, at the difficult border-line of science, they relate to wide-spread

---

[85] And even if a bloc universe ontology is accepted in the abstract world of physics, *time flow* remains a central concept in *human reality*. The denial of its reality blocks discussion of it in *real philosophy*. This is analogous to scientific determinists denying there is any real freedom at a fundamental physical level – and so blocking the *issues about human freedom in a social and political sense.*

[86] A series of academics and their students becoming academics in turn, all from the same institution.

[87] Actually what we have to say in most of these cases is that *we just don't know what the real answers and explanations are* – not any sense of a broad and convincing scientific consensus. But this is precisely what the conservative philosophy of science does not want to admit: that *they don't know something, that science is incomplete.* Instead of saying "we just don't know whether various UFO sightings may be alien spacecraft", for example, they have said for decades: "anyone who talks about UFOs is a crackpot". Yet surely the widespread *phenomenon* of UFO sightings, and lack of explanations for many phenomenon, makes this a genuinely interesting area for scientific investiga-tion. I have explored a logic for evaluating such claims in the context of *inference to best explanations*, in (WEBREF Holster 2014 (d)). I intend to apply this to practical exam-ples of controversial creatures, such as Big Foot, South Island Kokako, etc.

phenomenon and experiences that millions of people have and talk about[88], and a key problem in evaluating them is that they are excluded from 'scientific discussion'. Their ideological exclusion from the philosophy of science raises real issues about what its purpose is.

The controllers of *philsci-archive* have a narrow perspective on philosophy of science, and on philosophy in general. They specifically want to exclude areas outside their tradition, which is essentially an academic carry-over from positivism, and they take materialism, reductionism and atheism as the norms of scientific philosophy. Their agenda is to maintain the official cannon of philosophy of science, as taught in mainstream University courses. It is *only* through universities that philosophy of science is funded, and maintains its institutional power. It is nowhere as large as physics of course, but still a substantial academic industry in its own right, with academic positions, journals, conferences, research positions, text books, etc. (And to these academics within it, remember is their *whole world*.)

The fundamental mechanism for sustaining this industry lies in attracting students, and making them spend years jumping through academic hoops, studying the ideas of theorists in the official cannon, specialising in areas that are continuous with their Professors' interests, until, by their early adulthood, this is *all they know about,* and they come to believe that this set of ideas and concerns and topics defines the subject. The ambitious students who are suitably conformist to the academic culture, who want to imitate their professors and spend their life in an ivory tower, who seek fulfilment of their ambitions by climbing the institutional ladder, are taken in as the next generation of lecturers and professors. It all intimately depends upon *sustaining the official cannon and the official hierarchy of academic offices and reputations.*

This is what *philsci-archive* is really about in its preoccupations with censorship: setting up a form of institutional control to *maintain the status quo in philosophy of science,* to keep the blinkers on researchers and academics, to keep aspiring academics locked into a certain power hierarchy.

Now you might say in their defence: "*Well, these Professors were very enterprising, they did a good thing, their preprint site is really useful for the people who are in their tradition, and they are simply in the general competition for ideas and position and power that everyone else is. It is just that they have been more successful than others. If you don't like it, then go and set up your own preprint site!*"

That certainly has some truth: they have been enterprising, they done a good thing for their preferred ideology and tradition. But it is also simply to acknowledge that they have been the most successful in a *Capitalist-style competitive business model* for this intellectual domain. The larger question is *whether this is the model we want* for philosophy, or for science. Now they have

---

[88] C.f. few people talk about phlogiston, Ptolemaic celestial spheres, caloric theory or the flat earth.

set up their preprint site, like a corporate enterprise it has become a monopoly in the field, with the usual propaganda functions.

This points to a real problem: preprint sites take on a powerful influence, becoming corner-stone institutions. When they are privately owned and controlled by a small group of individuals with strongly vested interests, like *philsciarchice* or *arXiv*, they are inevitably prone to abuse. As they become more powerful, individuals who control such things become more arrogant. Placing your papers on such a site invests a lot of trust in the individuals who own and control it. Being in the monopoly position as the *authoritative preprint site*, as the primary research resource in the field, gives tiny nepotistic groups very substantial power of control and censorship. Does the whole world of philosophy of science wish to trust this corner-stone institution to such a group?

Proposal for a new journal publication system.

The *philpapers.org* site appears to have a diametrically opposite policy to *philsci-archive* in regard to censorship. It accepts any material within the broad scope of philosophy, without any apparent censorship for *content*. The editors do not try to decide who or what ideas they support or dislike *at all*. It is freely *self-censoring*. In fact, it tries hard to collect *all the work in philosophy that it can* – like a competition to accumulate a universal resource in philosophy.[89] (And if the odd paper does cross a subject boundary – well who on earth cares?!)

As a result it has a huge collection of material – now approaching 2 million papers - but representing so many different philosophical points of view and traditions that the problem is now to narrow your search for material that you want – in terms of both quality and subject. It has search tools and indexes, and you can follow writers of interest, etc, but there is so *much* material, it is hard to filter what you want, or know if you have missed something of real importance to your interest. Indeed, it is a full-time job keeping up with publications in any major area of philosophy, just as it is in science.

It is primarily for this reason that we need the function that journals are meant to provide – to sift through the mass of material, and bring the best work to attention. On this basis I propose the following innovation in the scientific and scholarly journal industry.

- Every domain – physics, chemistry, philosophy, history, etc – should have a global preprint server - somewhat like *philpapers.org* - that is freely accessible, and open to anyone who wants to post material.

---

[89] If we think of it as a music collection, *philpapers* is trying to collect every recording ever made, regardless of genre, as long as it is *music*, while *philsciarchive* is trying to collect a genre, say *Country and Western*, and thus legislate precise boundaries for the genre. Of course it is inevitable the latter has severe difficulties legislating this boundary.

- The preprint sites are owned *publicly and internationally* in some broad sense – not by individuals or university departments or journals (as with *philsci-archive* and *arXiv*). It is like an *intellectual commons* for the international community.

- Journals specialising in certain topics then *review and collect material to publish on their own specialised sites* from this large collection.

- Instead of individuals submitting papers to journals, the journals proactively filter and search for articles on the preprint site in their area of specialisation.

- When they identify an article they wish to publish, they contact authors directly, and negotiate agreements for publication in their journal.

- Journals can go through a process to improve quality and presentation, as they do now. Depending on the policies of each journal, the original article may be removed from the preprint site, and only the abstract left, with reference to the final publication journal. Or the full article may be left freely on the site.

- Journals decide their own policies. They may request exclusive rights to publication, or allow multiple publications; they may charge a subscription fee, or give free access; they may publish a hard-copy version, or only provide online versions.[90]

- The intent is to make journals act primarily as *quality and relevance filters* and do the real work of sorting out the vast amount of material, and bringing the best work to prominence.

- The intent is to provide a common forum for all publications, take the burden of finding publication outlets off researchers and peer-reviewers, and circumvent the hugely time-wasting and energy-sapping tasks of academic submissions, reformatting to journal specifications, etc.

- Authors can of course still submit work to traditional publishers if they wish – e.g. textbooks they want to sell commercially, as opposed

---

[90] In fact there are two print journals of essentially this nature found in NZ bookshops, *Nexus* and *New Dawn*, both published from Australia. They appear to publish articles gleaned primarily from web-sites, with some regular editorial columns and some commissioned articles. Both cover a similar range of controversial topics, including politics, alternative health, conspiracy theories, alternative science. They are both fun and interesting, sometimes enlightening, sometimes wacky. Their quality lies depends on their editors success in filtering the best articles they can find on these various topics from the vast pool of internet material. Both are produced on tiny budgets, but publish a wide range of material and points of view, which is only possible because of this method of sourcing material. Both have dedicated followings.

to research material. No one is forced to place work on a preprint server. But it serves to identify ownership and priority to ideas, and in this sense *counts as a publication.*

- A system of formal reference to preprint publications is used to allow referencing of work on the preprint servers. This is a major problem at present: you may have popular articles on a preprint server, but there is no official way for others to formally reference them as sources in their papers. Since preprint publication does not count officially as a publication, it does not count to establish priority to ideas and discoveries.

- A free software package is provided by the preprint site for anyone who wants to start an 'online journal' on a topic. This software automates transfers of material, manages agreements with authors, etc, and sets up a standard publication website for the 'new journal'.

- Thus journals that gain the trust of large numbers of readers to keep them abreast of the research will succeed, and those that do not will get few subscribers. Successful journals will be those with editors who show the best judgement, post the best article collections, give the best quality evaluations, and attract the most trust from followers in the guidance they give.

- The intention is to take away the excessive power monopolies of the leading journals, and make them compete on an equal footing with others.

- The intention is to redress the profiteering aspect of large journals, which charge significant amounts for individual downloads, with no compensation for the authors.

- The intention is to bring back some realistic competition between journals on the basis of quality, and undermine the monopolistic position of the gargantuan journals. Big journals like *Nature* or *Science* can currently behave as badly as they want, and remain invulnerable to any reaction from subscribers or audience. This is why the most prominent journals (e.g. JPS or BJPS in philosophy of science) tend to be the most arrogant, while secondary journals (like AJP) are actually better quality, but get little reward.

- The intention is to allow a better diversity and quality of niche topic publications, with journals becoming more like recommended article collections. This is specifically to address the issue of publishing *heterodox papers.* We need journals that *specialise in heterodox papers but still have reputations for real quality.*

- It also addresses a problem of *conformity to prescribed formats,* in terms of length and complexity. Almost all journals demand short articles, preferably of 5,000 words, sometimes allowing a maximum up to about 10,000 words. But complex ideas – especially new theories –

often cannot be given in this space. They are sometimes published as research monographs, but this is a tiny domain, reviewers often demand short papers to be published first, and it is not part of the primary *research communication domain*. By freeing up the journal publication format generally this problem is addressed.

I think that, although radical, this is a practical proposition. This proposal is aimed to break down the monopoly power of the academic journal industry, and return power and freedom to individuals. The fundamental problem with the journal industry is that it has two different functions: it serves two masters, with conflicting interests. The primary stakeholders it is meant to serve – researchers, academics and students – come second to executive stakeholders – with goals of maintaining power, prestige and profit.[91]

Implementing this kind of system would make it very cheap and easy for small groups to start specialised journals, following niche topics. The norms for publication should ideally be *free on-line sites* with *non-exclusive rights to publish*. But journals may give significant help to authors in revising manuscripts for example, and negotiate to retain greater rights. In other cases, hard-copy publications may be desired. Etc. The main criterion for success is that the hegemony of the most powerful publishers is diluted, and a larger number of smaller niche journals, with high quality filters, become established as authentic publications in the scholarly world.

For some further material on the rapid evolution of the journal industry, see Appendix 8 – 11.

---

[91] There is a parallel with the large public institutions that control our lives, ostensibly for our benefit. Government institutions – from Ministries and Departments to universities and public research institutes – have the explicit function of serving the population by providing services. But they are also industrial complexes in themselves, and have the function of sustaining themselves as organisations, providing an internal environment to benefit the career aspirations of their managers, and a propaganda face to serve their political masters. The fact that such organisations claim to serve the public interest (ditto: scholarly or scientific interest), and that such aims are specifically stated in policy and law, is effectively meaningless. This is explored further in Part 3.

PART 3. DEATH BY BUREAUCRACY

The social context: bureaucratisation of science and society.

The problems of science are not isolated to science: they are part of a much larger cultural pathology of our age. This Part explores this context in greater detail. Again I go back to López Corredoira for my starting point. He thinks science is in its twilight: that it is not yet dead, but on its death-bed, and will fade away into cultural irrelevance in succeeding decades. He has various reasons for this, which I briefly revisit as a starting point.

One is internal to science: he thinks the core physical sciences have already mined much of their ground, and there are not so many significant discoveries left – that the age of 'major scientific revolutions' is largely over. He thinks science has established the basic framework theories and explanations in many domains. On this point I disagree somewhat though. I think science – that is to say *real science,* not the conformist façade of 'scientific authority' that replaces it today – still has much further to go before it seriously comes to terms with the natural world. This is surely true of new domains of science – but I think it is true of traditional physical sciences like physics and chemistry too.

I think there is a new revolution in store for foundational physics, and this will be *the* major watershed in the scientific history of our age. It will dramatically revolutionise our *metaphysical beliefs* about nature at a deep level. It will overthrow foundational assumptions of various other sciences as well, and liberate them to explore new knowledge. Besides that, many sciences other than traditional physics are still in their infancy. We can see their forms emerging, they have some clear foundations, but they are really still in a stage of early development, like Enlightenment sciences in the 17th – 18th centuries. I have given a few examples from the recent science of water chemistry, the brain and consciousness, and information theory (including logical semantics and AI). These are already claimed to have mature paradigms by many orthodox experts, but in reality they have barely got off the ground yet.

López Corredoira is right that *establishment science* is bankrupt, and continuing to pour increasing amounts of money into its institutions will produce little more knowledge. The exclusion of heterodox thinkers and repression of challenging ideas from the scientific mainstream over the last few decades means that three or more decades of ideas and discoveries have been sublimated under the rule of scientific authoritarianism; but they lie in wait if creative, heterodox science is revived. This prospect is a rationale for Pollack's "Institute for Venture Science", which explicitly aims to mine *revolutionary ideas,* on the expectation that they lie dormant among heterodox thinkers, who have withdrawn or been excluded from the institutions.

Where I most agree with López Corredoira is that much of the science industry and research paradigms of our own day are exhausted: the vast bulk of scientific funding and effort is now spent on pointless bureaucracy and incompetent 'research' that has no future contribution to scientific value. It is time and money, talent and opportunity, flushed down the drain of history. It is sustained simply because our power society prefers the status quo of familiar institutional failure to the uncomfortable challenge of confronting real problems and meaningful questions. This is the main theme of this Part. Science is conditioned by the general bureaucratic culture of our age. Its modern form is sustained by a culture of greed and self-interest locked in a vicious circle with institutionalised power.

This is related to López Corredoira's theme that we have developed vast amounts of information, but little wisdom, and instead of the industrial-scale mining of more information, what is most relevant now is to go back to the search for wisdom, to leave science behind and pursue the synthesis of understanding; to pursue philosophy again. With this need I absolutely agree. And I share his view that such a goal will not be realised in the context of our present institutions. We might be able to genetically modify pigs to grow wings, but they ain't gonna fly. Public institutions of every kind have evolved into feudal bureaucracies, thriving on the mediocrity that institutionalised power allows, self-justifying, self-protecting, arrogant and spiteful to outsiders. I see this as an expression of human nature constantly lurking in the background of human history. We see today the reassertion of a fascist personality in bureaucratic-corporate management. We would be fools to expect real solutions to deep-seated social or scientific problems from the entrenched institutions of today. They are intent on protecting their own wealth and power, not providing solutions to real problems.

There is a more revolutionary prospect remaining: that social structures for comprehending knowledge and developing wisdom will be regenerated, as our present systems fail. But how can this happen? And this brings us back to López Corredoira's central critique, which is that science has irrevocably changed its organisational structures from those that originally empowered it, becoming institutionalised and bureaucratised, overtaken by greed and self-interest and mediocrity, to such an extent that it is *structurally doomed to degeneration and failure*. With this I agree, and in the next sections I explore some reasons in detail, before turning to consider what our realistic prospects are, and what our responses might be. I begin with a picture of the larger cultural setting - the "spirit of our age" – in which the problems of modern science and philosophy are set. It is about the interaction of cultural and psychological factors that create the character of the modern institution, illustrated with personal examples of how this works in practise.

Mediocrity and the stratification effect.

López Corredoira and I both complain of a plague of *mediocrity* in the 'science industry'. Having worked in many roles, on many projects, as scientific researcher, information systems developer, research analyst, business analyst and statistician, in various organisations of government, corporate business and academia, I can assure the reader that *mediocrity of alarming proportions is the norm throughout*. Incompetence is rampant in high-level executive roles – in senior management, and analytic, policy, research and development teams underpinning them. Anyone with a high level of scientific competence who works in these environments will know what I am talking about. If you do not have this experience you may think I am exaggerating. Perhaps you may think that my expectations of competence are too high. Perhaps you may think: *"Not everyone can be equally good at scientific analysis, statistics or mathematics, information system programming, or whatever - which are your forte, but not everyone else's. Lots of other people, some with lower technical skills, have to do these jobs too, and you can't criticise everyone for incompetence just because they don't have an advanced capability in these subjects that you might have. And compared to some others, maybe you are incompetent too!"*

But the levels of incompetence I am criticising are extreme: they involve people taking on tasks that are *so far beyond their competence their very roles become farcical.* I am incompetent when I take on roles beyond my own ability too – and this is precisely the question: *what levels of ability are required for competence in different occupations and tasks?* For instance, *what level of competence in statistics is required to be a competent research analyst?*[92]

When we talk of competence, it should be realised that this is *relative to a role, a task, a problem.* It is not a characteristic of individuals as such. We are all

---

[92] To give some answer to this question, which is pertinent to what follows, I would say: at very minimum you should know the concepts of *mean, standard deviation, t-test, correlation, effect size.* You need to know such concepts to be able to understand basic statistical analysis. That may seem extremely minimal, but few analysts in NZ government and business roles know the last three concepts – and probably half do not even understand the concept of a standard deviation! It is rarely used in 'business analysis'. The vast majority of research and business analysts do no more than report *averages*, with a few significant *ratios* – and managers rarely want anything more this, because they have even less analytic capability. To be competent to analyse and interpret original research data, of course you need to know far more than these basic high-school level concepts. You need to know what *statistical models* are, and a host of other concepts. Some are generic, some are specific to domains. It is impossible to make a list of statistical concepts to define competence: rather, you must identify *ability*. This is what university degrees used to mean: a demonstration of *ability to master a subject, to learn new concepts, to solve new problems.* This perspective is missing in our new technology-based and vocation-based education.

incompetent at various things – but most of us are good at some things, and it is important we are good at the things we do for our jobs, the things that others rely on us for, the things we get paid for and have responsibility for. Even more so in our intensely inter-dependant society, with hierarchies of specialists and grid-locks of dependencies, where everyone is reliant on others.

But the more demanding the job, of course the more difficult it is to be competent. Some demanding occupations nonetheless maintain high levels of competence – jobs with *critical performance levels*. Airline pilots seem to me to be very competent at their tasks – surgeons also generally seem very competent. This is because the outcomes are directly observable, and incompetence is life-threatening. I trust pilots to fly the plane I am in, and surgeons to do routine surgery on me.

By the same standards, *I trust hardly anything most research or business analysts say, and hardly anything most government or corporate managers say* – except in the rare situation where I know personally that I am dealing with someone capable. In the same way, *I trust little that peer reviewers in philosophy of science say – and little of what most physicists say about novel physics, and certainly nothing they say about philosophy.*

This is because these are *not critical performance cultures*. On the contrary, they have degenerated into dismal mediocrity, with a lazy arrogance towards making difficult judgements. The few highly competent practitioners are swamped by the general mediocrity that prevails.

And this is the same in the health system too – while surgeons are now generally very competent, being highly trained, with effective technologies to help them, and with patient outcomes closely monitored these days, your main of chance of being killed or significantly disabled by 'medical misadventure' is from *bureaucracy*. You are probably *100 times more likely to be killed or seriously disabled* through the administrative incompetence of the public health system and funding agencies than by *surgical error* – through bureaucratic denial of entitlements and treatment, postponements of surgery, mismanagement of surgical waiting lists, poor management of hospital and health system resources; or by poor initial diagnosis by GPs or consultants, the gatekeepers of medical treatment before you get to see a reliable specialist. It is only when you get into surgery that you can have confidence.

And we only have competent surgeons today because activists over the decades have pressed for accountability of surgeons – a profession that in eras past was notoriously prone to cultures of arrogance, greed and unaccountability, placing itself above criticism. Incompetent surgeons do still occasionally persist in the system, and they can cause great harm. But since their performance has been brought under objective scrutiny, surgical errors are relatively rare. Yet there is no similar medical accountability for *health bureaucracy administrators*, whose incompetence causes vastly more harm than surgical failings do today. There is no objective scrutiny of senior health system executives – they are business functionaries. Their 'accountability' is financial – measured by 'business productivity', rewarded by bonuses for achieving financial targets. At the very top of the system, senior executives, up to the Minister of Health, rarely have basic science knowledge, and almost never have direct experience working in the front line of health care. They are *professional managers and politicians,* not experienced doctors, nurses, surgeons, therapists, psychiatrists.

This reflects the modern business philosophy that *management is a generic function:* that hospital management is like factory management, business management, bank management, science management… That management is a separate professional specialisation, and all management roles are interchangeable, because in the end they are all *business management roles.* Indeed, as part of standard jargon these days, all NZ organisations talk of themselves as *businesses* – hospitals, schools and universities, govt departments, funding agencies, science institutes. (In IT and business projects, all staff in the organisation are referred to as *business users.*)

It is very disturbing that so many modern occupations with high levels of responsibility and high impact on our lives are now characterised by high levels of managerial incompetence and low accountability. The classic occupations in this respect are those requiring the highest-level managerial decision making, tasked with 'strategic planning', 'policy development', 'organisational capability' and 'research'. These 'leadership roles' really demand *exceptionally high capabilities to do well.* But such abilities are rare, while the number of roles proliferate in the thousands.[93] The peculiarity of such roles is that people can do these important jobs very poorly, and not be found out – unlike most skilled specialist professions. Senior management, policy, planning and research, (including organisational self-monitoring) in large organisations is dominated by chronic levels of incompetence.

How do individuals in these roles get away with it? They have achieved a position in a hierarchy with large salaries and important job titles, or *offices.*

---

[93] There is a delusion that a large number of average performers will bring capability as a group that is lacking in each individual. A large group can lift a heavy weight that an individual cannot lift. But will a large group be able to solve an intellectual problem that none of the individuals can solve – e.g. a mathematical or scientific problem? Cultures of mediocrity are often reflected by proliferating large numbers of 'specialists' with the expectation of compensating for low individual competence by force of numbers. Would the combined force of 30,000-odd NZ Govt bureaucratic workers be able to break the Enigma Code that Alan Turing and a handful of specialists cracked in WW2? Of course not. But if this was their project, they would spend years and decades *pretending* to make progress – making project timetables, business models, process diagrams, procedural specs, policy committees, oversight committees, managerial hierarchies, sign-off protocols, dissemination systems, proliferating an empire of sub-departments; just as they do now on more pragmatic problems they equally have no hope of solving.

They are in the *meta-system*, the upper hierarchy of trusted rule-makers that defines the system itself. They *only judge each other* – their performance is not visible to outsiders. The quality of their work is not visible outside their circle of fellow 'professionals'. They rely on minions within the organisation, their working staff and technical contractors, to implement the plans and decisions they sign off. Managerial failures are hidden to a large extent by competent lower-level staff working around their inadequacies (and bending or breaking the 'official rules' to do so). But that only goes so far. In the long term, on the large scale, the performance of these organisations is severely degraded. Yet senior management is rarely held to account: they are in the center of a self-protecting, self-serving system that makes them anonymous when performance is questioned.

It is also worth emphasising that a major difficulty in accountability is that performance really needs to be compared *counterfactually*, which is to say, through a comparison of actual outcomes with what *might have been achieved*, with *what should have been achieved under the circumstances*. We need to consider how well opportunities *might* have been taken, not merely count those that were taken. We must compare what an executive or analyst or scientist *might have achieved*, with what they *actually* achieved. For it is easy to keep a mediocre organisation performing at the same mediocre level – with the manager saying *"Look, I did my job just as well as the last guy"*. Recall here earlier comments about the difficulty of counterfactual logic. *Counterfactual reasoning* is the central part of human analytic and scientific rationality – but academic attempts to understand counterfactual logic has so far failed. Computer programs can do Boolean logic (the truth-functional logic of 'or', 'and', 'not')[94], and some quantificational logic (SQL implicitly embodies a first-order logic of quantifiers 'some' and 'all'). But these are logically trivial. *No computer*

---

[94] And a concept misleading called *implication*: *"if P then Q" (material implication)*. This is defined to be *true* whenever either (i) *P is false* or (ii) *Q is false*. On this reading, the statement: *"If the asteroid had not hit Earth 65 million years ago, then the Dinosaurs would still have died out"* is *true* – simply because the precedent is false. Duh? Indeed, the statement *"If the asteroid had not hit Earth 65 million years ago, then the Martians would wear blue pyjamas"* would be true. Yikes! A long-serving senior lecturer in philosophy at one NZ university taught a popular introductory 'critical thinking' course (a soft alternative to studying logic), by making up hundreds of examples of intuitive commonplace statements of *counterfactual implication*, and forcing students to evaluate them as *material implications!* (This included his own deviant interpretation of negation, which he interpreted as a 'negative inference', instead of propositional negation.) It was very bad for students, but nothing could stop him teaching this nonsense – as a 'philosopher' in NZ he had freedom to teach as he chose. He had never studied logic, and did not know the difference between *material implication*, *logical implication* and *counterfactual implication*. But neither do many teachers of propositional (truth-functional) logic in either philosophy or computer science.

program can do counterfactual logic. There is no reliable rule-based system known for applying such logic to real situations.

Counterfactual comparisons of performance are notoriously difficult to make and to prove. To give the bureaucrats a break for a moment, academics are also very familiar with this problem. University departments are typically undermined by incompetent professors – the academic executive leader, appointed for life, and impossible to remove until they choose to retire. Research and teaching under incompetent professors is often paralysed for decades – falling years behind the changing research frontier, locked into failed paradigms that have become irrelevant, and driving away the brightest students. Bright ambitious junior academics can see that their department is *failing the opportunity to keep up with modern progress in the subject.* But as long as the department continues its legacy syllabus, filling its quota of students, and the professor doesn't commit some moral travesty in public, nothing can be done. This is partly because there is no official way to evaluate academic performance *counterfactually.* (I.e. how would the department have performed under *another* professor?) The bright academics leave and the department spirals into greater dilapidation. This is hardly exceptional: it is the *typical state of average academic departments found in* NZ. In NZ, bright academics with re-search ambitions almost invariably move overseas, because NZ cannot sustain any serious research university: a phenomenon of intellectual medioc-rity that has swept over a small country.[95]

Lazy executives, like lazy professors, get by simply by *doing nothing* – let-ting all opportunities and challenges equally pass by. This is how the NZ public service works. It is an established part of its character to have medio-cre performance, low morale, low capability – and to be paralysed by its own red tape if anyone does want change. There are huge opportunities to im-prove, but *improvements threaten those already in the system.* Realistically, to im-prove a bureaucratic department requires a revolution: the first and critical thing you must do is to dump the dead-wood that already chokes up the management and senior roles, along with their pets. But this is practically impossible in such a stratified power-based system, developed and evolved to maintain power. Managers progress to the top jobs after decades in the system, when they have established the conspiratorial trust of others that *they will not question shabby performance of their managerial colleagues (although they may be personally spiteful), they will not threaten the positions of other executives (although they may prevent their advancement); they will not rock the boat.* The established culture is

---

[95] Thus we also find the phenomenon of provincialism: a small number of leading international universities gathers most of the research power, and without intellectual leadership and creativity, 'provincial' universities descend into mediocrity. They often have isolated intellectuals with talent: but power hierarchies mean no one can achieve much in isolation – and mediocre academics resent their bright colleagues. In a small country like NZ, this 'brain drain' has become a terminal condition.

to maintain a succession of shabby but loyal managers, waiting in line for their turn to inherit privilege. This is the structure of a feudal aristocracy.

There is a specific psycho-social mechanism at work here, as noted earlier, which I refer to as the *stratification effect*. It is an adaptive interaction between *personality* and *culture*. Individuals with certain personalities and capabilities enter an organisation, and get churned over in its culture. Some of them are retained and rise to successful positions in this *ecological niche,* while others leave for other ecological niches. It is like an *evolutionary adaptation* effect, seen in natural ecosystems, because the organisations provide environmental niches in which some thrive and some do not. It is a *cultural adaptation* mechanism because the environment itself is a human artefact created and shaped by the individuals retained within it.

Bureaucratic systems – formalised rule-bound institutions of all kinds - stabilise around a culture of mediocrity, retaining those with a strong drive to conventional career ambitions, a strong drive to conspiracies of self-interest; and ejecting those with real capability who seek an environment to express themselves in a natural *vocation*. (It is very noticeable that few creative intellectuals feel a natural vocation to be professional bureaucrats or managers; just as few compassionate individuals feel a natural vocation to be torturers or executioners. These intrinsic qualities pre-determine the pool of personalities available to become bureaucrats or torturers.)

I note, as a theoretical point, that an interaction between *psychology* and *culture* is necessary to explain how social institutions form and stabilise. For institutions are composed of individuals, and their cultures and organisational behaviours are ultimately driven through patterns of individual *human* psychology. The bridge between these two levels of description – the personal and the institutional – is poorly understood in social science, and frequently taken for granted. Nonetheless all social theories call on assumptions about human psychology and personality. For instance, that *humans act rationally in their own selfish interest* is an idealising assumption of classical economics. That *humans act politically through a sense of class identity* is a similar assumption of Marxist theories. Social philosophy always takes some such assumptions for granted. I am not proposing any theory of this, just calling on commonplace observations, made as a participant-observer in various social roles.[96]

---

[96] Most sociological treatments start with some large 'methodological' perspective, and typically declare their 'methodological paradigm' as belonging to some tradition, like Capitalist economic analysis, or Marxist social theory, etc. These are usually very boring and unenlightening. Writers like Lopez Corredoira, Woit (2006) or Gleick (1987) are not trying to be sociological theorists, nor ideological propagandists; they are akin to informal cultural anthropologists, making intuitive observations, as participant-observers in the cultures they describe, based on their own perceptions and judgements. My account is similar in kind.

Personality can be seen to enter in this way. Highly self-serving people (a form of ego-centricity) are a *personality type* – with a stable, life-long personality trait – that makes up a sizeable percentage of the species. Some 5%-10% of people appear chronically and profoundly egocentric, dominating the pool of the most materially successful – and most destructively greedy - people on the planet.[97] They have little self-criticism, little empathy for others, and strong drives for social position, status, the trappings of external recognition. In the extreme, it is often estimated that one percent or so are fully-fledged narcissists, sociopaths and psychopaths. They are demanding, assuming entitlements to everything. They think they deserve far more than others. They are ruthless. They rate their own performances and capabilities far higher than is realistic. They readily take on roles far above their abilities – desperately wanting the *role and its rewards,* but lacking the self-perception and self-criticism of their own performance that most people have.

Bureaucracies (both government and corporate) give these people the ideal culture for their ambitions, by setting up a formal power structure, in terms of official *offices,* power-roles, inhabited by other egocentric personalities all playing the same game, so that once in these roles *objective performance is no longer questioned.* Once the ambitious personalities cross the threshold from *workers* to *executives,* their focus turns to their personal relationships in the elite power network. The workers and systems they manage are secondary, mere pawns in their game. Not all executives are like this of course: some are very good, and to be good in these demanding roles requires people of really exceptional talent. But stratification into *mediocrity,* through aspirational, grasping self-promotion, is a dominant effect, not a minor one. As a result, the 5%-10% of extreme egoists in the general population is concentrated by the *stratification process* to 50%, 60%, ... 100%? in the social elite. I.e. eventually to the level where it fully overtakes the institutional culture.

The same syndrome appears in the teams these executives and managers run. Teams can sustain performance with 10%, 20%, 30% of incompetent workers in core tasks, who are carried by others. This is normal – in fact, a healthy reflection of the natural diversity of ability – those less competent in

---

[97] We should remember that the history of human civilisation is *really* dominated by violence, murder, war, slavery, genocide: the brutal exploitation of humans by each other, enabled by forming into powerful groups. The human race is *not a nice species in general:* it contains a range of distinctive personality types, some benevolent and even saintly; some malevolent and psychopathic. We have intricate rules to suppress direct interpersonal violence, but the most aggressive individuals, the wolves, *remain among us in large numbers,* and dominate the power structures of society. The human race seen in terms of divergent personality types is in many ways more like multiple different species in competition, rather than a single species – and the dynamic this sets up makes our power structures intrinsically unstable. As a result, the institutions of society in any period have only a limited lifetime before being overwhelmed by internal corruption. The modern institutions of science are an example.

some tasks compensate in others; and human activities, being as complex as they are, really requires groups with a diversity of ability, not a simple one-dimensional standard of competence. But when teams are degraded to 50% incompetence in core abilities, they begin to go into a tail-spin, the remaining competent workers begin to abandon ship, and incompetence degenerates to 60%, 70%, ... 100%.

This is the situation observable *throughout most of the senior administration, and subsequently operational team workers, in the NZ public service* – and I am quite sure in other countries too. For this is a universal effect, pertaining to the nature of bureaucratic psychology itself. It could only be limited by being *actively* countered – just as bullying and violence by police, for example, can only be limited by being *actively* countered. Bullies, people who enjoy dealing out physical violence, are naturally attracted to organisations like the police: it promises them a safe and legal outlet for their natural inclinations. Without this inclination being *actively countered*, a police force quickly descends into a gang of thugs – as shown repeatedly in practice throughout history, and still rampant today. There are well-intentioned people who want careers in the police too: but when bad cops dominate, the good cops leave. Equally, without the egocentric stratification effect being *actively countered*, bureaucracy and management quickly descends into a self-serving mediocrity: a gang of middle-class thugs. "The price of freedom is eternal vigilance", as the saying goes.

But there is little to counter this tendency within bureaucracies today, and this is what has happened in NZ: the public services have become increasingly Stalinistic organs of control, seeking to regulate every aspect of the lives of individual citizens, dispensing threats, punishment and fear. Ordinary people, ordinary businesses and professionals in NZ are now deeply afraid of state-corporate bureaucracies of all kinds – of IRD, ACC, Health and Safety, Housing, Building, Environment, Industry, Social Development, Welfare, Justice, Police ... and for teachers and school principals, the Ministry of Education... and for scientists, the various shape-shifting Ministries that control and dispense science funding. These bureaucracies can and do threaten and destroy your business, your job, your family, your home, your entitlement to health or welfare or retirement benefits - your school – your research institute – your scientific project - at the stroke of a computer key and the whim of a malicious manager.

Bureaucracies are invariably malicious against anyone who *flouts their rules,* for their rules are their touchstone of power – and their deepest fear is of non-conformity. Flouting their rules is a *threat.* It is important to realise that this provokes a deeply psychological reaction, not a rational or logical response. It is a reaction of fear, anger and reprisal. It is conveyed to the public through the actions of individuals – in the first instance by case managers, front-line officers; dictated to at one remove by second-tier managers. They project the spirit of the bureaucracy by interpreting its rules and processes.

They attack non-conformity to *enforce their power and remove the threat*. Rules and regulations are the bedrock of their reality, and without it they are empty shells.

Bureaucracies have a similar character, everywhere and in all ages. It is one of the universals of civilisation. But material conditions change, and modern bureaucrats now have a huge technological advantage over the previous generation of authoritarian bureaucracies (exemplified in the unfettered drive for state control by Stalinist Russia and Nazi Germany), provided through scientific technology. They are equipped with the modern tools of electronic surveillance and computerised information and control systems. This increasingly allows them to dispense with human capability and human judgements, which in the past was a natural limitation, and had a certain humanising influence even in the most totalitarian state control.[98] For the vast number of victims of crimes by the state-corporate estate now there is no redress; bureaucratic injustices, errors and stupidities are perpetrated by computerised control systems. Computers allow bureaucrats to expand and multiply *rules of control* endlessly, and thus *objectify* their power.

This is also the general cultural context of modern science: for the public science institutions are in a continuum with the bureaucratic sector, run by similar executives, swapping positions of privilege with each other, bullying and intimidating the front-line scientists they are really supposed to enable and serve. The driver for this behaviour comes back the psychology of those who thrive within bureaucracies or state-corporates. They pool people with a particular psychological profile: egocentric personalities, self-important, rule-based paradigms of morality, strong sense of social hierarchy, conformist to those above them, tyrannical to those below them, lacking intellectual or creative spirit. The organisational mediocrity of their culture magnifies these negative personal qualities of their staff. The dictatorial behaviour of bureaucracies is not an abstract property: it is a projection of the diminished human spirit we find within them.

The scarce resource driver of mediocrity.

It is also important to recognise an intrinsic *scarce resource* driver behind the chronic levels of mediocrity – the 'scarce resource' being human intellectual talent. Again it applies equally to science and research, technology and bureaucracy. In the modern context, we must question whether the *human*

---

[98] As a paradigm of its evolution, we see the rise of US military attack drones, killing 'terrorist suspects' at a distance, in foreign countries, at the press of a button; part also of the intense technological militarisation of domestic police forces. This makes the 'projection of power' into a kind of game.

intellectual resources are available for the challenging tasks demanded by the complexity of our systems.

The first step in optimising the competence of a work force lies in *matching workers to suitable tasks*. If we *randomly* assigned people to roles, or by criteria unrelated to their merit in that role, we would get high levels of incompetence. Instead, of course, we try to match people to roles on merit – which obviously helps. (Although the idea we have a *meritocracy* is a delusion: success in matching talent to roles is probably not even 50% effective.) But assuming we can match people to tasks on merit, what if the tasks are *simply beyond the competence of the available work-force force?* Suppose we really need a thousand talented information system developers, but *there are only a 100 in the country.* (Which is probably about the case in NZ). Suppose we really need 20 talented philosophers and 200 talented physicists to teach in university departments, but *there are only two competent philosophers and twenty talented physicists in the country.* (Which again I would say is probably about the case in NZ). Suppose we need a thousand competent research analysts for government departments *but there are only a hundred such people available.* (Again the situation in NZ). We have an in-built talent short-fall of 90% in high-level intellectual roles, *especially those requiring scientific, mathematical and technical capability.* What happens? The roles have to be filled, and they are filled – by incompetents.

The phoney egalitarian ideology (chanted in our educational propaganda) holds to a fantasy that it is all just a matter of training and education. *Just send staff on more training courses to develop high-level capability.* But this ignores our intrinsic limitations. If you think of it as *physical abilities,* our intrinsic limitations are obvious. It doesn't matter how much training I do, *I will never become a weight-lifter.* It is beyond my intrinsic physical potential. I simply do not have the right physique. You can tell that at a glance. And it is not something I am likely to delude myself about. I simply wouldn't apply for a job where I had to stand on a stage in front of an audience and try to lift large weights.

But when it comes to intellectual-based abilities, we are much more self-deluded and vain about our capabilities. This is a particularly virulent syndrome among aspirational middle-class people with above average but still modest abilities – one to two standard deviations above average, say. (This may put them in the top decile: but remember there are 400,000 people in the 'top decile' in NZ; 30 million in the US; over 700 million world-wide). With intense ambition and striving, averagely intelligent people can pass exams and *over-achieve* in the education system. Combined with their social advantages and sense of entitlement this lets them take up high-level roles. But they are still two standard deviations in real natural ability below what is needed for the roles of *real intellectual leadership.*

Only a tiny number of people are really capable of becoming leading software developers or mathematicians or statisticians or business executives or scientists - capable of doing high-quality *innovative work.* And science, analysis, software and business today are dominated by *advanced innovation.*

The average talent in these fields cannot come grips with the innovations of the last two generations. This talent scarcity is inevitable, the same way only a tiny number of people are capable of becoming professional athletes. Because these are highly competitive activities, the standards for high performance are set by people with *freakish* abilities; and the standards for competent performance require very high ability - just to stay on the same page with the leading performers.

Bureaucracies – extending to all large institutions, including universities - are especially in denial about the intrinsic impossibility of turning ordinary intelligent people into intellectual acrobats and innovators. They spend lots of money sending ordinary intelligent staff (who have already forgotten high-school level maths or science) to IT courses, management courses, statistic courses, etc, expecting them to master these difficult domains at a creative intellectual level they simply *cannot* perform at – because the standards are set by intellectual freaks. Our education system cranks out quotas of professionally qualified drones to order. Modern professions are *swamped with mediocre over-achievers*. They have skills for basic operational tasks, with academic qualifications and impressive sounding roles in their CVs; they look good on paper to HR staff. But they are incompetent when faced with the real challenges of their roles: challenges of innovation. They are like mountaineers with basic skills to climb easy routes to a certain level, but *who get stuck as soon as they hit an unfamiliar rock-face*. They become a serious liability when over-promoted to expedition leaders.

The situation is in chronic failure in business research and information system development roles, where large institutions, with huge system problems, are swamped with incompetence. Most CIO's and IT system architects in NZ state-corporates would be found incompetent even as basic *programmers*, let alone as overseers of mega-million-dollar software empires.[99] Similarly with senior research and policy managers: most are incompetent at senior *high-school statistics*.

At the rock face, business workers sent on software training courses come away able to press a few buttons in an STK (software toolkit), but without the basic capability to *organise information systems* effectively. In fact most computer science graduates have little idea of this either. Their training gives them only superficial 'technology' skills – leaving them completely ignorant of how to develop an information *system*. It is like builders learning to use a hammer and saw, but having no idea of how to build a house as a

---

[99] *Chief Information Officers*, the new modern executive business role, termed after *CEO* and *CFO*. Of course they will tell you that *programming or developing systems* is beneath them, a job for their techies; they are on a higher level, of organising large system purchases – do we go for Microsoft or Oracle or SAP or IBM? It makes no difference: *if you don't know how to develop an information system in the first place you are never going to succeed by throwing more money at IT platforms!*

structure. They have no ability to create a *structured information system*, but are able to use certain programs, talk in IT jargon, and they have a low-level qualification for credibility. HR and business people take *Microsoft Technical Certificates* seriously as IT qualifications! This is like taking a two-day first-aid course as a medical qualification. "OK lad, now you've got your CPR certificate and you know how to apply bandages to stem haemorrhaging, you'll be starting in the MASH unit in the Eastern War Zone next week."

In fact even a three-year *BComSci* degree is only a low-level qualification, but it is at least serious start *if you have an aptitude for development*. (If you don't have a natural aptitude for it, and 90% of computer graduates don't, then forget about trying to *develop information or program or architect systems!* Stick to system admin and technical button-pushing jobs.)

Such 'vocational qualifications' allow people to become in-house leaders for major system developments and innovation projects. The result is a plague of incompetent information system and software development. Like the vast sprawling slums found in third-world cities, they have created vast degraded information environments that are extremely difficult to reform once they are established.

The same goes for statistical analysis courses. Large numbers of 'business analysts' and 'research analysts' and 'applied scientists' really require sophisticated quantitative analysis these days, and they are paid by their institutions to go on training courses. These are usually expensive exercises in using basic software functions in commercial statistical software packages. Trainees gain no more realistic ability to understand statistics as a science than when they started. Their trainers have no idea either – that is why they are *software trainers*, teaching the use of software functions provided by systems they are marketing, rather than practitioners of the art themselves. Few industry 'statistical analysis courses' teach how to do meaningful statistical analysis, which requires a holistic grasp of many concepts, an integrated knowledge of a science. For example, for almost all real statistical analysis, *statistical models are essential*: and identifying the appropriate types of models to use is the first step. But it requires quite sophisticated mathematics to understand what statistical model equations *mean*. You can learn to press some buttons and generate models – but if you don't know what the *models mean*, the research results are – surprise, surprise – *meaningless*.

The only realistic solution for most applied scientists is to *work with an expert statistician with some advanced mathematics*, not to try to become an expert in these areas themselves. But there are very few such experts in circulation. For most real mathematical talent is now locked away in academic specialisations. In the modern business world, real experts are generally only available through expensive consultancies – an industry itself fraught with greed, incompetence and erratic quality standards. The traditional solution, of forcing analysts and applied scientists to become jack-of-all-trade statisticians, is an absolute failure. It is a major reason most applied scientists today produce mush.

The simple demographic fact is that the number of *very high-skilled specialist* roles has sky-rocketed – and it has vastly outstripped the intellectual resources available to provide capability or competence in these areas, across the vast domain of bureaucracies, corporates, scientific research institutes, educational institutions, professional businesses, etc, that require genuine intellectual talent to keep up with the frontiers of knowledge and organisational complexity.[100] We try to expand the talent pool beyond its natural limit by dumbing-down and automating tasks; by defining rule-based methodologies as recipes for drones to follow; by creating software toolkits to perform technical tasks without understanding; by breaking complex holistic tasks into specialised technical parts – and end up with a flood of incompetence in high-level roles: people who *cannot think at the level required to make meaningful judgments in their fields.* Drones who follow pre-defined methodologies they do not understand. This is an epidemic throughout high-level management, planning, policy, analysis and research – and a plague in science. In the state-corporate sector, we have given over the controls to incompetents – while a small army of pure academics, on whom we spend millions each in higher education and salaries, live in private delusions of grandeur, working on pretentious projects that are far beyond *their* capabilities.

A real example: NZ Ministry of Education Research.

To illustrate this in the real world, I will give a real example in some detail. For it may be wondered, for instance, if *there really are teams of analysts who cannot do t-tests or simple correlation analysis* as I have claimed? Am I not exaggerating? But it is very common: indeed, *it is rare to find an analyst in the NZ public service who knows how to do a t-test - or who does any meaningful hypothesis testing in their work.* The following is just one of many examples I could give, and chosen because I was a full-time permanent analyst in this situation – and

---

[100] Just think of all the modern domains of science and technology, professional and business sectors – there are a *thousand of them,* from Accounting to Zoology. There are a dozen major branches in each of them from Accounting to Zoology - each a large science in itself. There are dozens of major specialisations in each branch. We end up with *tens of thousands* of major specialisations – not just across sciences but all professions, because they all have serious intellectual dimensions. As Lopez Corredoira notes, the number of professional scientists has dramatically increased in 100 years or so – perhaps by 100-fold – but the number of specialisations has exploded even more – by 1,000-fold! How do you split the intellectual resources of humanity across all these fields, to maintain intellectual quality? You can't. Instead of vast numbers *improving* standards, a swamp of mediocrity prevails.

because this organisation has an appalling record of failure that has real impacts on science education.[101]

In 2000 - 01, I worked for a year as a research analyst, with a permanent full-time role, for the NZ Ministry of Education's *Comparative Education Research Unit*. This is the premier research unit in the organisation, with a multi-million dollar budget to participate in international OECD research projects, including large sums for international travel to project conferences, etc.[102] This is the *flag-ship research project* for the NZ Ministry of Education, a *multi-billion dollar bureaucracy.*

I immediately discovered that no one in this *dedicated research unit understood how to do a t-test or simple ANOVA, or simple correlation analysis* - concepts 16-18 year olds are expected to learn in high school statistics. One analyst struggled for six months to do a *single t-test* (without success), before I was asked to step in and help him. This was his sole task. He sat quietly in a corner, month after month, endlessly reworking the same set of figures. He was a particularly unfortunate case, having a degenerative brain disease; but he was only in the early stages of impairment, and in fact had clearly never understood how to a t-test for the last two decades. Other senior analysts, in their forties, were no more competent. The unit manager thought he knew how to do a t-test, and held a special seminar after I had been there about 6 months, to show everyone in the unit (except me[103]). After floundering for a couple of hours, he finally gave up the lesson, with the white board covered in a scrawled mess of failed calculations. The analysts stirred from their afternoon slumber and stumbled back to their crowded overladen desks no wiser than when they came.[104]

Equally, when I arrived, no one in the unit knew what an *effect size* was. My first task was to write up a paper explaining the most basic of concept of effect size, Cohen's d statistic: *(Mean$_2$ − Mean$_1$)/Standard Deviation* to this unit of specialist education research analysts. An influential professor of education had recommended the MOE use effect sizes to present results, and had done

---

[101] I would not want to give such examples from subsequent consulting roles, because there is a genuine need for confidentiality by consultants, especially when providing specialised advice to organisations in problematic situations; but this example is from a routine BAU job, and I can fairly call on whistle blower protection if the organisation objects to being truthed.

[102] I should add that I had no international travel in this role: it was a perk of seniority, with the senior manager going on regular European and American junkets, followed by the Manager, and then the senior analysts, who often only made it to Australia.

[103] The unit manager secretly arranged this seminar when I was meant to be at another meeting; but this was cancelled. When I returned early, I found all the analysts in a seminar being shown how (not) to do a t-test.

[104] I should add that t-tests were not satisfactory anyway, as multi-variable analysis was needed, but this was going *way* over their heads.

extensive meta-analyses; but no one knew what he was talking about. 'Effect sizes' are universal in social research, used for reporting group differences in a standard form. After long discussions I realised they struggled with the prior concept of a *standard deviation*. (They did have a qualitative understanding of a standard deviation, although they did not know what the *calculation* meant, nor did they understand that a *standard error* is the standard deviation of the estimate of a mean.)

Equally, no one knew what a *correlation* was, as a technical concept, and correlations were not used in any analysis before I came – despite being the most powerful simple tool available for many purposes. The term was used informally, as ordinary people everywhere speak of 'correlations' among effects; but no one could calculate or interpret correlations as a technical concept. All but one of these researchers had been in the MOE for decades; most would remain in their roles until they retire, choking the life out of the organisation. They were incompetent as a research unit and largely self-serving as employees. At the same time, it wasn't that they didn't *try*. Some were neurotic workaholics, and *wanted* to perform. The demands of the job were simply beyond their competence. I do have considerable human sympathy for them. They would struggle to perform in any professional job[105], but have found a niche in the urban jungle where they can survive. But they have found a niche at the expense of the NZ education system and the NZ public. Management is to blame for allowing these key roles and functions to be so degraded, to the detriment of the public interest they pretend to serve. But management was even more mediocre in their roles.

How did this situation of terminal incompetence arise? The unit had been carried for some years previously by a Senior Manager who was a competent statistician, and had left shortly before I arrived (to move overseas). They had appointed me to get some technical capability. The new Senior Manager (directing this unit and a couple of others) had herself been promoted from a senior analyst in the unit. She was ignorant of statistics, and could give no meaningful guidance on anything technical. She was domineering and dedicated to in-house politics: the perfect candidate for NZ public service middle management. She spent much of her time flying around the world to international education conferences or flying around the country to chardonnay lunches with cronies from the NZ Education research communi-

---

[105] With one exception, a recent 'policy' graduate, in his third year at the MoE, who was well on the way to becoming a capable analyst. But he left shortly after being promoted to senior analyst, moving to a policy role in the Australian MOE. He had little statistical competence, but was intelligent, ambitious, and capable because he had the ability to learn new concepts. He could have learnt how to do quantitative analysis adequately, but couldn't be bothered with the effort required, preferring to become a policy analyst instead. He realised that scientific analysis is not valued by the hierarchy, while policy analysis is the key to power and promotion.

ty. These are academic-bureaucrats from universities, teachers colleges and research institutes whose critical function is to suck up to MoE management to help get research grants and influence to keep their organisations afloat. It was very much a *life-style job* for her. She was a keen wine buff, and deftly coordinated her job with opportunities to socialise at free alcohol events ('professional networking' as it is called). The MOE, and our unit particularly, hosted regular drinking sessions. At the financial year-end, her research unit budget conveniently had a surplus of some four thousand dollars (out of a total budget of a few million). This was spent purchasing a bulk supply of wine and beer to stock the in-house cellar. (Otherwise the money would be wasted, she explained: any surplus from this year's research budget that was not spent would just go back to the MoE funding pool.)

The task of routine *data collection* by the unit was its main function, and this was done reasonably effectively. It was hardly demanding – the OECD studies were usually run in three-year cycles, and the scale of data collection was quite small. Part-time casual staff were hired to do all the donkey work – entering test records into the computer, organising school visits, etc. This test administration and data collection was routine work they could manage – it was determined by a 200-odd page rule book. However the quality of *analytic work* done in the unit – analyses of specific interest for NZ education, interpreting results in the NZ context - was pitiful. It caused constant performance crises, as analysis and publication of results was repeatedly delayed, for months that turned into years. The studies conducted when I was there were published about two years late. The published booklets were copies of previous formats with new data tables inserted; expensive, glossy, pompous marketing exercises; meaningless to inform any aspect of NZ teaching practice; meaningless for NZ teachers themselves; meaningless for the NZ pubic; a marketing exercise for the MOE *to its own political hierarchy.*

While there were questions raised about performance, and some hostility from other units, the ongoing performance crisis was hidden from outsiders. Other units in the MoE had similar character. The pattern was basic competence at routine administration tasks that they had been performing for decades – like property management, teacher salaries, accounting and HR - but incompetence at anything analytic or intellectual, challenging or innovative. Research and policy suffered most.

How did the analysts in this unit – five FTE analysts in a team unit with a total of about eight FTE staff - carry off their jobs? Well first of all they did practically nothing as *research analysts.* Rather, they were primarily engaged in *collecting data,* for international research programs that provided strict methodologies defined in rule books. Country-level teams were provided with comprehensive manuals detailing all the rules to follow, specifying how to choose samples and administer tests, and provided with all the software needed to record results. Data was returned to central HQ in Germany or the US for analysis, and results were provided back to country-level teams to write up in their own way. The unit had to do only a very small amount of analysis of their own, of special interest topics for NZ. The previous senior manager had spent years babying the unit, doing everything technically challenging, so their incompetence was highly exposed when he left. But as far as their core tasks went, they could simply copy previous research reports, replacing graphs and tweaking a bit of jargon – painting by numbers.

Their incompetence was exposed however when they were expected to contribute to wider MOE research. The OECD studies are for one narrow and specific purpose: to compare 'achieved educational performance' internationally, to make rankings of countries. This is the major political outcome for the Minister of Education, and becomes national news, especially as NZ slips down international rankings for high school language, maths and science education over the years. The primary goal of the unit was to keep this high-prestige research program going – this gave them a large budget, international travel, prestige and resources, and kept them safely away from real analysis tasks. They could simply 'follow the manual' without understanding any concepts. Their biggest concern was the Minister's reaction to poor national results – which of course makes bad political news.

But the cost to devoting practically all the MoE's research effort to these 'high prestige' international projects was to ignore domestic education research, that should be used to address the specific and glaring educational failures of the NZ system itself. The data obtained from the international surveys was specifically and narrowly designed to allow international comparisons between countries. Samples are small (on the scale of about 100 classroom samples per study), and it is poor quality data for assessing domestic education. It is not designed for this. There are far richer domestic NZ data sources available – from research studies and standard assessment tools that are widely used across the country, with tens of thousands of individual assessment records.

The MoE comparative education research unit was expected to contribute to wider research analysis, but continuously resisted this, for two main reasons. First, the domestic data sets had barely ever been analysed before – with only superficial reporting of averages, usually by the obvious factor of decile – so there was no *predefined method of analysis* to copy. The research analysts had no idea of how to do real, original *data analysis* for themselves. They were completely incompetent to develop a *secondary data analysis program* for themselves. This prospect threatened to expose their incompetence. Secondly, the international studies had high prestige and large budgets, lots of perks, lots of travel, while the domestic studies were low prestige with shoestring budgets, and demanded real work and real analyst capability.

A good illustration of this happened one day at a meeting with another unit we were supposed to support with data and analysis. The senior manager for Early Childhood Education – one of the rare managers in the organisation who was dedicated and competent – asked if we could provide data and analysis to show how key differences in achievement levels, analysed by decile, ethnicity and gender, carried through school life, over time. She was well aware from experience that large class-based disparities already established when children first come to school at 5 years are carried through middle and senior primary school and secondary school. She wanted some real data analysis to confirm and quantify the effects. She was trying to get support to provide for free early childhood centres, as a realistic way of addressing large class-based inequalities that are ingrained in NZ education.

At this request, my manager hummed and harrumphed and said it would be very difficult. Pressed to estimate a time to produce some data, he said it would take at least three months of analysis work – even if it was possible to start on it immediately, which was impossible. The Early Childhood manager looked dismayed: everyone knows that three months means six months, then a year, or more exactly, in an degraded environment like that: *never*.

Yet I was working daily with the data she needed – having been assigned to take care of troublesome domestic data collections that management wanted to ignore - and had key data sources at my finger-tips. She was perfectly entitled to have the data. I told her I could get her the data and basic analysis within a day or so. By the next day, I had extracted and compiled the key results she needed, from about 5 different national assessment sources, with the essential analyses she wanted. I had the results back to her in a *day or two* – rather than *three months*. Results showed that sub-groups coming to school already disadvantaged at five years old (with low-achieving groups being already 1-2 school years behind high-achieving groups) maintain the same disadvantage throughout their school life. This was of essential importance for her case to get resources and support for early childhood education. She knew from her own experience that educational disadvantages established in early childhood carry over for the rest of school life. Her mission was actually to improve early childhood education. But this was of no interest at all to my manager. He was resentful that I provided her with the data she needed.

This is also an example of what I meant earlier when I said that *in our highly interdependent world, it is important that we are competent in the jobs that others depend on us for.* Dependence on my unit manager undermined the Early Childhood Manager's capacity to do her job. It was hardly the first or last time. A couple of years later, I saw her on national TV, being interviewed about why she had angrily resigned from the Ministry of Education. She was bitterly critical of the organisation, where she had just wasted five years of her life fighting against a smug bureaucracy for support for Early Childhood resources. In fact, she was the *only* manager I met in the entire organisation, in the course of a year, who impressed me as serious and competent in her role, with a commitment to actually trying to improve educational outcomes.

I think she was also the *only* person I worked with in the Ministry of Education who had actually been a teacher.[106]

I will add another anecdote about my MOE unit manager, which connects with a theme I will also return to: the failure of information systems in such organisations. Information systems are integral to all modern enterprises, and their failure is central to organisational failure. The unit manager had one special task: the most challenging technical task ever undertaken in the research unit. A task of breath-taking ambition. A software program. He spent most of his time on this in my first months there – and as the months went by, progress was slow. At weekly unit meetings, he always reported he was on the verge of getting it to work, but always needed another week or two, to iron out the bugs…

The task was to write a program to produce what are called *jack-knife standard errors,* generated from our small trial data sets This is a well-known technique to estimate standard errors for stratified samples, where stratification or clustering effects make the assumption of random normal distributions invalid. (Key stratification of samples was by ethnicity and decile, with clustering by classroom-room level sampling.) Eventually, after several months, I was asked by the increasingly impatient senior manager to help him. (He was falling months behind in his core admin tasks, and this was becoming visible to *her* boss).

After spending a few hours discovering what the problem was (it had been kept rather secretive up to then), I looked up the standard formulae for producing jack-knife standard errors, and wrote a program to implement it and generate the jack-knife estimates. It requires iterating a procedure over multi random sub-samples, so it has a moderate complexity; it is not just plugging numbers into a simple formulae. But it is not very challenging, and took about a day to write the program, and a morning to test it. It worked perfectly well, produced its results in a few minutes, and I subsequently used it for analysing some domestic studies. It is not a challenging programming task *for an experienced programmer with some mathematical background.* On the other hand, it would be very challenging for an amateur programmer without techniques to iterate procedures effectively and without the ability to read mathematical equations. The manager unfortunately was an amateur programmer, with no mathematical background. His attempts to program it were doomed to frustration from the start.

---

[106] The MOE policy analysts, ostensibly responsible for such concerns, lived in a supercilious fantasy world of political agendas. They regarded the education system not as an environment for *learning,* but as a social engineering tool. They were intent on a politically correct syllabus and politically correct organisational processes, conforming to a neo-liberal bourgeois moralism.

But the manager disdained to use my program, and continued with his own attempt. In fact he was dismayed that I had *developed a program from scratch*, instead of trying to fix the software mess he had created. If I had fixed his mess, of course he could have claimed success. But it would have taken me a week or more to unravel the mess he had made, and it would always be compromised because if its underlying poor design. Far easier just to build a proper version from scratch. But how could he abandon all those months of work he had done? He ignored my solution, and asked programmers from the OECD project HQ to fix his problem. It turned out he been trying to modify a more complex program he had got from them (he had no idea how to program it from scratch himself), and he had wrecked it. They subsequently fixed him a workable version, and the topic disappeared from conversation.

This is a typical example of where *technical programming incompetence* can be a killer. The line between being capable of doing something technically complex, and *not being able to do it at all* is typically stark. You can fluff around some policy analysis by putting in a few catch-phrases you know people want to hear, and no one will really be able to tell that you don't know what you are talking about. But consider some standard maths or programming problem, say from a university exam, where you need to prove a result or write a short program. If you do not know the subject it will be *completely* opaque. You can look at it, not just for hours or days but *forever*, and not be able to solve it – or even understand what it means. You can't *pretend to solve it*. Yet once you learn the concepts and technique it becomes transparent, and you can solve it in ten minutes, and *get it right*.

The same goes for many programming and analysis problems. Amateur programmers, with no idea of proper techniques for recursion, functional programming, automation, or no idea of the scope and suitability of different programming tools for different tasks, will spend months and years cobbling up ugly, over-complex 'work-arounds', kludges, that are unsustainable and unmanageable. They require constant user intervention, take large amount of time, a prone to bugs – and have no future. They become legacy problems that someone in the future is going to have to unravel. When you have layers of inter-dependent programs of this kind, information systems turn to mush. This kind of incompetent programming *overwhelms* large organisations. I have wasted years of the time of my life fixing this kind of crap, often for arrogant and resentful managers who caused the mess in the first place and now *demand* solutions; until I finally decided to cease working in this sector altogether.

The unit manager in this case was regarded as a technical wizard by the senior manager, because he could work out how to program *something*, he could follow technical instructions in manuals, and get programs supplied by the project HQ to run, and so on. The unit was absolutely reliant on him for this capability to do anything – others in the unit were technically helpless.

He sustained himself in this role of 'technical guru'. Of course he was not comfortable when someone with real programming experience came along and saw these technical challenges as routine. For such a personality – a very common personality in bureaucracy - this creates an impossible position. An open, honest, confident manager with the same technical abilities and limitations would easily navigated this situation, and simply put the new analyst to work on technical problems suited to their talent and to the needs of the organisation. For this MOE manager it was a crisis of ego and control.

As this uncomfortable relationship unfolded, after some eight months at the job, the manager responded by giving me a poor job review – rated as *just meeting expectations* in areas of programming and statistics; and *failing to meet expectations* in other categories, such as organisational knowledge and following orders (areas that have no objective measures). This is equivalent to getting C and D grades. In terms of technical capability and conceptual knowledge, my performance was well ahead of all the analysts in the unit; and I had a proven work record and much wider experience in many previous projects from which to judge my own performance. The manager took an obvious pleasure in giving me the poorest job performance review he could possibly muster. I made little comment, but realised immediately that working under such a manager was poisonous.

Ironically, I was subsequently compelled (as part of a new management innovation to get a '360-degree' picture of managerial performance) to give a job review of *his* performance as my manager. I gave an honest assessment, with low ratings in almost every area. This was not exceptional in most respects: his performance as a *manager* was criticised by all his staff. Well, he was quite simply one of the worst managers I have ever encountered. Other staff complained that he was a year behind in their performance appraisals and other routine managerial tasks, which affected their potential pay-grades. But it was my low ratings for his technical competence that hit a nerve. This conflicted with the fantasy picture that my senior manager had built up, and had sold to *her* boss.

I was then bitterly attacked for this assessment by the senior manager, who angrily pressured me to withdraw my assessment. She was afraid her boss (the group manager) would see criticism that reflected badly on her. I subsequently resigned. She subsequently engaged her networks to attack me – something I had seen her do to others with great vindictiveness. (In fact, she confided to another analyst that the reason I made her so *angry* was because I reminded her of her ex-husband, apparently a talented analyst, who she hatefully resented out loud on many occasions.) This stymied my career prospects for some time. But I moved on and became a successful private sector programmer/analyst contractor. I avoided work for any government-related organisations for over ten years. For any creative and constructive individual, the experience of working in such an environment is soul-destroying.

As an epilogue, I was finally talked into doing some work for a couple of other government organisations many years later - now as a senior contractor being charged out by multiple layers of expensive consultancies – and found managerial incompetence as dismal as ever. My last role, developing a 'business intelligence' system for another core government department in 2013-2014, finally convinced me to avoid any further participation in the NZ state-corporate sector. There are a lot of nice, sincere and dedicated *workers* in the public sector – just as much as anywhere else. But *project and innovation and policy management* is particularly poor, along with executive management. Policy and research capability is appalling.

To illustrate the incestuous nature of the public service, my manager in my last role had come from the Ministry of Education, recruited as a specialist 'business intelligence manager' to develop capability, primarily software systems, for another degraded Government Department. But she had no idea of modern 'business intelligence' systems: she had never used one! She had never developed *any software*. She had never worked on *any* development team. She had never heard of *dev, test* and *prod* servers. She had never heard of conceptual models and data models. She knew nothing of database design, warehouse design, datamart design. Her idea of a BI information system was the typical point-to-point data extraction always visualised by amateurs: the childish design that leads straight to BI hell.

On the statistics side, she did not know what a *t-test or a correlation* was. She had absolutely no quantitative analysis skills. She claimed to be a SAS programmer, but no one had ever checked - her senior manager couldn't tell a programmer from a jellyfish, and neither could anyone else in the Department. I soon discovered she had no programming skills. She did not know what a SAS variable is. She had no concept of a programming loop, with iterated variables. She had no concept of automation. She could execute one-line SAS commands – that was what 'SAS programming' meant to her.

We are not just talking about some low-level analyst: she had been given control of developing the whole analyst environment and directing in-house education, for about 25 analysts, in control of *national priority databases* - passports, citizenship, births deaths and marriages, call centres, finance, etc. When I had to hand over managerial control of various BI systems I had built, which extracted information from these on a daily basis, she was unable to follow simple database administration instructions, and she did not think that the rules of the system applied to *her as a user*. Why, she was the *manager!*

When she needed to hire an analyst-programmer in a permanent role to do the technical part of her job, she turned down all candidates with any serious programming or analysis skills – the capability the Department most desperately lacked - on the basis that *the candidates were not from the public sector.* They would not understand the public service culture.[107] She would only employ a career public service drone, like herself, *who conformed to her public service expectations, and who would not be a threat to her position.* The lesson I take is exactly the opposite: *do not employ anyone who has had a career of any length (say more than 3 years) in the NZ MOE, or any other NZ Government department. Especially* not in any programming or analysis role.

The state-corporate context of science.

A few petty, isolated examples? you might wonder. Everyone becomes disillusioned with their managers and colleagues at times, everyone knows that office politics is a universal pain in the butt, and everyone goes through a job or two that they hate. Why even bother to complain? Well, because these examples are not isolated: this level of incompetence, of systemic mediocrity, is *typical and universal* throughout NZ state-corporate organisations. It is especially the curse of anyone with real scientific or intellectual ability who wants to engage with life outside academia. In a wider sense, it is the curse of everyone who wants to participate in any business or enterprise in NZ. It represents a grim outlook for NZ public sector capability. It represents a systemic failure in every aspect of governance, as systems slowly spiral into decay, mismanagement and corruption from within. These examples show all the features that typify the NZ culture of mediocrity:

- Systemic lack of capability in analysis
- Systemic failure of information systems
- Self-serving and bullying management
- Destroying real opportunities for achievements
- Isolation and exclusion of serious performers
- Incestuous bureaucratic employment practices
- Feudal empire-building culture
- Failure of neo-liberal management to have any effect

The Ministry of Education determines the future of the NZ education system, and dominates the education industry. Yet its senior research analysts are not even competent to teach senior high-school statistics. Its policy analysts and bureaucrats generally have *no* scientific competence – and most certainly do not value science. Almost no one in this huge organisation that governs the teaching profession has ever been a teacher. The MOE has a

---

[107] Actually I wonder if this discrimination is against employment laws?

dismal performance over many years, some on public record: disastrous software scandals, arrogant treatment of schools, principals and teachers; disastrous national curriculum and assessment reforms; a disastrous school mathematics reform recently abandoned after years of failure; etc, etc, etc.

This is not exceptional or isolated or petty: it is the typical state of degradation of public sector institutions in NZ, and it is the *major* problem for the future of NZ as a nation, and certainly as a scientific nation. For we surely depend on leadership from a public sector that controls some 40% of our GNP, dictates vast tracts of rules, regulations and policies, and determines practically all our scientific and social research spending. NZ has been in steady decline for about 40 years now – since the floodgates opened to mass emigration to Australia and other developed Western nations in the 1970's, the cream of intellectual talent has been streaming off-shore to countries that offer at least some opportunities for intellectuals. NZ remains a back-water, where middle-class professionals and bureaucrats can have an easy life, riding on the back of the wealth produced by a large primary produce sector – but *it has no sustainable future as a first-world nation. Don't move here if you want a career that values intellectual or creative abilities.* There is no R&D sector, no corporate sector, practically no technology or scientific opportunities outside the public sector, which dominates all employment for researchers, scientists, analysts, and computer system developers. In this respect, NZ is a special example of what happens when a country loses its intellectual and scientific brain. There is no longer the intellectual horse-power available to empower a modern developed economy. It is stuck in a rut as a farming province, providing cheap primary produce for developed nations.

Politicians are not unaware of this. They despair at public service capability too. Successive governments propose public sector reforms, with seemingly endless 'restructuring', and have tried repeatedly to introduce neo-liberal corporate-style management practices – to no avail. The problems cannot be fixed by *managerial reforms within the established system* because the levels of incompetence are too extreme. Mediocrity is too systemic. The public sector cannot attract or retain intellectual talent, and *you can't make a silk purse out of a sow's ear.* The human capability to navigate the challenging tasks of the modern world simply does not exist within the NZ public sector, or in NZ's stymied private sector either. It does not exist *at any level of leadership, middle-management, analytic, research, technology or policy.*

NZ Public sector scandals are found in the media on a weekly basis – and those are only the ones journalists discover, and that their editors figure will make popular copy between the staple of celebrity scandals. Deeper, chronic, ongoing failures, failures of core capability, remain hidden under layers of feel-good image marketing. The mainstream media rarely ventures into such territory: it is intent on a positive nationalistic spin of NZ as a progressive, innovative country. Stories about the real intellectual impoverishment would hurt the popular national ego.[108] On the other hand, the public is so used to the systemic failures in the public bureaucracies that scandals of incompetence barely draw comment beyond a resigned cynicism. People just shrug and say: *So what's new?* If provoked, practically every adult in NZ will tell their own never-ending stories of bureaucratic incompetence, ranging from routine failures of basic services to serious abuse by government agencies – stories repeated over decades past, but more common than ever today, as people are forced into more and more bureaucratic relationships, governed by more and more complex regulations, with more and more opportunities for failure. (Try asking any small business, farmer, or self-employed contractor or consultant about IRD, WorkSafe, or resource consent processes. Large numbers of these people are openly contemplating *business suicide: selling up and getting out.*)

I also use the example of the education bureaucracy because the failure of the education system has an intimate connection with the degraded scientific capability in NZ. Education policy is primarily concerned with *social engineering*, not with supporting educational goals in schools or tertiary institutions. Its operation is dominated by policy ideologies, dedicated to imposing the middle-class political conformity of its smug bourgeoisie management. All public institutions in NZ have a similar character. Welfare, health, economic, justice, business, environment, social, education – and science - are dominated by top-down *political agendas,* not by goals benefiting the stakeholders they are meant to serve. Consequently, loyalty to an organisational hierarchy and ideology is paramount at all levels of management and leadership. No one who challenges an organisational ideology from within is going to last long. The dominant mode of thought within management is a fuzzy blend of self-serving *political, moralistic and emotionalised ideology,* with objectivity, science, or factual knowledge subservient to these. It is precisely the antithesis of scientific thinking, which seeks to separate objective knowledge from religious or political belief systems, to allow impartiality from our emotional reactions towards consequences of facts and theories, to apply analytic intelligence to problems.

---

[108] Bizarrely, NZ retains a popular assumption that it has one of the worlds *leading education systems (!?!)*, that it is *highly innovative* - and even that it has *world-leading scientific-technology innovation (!?!)* These myths were inherited from the 1950's, when NZ did have one of the leading *per capita* incomes in the world, based on an innovative farming sector, and a privileged economic relationship to sell produce to Britain. Since then, it has declined steadily to a second-world capability, with a dismal scientific and innovation performance, mediocre education, a large underclass of poor and unemployed, the highest imprisonment rate among developed countries outside the US, and a popular-political culture derivative from the US.

Corporatisation of NZ public science.

The same degraded culture also dominates the public science sector in NZ. It is reflected in the *corporatisation* of public science and science funding institutions. The bureaucrats who control science and research funding are *business managers* who are grossly incompetent at science themselves – just as MOE bureaucrats are themselves grossly incompetent as teachers or educators. There are many public science research funds, administered by government bureaucracies. Funds are advertised as 'contestable', but they are realistically only available to other bureaucracies – CRIs (Crown Research Institutes) or SOEs (state-owned enterprises) and their spin-off companies, masquerading as 'research institutes'; and to public universities - a sector also seen by government primarily as a *corporate profit center.*

Public research money mainly just sloshes around in a circle between government departments, CRIs, SOEs. There are multiple agencies posing as *scientific research agencies,* such as *AgResearch, HortResearch, ESR,* etc, who receive the vast bulk of science research funding. But they are essentially tasked with making profits, with acting as businesses. 'Profitability' – by whatever monetary measures the bureaucrats wish to impose - is the performance measure. They do no pure science research worth the name; and little applied research of any quality. Three or four decades ago, their predecessors (such as the DSIR) were real science research agencies, doing pubic-good applied science; mostly routine science, occasionally inspired, effective as a scientific platform in the context of an agricultural nation. But their capability has been degraded over the last decades, as successive governments have restructured them on a neo-liberal corporate model.

There is now a flood of complaints from senior scientists who have left these institutes, about the degradation of their capability, and the failure of their corporate-style management. CRIs or SOEs now provide commercial services that compete with private sector companies. They cannot support 'public good science' because that would contradict their commercial imperative. Instead they increasingly try to use their Government muscle to create *GPPs* - government-private partnerships. Of course, supporting their corporate partners means locking all other enterprises out, and denying support to industry as a whole. For example, ESR ("environmental and scientific research"), effusively claims in its own marketing to be NZ's 'premier scientific research institute', but it really does *no scientific research* at all. It provides technology services, primarily chemical analysis of samples (clinical hospital samples, forensic samples, food and drug samples, chemical samples, environmental samples, etc). The key clients are mainly other public organisations (police, customs, hospitals, etc), with whom they have long-held monopolies – although these are unravelling now, as these clients are now being allowed to look for competitive alternative services. They compete against commercial companies (like Hill Laboratories) for private work – in a losing battle, despite the advantage of large government subsidies for infrastructure and salaries. Without huge subsidies they could not survive. They really don't know if they are a bureaucracy or a business. Their senior executives and directors are third-rate business management clones and quasi-political wannabes.

This State-Corporate identity crisis appears to most commentators to be a direct result of neo-liberal government policies over the last several decades, that sees every government institution acting as a *corporate profit center*. This demand to be profitable now extends to most core government services. All regulatory bodies fix their prices for services (i.e. for providing licenses for activities) as monopolies, and charge far more than any private companies with equivalent kinds of service complexity do. They legislate to create vast client bases, and legislate their own business environment in an *ad hoc* manner as they please. Their fees are officially justified to cover their own inflated bureaucratic costs.

Neo-liberal reform of NZ science.

López Corredoira complains of the corruption of greed and money as a primary driver of wider corruption of science and society. The NZ public science sector has become a perfect illustration. The failure of AgResearch, NZ's premier agricultural research institute, provides a popular target for such criticisms. Condemnatory criticisms are rife. Professor Dick Wilkins, now at Waikato University, in a recent article in the NZ Herald concludes:

> "To be blunt, we are fast approaching the stage where, if AgResearch were to disappear from the face of the earth, there would hardly be a ripple in New Zealand agriculture. This is a tragedy, and the only solution is to disband AgResearch as we know it, and reinvent it as a science led organisation." (Willis, 2015) [109].

Another trenchant critic is Dr. Doug Edmeades, an ex-AgResearch scientist now working as an independent consultant. Edmeades is no left-wing radical: he is a conservative, orthodox agricultural scientist, who puts forward forthright opinions. He has an interesting 2004 study in which he identifies the core failure of NZ's public science institutions as a failure of the organisational model, specifically, the dumping of the *normative, public good model* for the *neo-liberal corporate profit model,* in the ideological reforms that swept through NZ in the 1980's.

---

[109] "Emeritus Professor Dick Wilkins worked at Ruakura from 1989 to 1995 before moving to the Department of Biological Sciences at the University of Waikato."

"The reforms have moved New Zealand's government funded science from a normative organisation into a performance-based organisation, managed by measurement, particularly financial measurement". (Edmeades 2004, p. 88.)

He recommends returning to a normative, public good model. He understands that science is based on values outside commercial ones, and its quality is reflected by its human capital.

"Ziman (1994) has identified some of the necessary conditions for science to flourish as follows: 'Any research [science] organisation requires generous measure of the following: a) social space for personal initiative and creativity, b) time for ideas to grow to maturity, c) openness to debate and criticism, d) hospitality towards movelty and e) respect for specialised expertise'. These very general requirements emphasise that science is a creative, human activity. Devine (2003) takes this further when discussing the role of science in economic development: 'Human capital, embodied in the people with the skills and 'knowhow' is the key to economic development'. The key resource, in any science organisation, is the human resource – the pool of suitably trained and experienced scientists – not bricks and mortar, typically the major resource in a commercial operation." (Edmeades 2004, p. 89.)

I think probably every *natural scientist* I have ever met would agree with this.[110] But this opinion falls entirely on deaf ears. Ten years later, his criticism grows increasingly trenchant, and his popular articles increasingly polemical, as he sees organisational failure in NZ science rife, and worsening, with no redress in sight. The following is an example of his present view, expressed in a series of recent articles in a popular farming sector magazine. The following article is entitled *"McScience diet is hard to swallow."* Edmeades, *NZ Farmer,* June 15, 2015.

---

[110] I was going to say 'every scientist', but I remembered that many (so-called) *management and business 'scientists'* would disagree with it, supporting a neo-liberal economic fundamentalism instead.

Agricultural research is a shambles, says **Doug Edmeades**. How did it come to this?

MY spade has hit a wellspring of agreement – agricultural research in New Zealand is a shambles and it appears from feedback that many agree with my assertion.

So let's clean and sharpen the spade.

How and why has this happened?

The Labour government started the process in the mid-1980s to commercialise agricultural research and technology transfer – remember "users pays"?

This policy bud blossomed with the passing of the CRI Act (1992) and the establishment of the Crown research institutes.

Scientists were now required to do research for the public good and make a dollar – a return on investment.

The reforms, we were told, were intended to: Increase the efficiency, accountability and flexibility; achieve better alignment between science and industry and enhance technology transfer.

It is beautifully ironic that I left AgResearch Ruakura in 1997 because I could see that gap between science and the farmer was becoming increasingly wider and I was told, "Your job is to make money for the CRI, not inform the farmer".

Truth be told, I was depressed, distraught and confused.

Was I going mad or was the new science system insane?

I embarked on a post-graduate diploma in management to ease my troubled mind.

I got to read erudite papers on organisational models – why are there so many different models and why different models suit different endeavours.

The commercial model is good for delivering tangible goods to the public.

You can buy a loaf of bread or a hunk of 4x2 anywhere in New Zealand at about the same price.

The unseen hand of the open market applies.

But it is not the best model for delivering services that have a large intangible component.

Hence the distribution of education and health services to society is achieved largely via a public service model.

It should be no surprise that most of our agricultural industries operate within the co-operative model – specific goods and services for the benefit of a specific sector of society.

In this context, it is fascinating to realise that most of our humble human activities are performed within the not-for-profit model – sports clubs, social clubs, incorporated societies, charitable trusts, churches etc.

The focus is on community good for the good of community.

The "goods and services" are largely intangible.

Importantly, the absence of the profit motive does not mean inefficiency – it simply means that any money left over (profit) cannot be appropriated away from the organisation.

Where does science fit along this continuum?

To answer this we must explore the question – what are the values implicit in science?

Is there a set of operational conditions that would optimise the conduct and delivery of science for the public good?

Here are some of the important ones:

▊ The integrity and objectivity of science and scientists must be protected.

▊ The motivation for doing science and speaking about science must not be compromised.

▊ The organisation must be open internally and externally to discussion and debate. This is essential for defining problems of public importance and the development of scientific ideas.

▊ Science is a creative activity and hence scientists must have time and space. The funding, planning and reporting processes need to reflect the long-term nature (five-15 years) inherent in science activity.

Without any objective analysis of the values and requirements of science, the Rogernomes thoughtlessly assumed that the one-size-fits-all commercial model could be applied to science.

They could not have done worse.

The purpose of science now, operating under the CRI Act, is primarily focused on making money, hitting revenue targets and gawking at the bottom line.

Welcome to what is now being called Post Normal Science. Science is no longer about seeking to understand the world around us for the betterment of the public.

The role of science is now to serve the machine – the CRI.

The age of McScience is upon us.

And spare a thought for Dr McScience.

We know from surveys that he is distrustful of management and disillusioned with his career, to the point where he would not recommend a science career to his children.

We know he spends little time doing science. He is too busy being "accountable" to be accountable, preparing budgets, reporting on projects and developing new funding proposals.

He must rush to and fro, greasing the many palms of the many "shareholders" that feed him to ensure his begging bowl is never empty – a never-ending rush to stave off the day when the wolf arrives and claims more colleagues who "regretfully" did not make it into next year's budget.

Dr McScience tries hard to smile when he hears his senior management repeat the mantra "Our people are our most important asset".

It is a pity that there is no heading under "Assets" in the balance sheet called "Scientists". Its absence allows an otherwise perfectly sane board chairman to explain that more science staff have been laid off to protect the company's assets.

My point should be obvious: When the values and needs of science are considered, the last organisational model you would choose for science is a commercial model.

That is the tragedy of the CRI Act and it is the single most important reason why agricultural research in New Zealand is a shambles.

Is there a solution? I believe so, but that will have to wait for the next column.

▊ Dr Doug Edmeades, MNZM, is an independent soil scientist and managing director of agKnowledge. He was Federated Farmers' Agriculture Personality of the Year in 2012 and is a former Landcorp Agricultural Communicator of the Year.

Criticism of NZ's public science sector by senior scientists who have withdrawn from the bureaucracies is rampant. The most universal theme is the destruction of scientific culture by commercial models, and bureaucratic incompetence to manage science. The corruption of scientific values by corporate profiteers has of course long been notorious, through the machinations of Big Tobacco, Big Pharmacy, Big Chemical, Big Oil, Big IT, Big Genetics, etc, with a long history of concealment and corruption of scientific research for commercial gain, and destruction of competing companies, products, ideas or critics to maintain power. Observers since the 1960's are deeply cynical of scientific research or opinion from the corporate sector. In fact, you can predict private-sector scientific opinion on almost any controversial topic *from the industry the scientist comes from.*[111] Well, we all know that 'scientific opinion' issuing from *any* sector with self-interest at stake will be calculated as a marketing exercise. But the same force for corruption and mediocrity was *deliberately* introduced into NZ's public sector science in the 1980's through ideologically-driven government policy.

This refers to a dramatic change in management of NZ public institutions in the late 1980's, when a radicalised new Labor government ushered in sweeping neo-liberal reforms, based on an extreme version of free-market economic policy, applied to all goods and services, extending this to what were traditionally public goods – public-good science, as well as energy, communications, transport, health, education, justice. This brought a dramatic change in state management philosophy. Corporate-style feudal management based on profit-making was grafted onto the traditional state-sector bureaucratic function, to become the norm. We have since seen the dominance of corporate-style elite power-cliques across the new State-Corporates, including the new science corporates.

The dynamics of these elite power cliques is now definitive of the modern science industry, as an extension of State-Corporate bureaucracy. I now turn back to some of the psycho-social mechanisms that drive the behaviour of such organisations.

---

[111] Scientists from Big Greenhouse Gas industries (oil, cement, autos, dairy farming) will invariably belittle climate change and environmental issues. Scientists from Big Genetics will belittle concerns about GMOs. Scientists from Big Chemical will belittle effects of pesticides, herbicides, chemical additives. Etc. If you can predict opinions from self-interest factors, this shows they are not scientifically objective.

The dominance of elite power-cliques.

The dominance of power-cliques is so universal in large institutions, government and corporate alike, that its rationale is one of those invisible cultural assumptions that people seldom question – like the role of aristocratic title in a feudal system. The people in these cliques may be incompetent in their roles, but they are not stupid. Along with strong personal ambition, their common feature is high *social intelligence:* meaning social manipulative ability. They are good at maintaining their power networks, and at talking, presenting, acting out roles, manipulating perceptions. They are thespian managers, focused on managing relationships with the other power players in the executive circle - as opposed to the real managers that you get in front-line services or in small free-enterprise type businesses[112], who are focussed on the business services, and know the detail of what their business and their workers do.

Social intelligence is the main type of intelligence of the human race, the most universal. It is what gives us capacity to participate as social and moral beings. But the people who succeed in the power elite generally lack interest in the second main sort of intelligence – *rational or scientific or technical intelligence,* the intelligence that mathematicians and scientists and analysts and researchers and technologists and planners need. This is the ability for rational planning, calculation of complex cause and effect, counterfactual reasoning, conceptual vision, creative problem-solving. This is the capability to *concentrate on and solve intellectually demanding problems of causality.* This is the more recent type of intelligence in human evolution, and it has the most extreme variations among individuals – with our higher scientific and technology innovations powered by a tiny group having this mode of intelligence in an extreme form.

As López Corredoira observes, ability in science is not distributed *democratically.* Yet it is generally felt that individuals with such gifts owe it to society to use them 'democratically', which is popular parlance meaning *for the benefit of the group; or for the whim of the masses.* Thus while it is expected as part of the natural social order that those with high social intelligence – sales people, executives, marketers, merchants – will exploit their abilities to make money, those with high scientific intelligence are expected to offer their intellectual gifts as servants on the altar of the 'public good'. The invention of medicines, machines, power plants or weapons is considered too *important* to be left to purely commercial exploitation by individuals. This is part of the popular view that *scientific products and power* needs to be controlled by social leaders. Science must be carefully *managed* so its products are used properly. Scientists (and other advanced intellectuals) are expected to be *grateful* to be

---

[112] The ideal of small *free enterprise* business being the antithesis of state-corporate *free-market* ideology.

allowed the opportunity to work, grateful for the opportunity to be published, grateful to anyone taking an interest in their work.

Executives often dislike people with this kind of intelligence that they do not have themselves – it makes them feel intellectually threatened, which is the *worst* feeling they can have, being the biggest egos in the building. They are not aware of how anti-intellectual they are, for by their own standards they are outstanding thinkers. They are such brilliant social-business thinkers that they do not need the pedestrian analytic-scientific intelligence of their intellectual servant-class ('geeks', 'egg-heads', 'propeller-heads', etc) to work out what is best when it comes to *decision making*, to policy or strategy or management – they only need experts to implement their decisions. They like technical reports to be simplified, dumbed-down to a presentation they can take in at a glance. They do not like intellectual complexity. They like elevator speeches – 30 second encapsulations of ideas. They like simple statements that capture consent. They like rhetoric. They like the sound of their own speeches. And they like their own society: the society of *other power net-workers, social game-players. Important people in the world.* For theirs is a world of people like themselves; a world of actions and acquisitions and legal certainties; not a world of ideas and possibilities and mysteries.

These people have a monopoly on the roles of the *upper management and executive,* they control higher-level decision making, over the heads of creative intellectuals who can think through problems carefully. They usually have little objective idea of what their decisions mean in any detail. In management meetings with staff, the leading analytic or technical staff often politely pretend that the executive is talking sense – knowing full well they are just making speeches, blabbering generalities, repeating clichés. It is this mediocrity, structurally built into our institutional systems at the highest levels of power and decision making, that is the determining feature. The competence of individuals below them becomes a problem too, because they undermine the technically and intellectually talented workers below them in the hierarchy, driving away the talent, and retaining the mediocre conformists that they can easily control. This is part of the adaptive social feed-back mechanism that sustains them.

The problem is also seen in the career structures in these organisations. To progress beyond a limited point, you *must take a managerial role.* High-level intellectual staff – leaders of research, analysis or innovation projects – can only progress to the salary and influence levels of lower middle-management staff. To go any further in these organisations, you *must take a managerial and administration role.* The most pitiful drones in middle management have more social and material rewards than the most talented and hard-working intellectual staff. *IT project managers* for instance rarely have any programming knowledge – but control programming teams. *Policy project managers* rarely have any analytic knowledge – but control analysis teams. This is the expression of the neo-liberal corporate business model, instinctively designed to give opportunities to an executive class. Their attitude is: *We are the important people, the aristocrats of the enterprise, with the vision and social intelligence to make the important decisions. Technical staff are there as slaves to provide the capability to put our important plans into action.* This is very much the modern public perception of science: *scientists are there to provide the technology for us to use: they are accountable to our public purposes, not to their own judgements.*

Of course the executives can rightly say: "*We can't just put intellectuals into executive power-playing roles, it is a full time job managing these power relationships, it is all about your network with other power players, controlling the propaganda channels, intimidating others in the network to get what you need for your own projects, your department, your organisation, getting their confidence that you know how to play the game with them. For this is a feudal competition for resources and power, between departments within your organisation, and between organisations. You cannot do this as a part-time job while concentrating on complex technical matters. You must devote your entire energies to your social performance in meetings – you will essentially spend your entire time in meetings. You do not have time to think. This why we have to have this specialisation. And rewards for individuals in these power-roles must be commensurate with their responsibilities and their place in the power hierarchy: they are dishing out huge budgets, making the big decisions.*"

This true too: as the system is organised now, it *is* impossible for intellectuals, research or analytic or technical leaders, with their minds full of all the detailed content of their domains, to be effective in the power structure. But this is exactly the *problem. As the system is organised now,* intellectuals are excluded, and only those who have developed power networks, who have mastered organisational propaganda, who understand how to bully and manipulate the others in the power networks, can succeed there. We are locked into this stratified world, with control of power and money at the top, pursued for its own agendas. The world of science has now been stratified into this neo-liberal business model too. The question is not whether intellectuals can succeed in the present system – they cannot. The question is whether the system itself can succeed without intellectuals.

The voodoo science of neo-liberal business management.

The neo-liberal business management ideology must be identified as a pillar of failure. Not just as applied to the management of science, but as a *scientific theory of management.* Academic 'business science' theorists have a model of business as a kind of machine, with people as cogs. The model tells them that by putting 'metrics' on 'performance' of workers they can ensure efficiency – weed out poorly performing workers, enforce conformity to unit standards, etc. It is all about *optimum productivity at lowest cost.* It has a reductionist metaphysics of 'inputs' and 'outputs'. It uses a simplistic one-dimensional model of value: *money is the measure of all things.* This is the fundamental axiom of conventional economic theory of course, which cannot be questioned within the orthodox science of economics. Thus the criterion for success is *profit for the business owner.* Science is regarded like any other factory, as a production line with inputs and outputs.

This focus on 'performance metrics' and 'financial profit' is a delusionary idea of how people really work, of how businesses and organisations really work, as *ecologies and cultures.* It blinds us to our larger natural goals, the creation of value in a much larger sense than money. And it blinds us to the fact that failure originates primarily from *mediocrity and delusion at the top,* not in the enslaved work force that has to work in degraded systems forced on them. (There is an assumption that the business owners or executive level of management *define* the goals and strategies for the organisation by some natural right, and the 'problems' lie in optimising the performance of the 'work force' and its systems for these goals. If you *define* the goals as *whatever the business owner or executives want,* then they can never be wrong – even though they destroy their organisations by their mistakes.) The focus on profitability is a delusionary idea of value at the deepest level; and it distracts us from the fact that the 'science factory' of recent times is producing little of *real value,* but is being paid increasing amounts of money.

This goes hand-in-hand with another omnipresent theme of neo-liberal business management: an obsession with 'systems', 'processes', 'methodologies'; wrapped up in jargon endlessly babbled at business meetings. In any area of large business organisation nowadays, you will find people earnestly discussing *process,* analysing every difficulty as a 'failure of processes', recommending solutions to every problem by 'defining new organisational processes'. It is taken as an axiom that you only have to get the *organisational processes right, the system right, the methodology right,* and the business will automatically perform. Analogously in the scientific context, you only have to get the *scientific method* right, the *scientific process* right – and the scientific results will automatically come. Or in the social context, you only have to get the *rules right – the laws and regulations and control systems* – and the social utopia will follow.

'Business management science', which generates this jargon and vision, is itself one of the archetypal voodoo-sciences in history. Its language it not about the real world, but the projection of imaginary mechanical models of human organisation. Its theorists and practitioners live in an artificial reality of their own constructs. People with real *primary* jobs - doctors, nurses, teachers, engineers, scientists, chemists, programmers, statisticians, psychologists, social workers, electricians, mechanics, tradespeople, farmers, etc, etc - all study subjects focused on an *objective external reality.* They all know something about the real world: a world anchored in nature. Their technical language may become entwined in jargon, but it still refers to real things.

Business management science focuses on an artificial world of organisational structures, a *construct of artificial rules,* analysed through arm-chair theories. This is called a 'science', but it is not about *knowledge of reality,* it is about arbitrary, legalistic, *rule-defined* methods for domination and control of human activity. 'Business management' is a prime example in our culture of a meta-level activity: an activity seeking to organise and control the *primary, creative* activities of human beings, from a second level of power: the conventional power of the *boss*. It easily becomes a world of fantasies: fantasies of control, power and wealth. Business theorists and managers, financiers, economists and bureaucrats, indulge in this artificial world, and indeed convince themselves that *they create* the wealth of the world, that *they own* the wealth created by their workers; that *they have a natural right to it* because they rule it. In reality, their wealth is the result of enslaving the population, stealing the time of our lives for their purposes.

What does all this have to do with the failures of pure science that López Corredoira is concerned about? Many things. First, this kind of *hierarchically stratified power environment,* with its feed-back loops that *maintain control in the hands of an ambitious but intellectually mediocre 'executive business elite',* has swept through the scientific institutions themselves; justified by a voodoo-science called 'neo-liberal economic management theory'. Secondly, the capacity for *scientific and intellectual thought* within bureaucracies themselves, the capacity to provide insight and solutions to the serious problems of society, is severely degraded – intellectuals being used as techno-slaves serving the purposes of managerial masters. Third, *the same levels of general mediocrity* found in bureaucracies and corporate management have come through into the leadership scientific and academic institutions, and then percolated through the research teams, as they have developed the same stratification. Why are these institutions so increasingly bereft of real talent – now become so dominated by technicians, or specialist geeks, instead of creative intellectuals? Because the *stratification feedback loop* now dominates their structure. Just as it allows the mediocre egoists to take on power positions, it drives out the creative intellectuals from the research teams below them.

The rule-based world view and moral philosophy.

I now emphasise another feature that dominates our modern culture: an intensely *rule-based world view*, which appears as the paradigm of social and moral organisation to people today, especially those most intent on maintaining control through the social hierarchies of power. Like the stratification effect, it has its roots deep in human psychology. It has become a dominant *social metaphysics* – th analogue of materialist reductionism in natural science. It is expressed the conflation of *morality* with *rule-following*. It exploits the ambiguous moral status of regulatory rules – rules often having little to do with morality or justice, but applied through *laws of the justice system* – which is meant to legislate on morality – and thus acquiring a false moral status. It is reflected in the modern world in an obsession with the automation of rules and regulations, now implemented by a vast technology of information systems, becoming an attempt to *program human reality as rationalised organisational structures.* Psychologically, it reflects a need of many people to have a foundation for their beliefs in some form of *authority* – whether the government, the church, or the scientific establishment. (Yes: it starts with the imprint of the authority of Mommy and Daddy). This is a very real feature of human psychology: many people need a sense of external authority, as an anchor for their sense of social reality. It is a kind of a security blanket for them. It has its pathological extension in the idea of an authoritative system of rules to control all human behaviour, to judge all human decisions against.

The larger Utopian vision among secular technophiles is of an artificially designed reality, with every aspect of life controlled by human design, for human purposes. This embodies a kind of rationalist metaphysics, with humans at the center of the world. A related metaphysical paradigm is found in the (industrialists') obsession with a *reductionist materialist* exploitation of the natural resources of our planet: breaking complex substances down into their elements, and then reassembling this meaningless 'dead matter' into meaningful 'useful' commercial products. Breaking complex natural eco-systems including life-forms down into raw materials, and reassembling their materials into products for human consumption. This is the Cartesian nightmare of reducing life to a purely mechanical system – but extended in the modern materialist age to all human life, and to the human spirit.[113]

Ultimately I think this reflects a delusionary metaphysical vision about the nature of the world and our relationships within it as human creatures. It

---

[113] Descartes, at the start of the C17[th] Scientific Age, held that animals are 'mechanical' and non-rational, but humans have non-material souls; hence *Cartesian dualism* proposes dual substances, matter and mind, that mutually interact with each other through the pineal gland. He also believed in God, as a third 'substance'. So he was far from a Materialist. But his 'mechanical-rationalist' philosophy initiated a movement towards materialism, and is sometimes blamed for the anthropocentric view that the material-biological world has no intrinsic value in itself, and is to be used for man's benefit alone. Descartes is certainly one of the 'classical philosophers' most worth reading, whether you agree with his views or not. He was pivotal in initiating the age of 'modern philosophy' (post-medieval or post-scholastic philosophy), by turning the primary method of philosophy to the analysis of *epistemology,* i.e. theories of *knowledge,* instead of starting with *metaphysics,* theories of *existence and ontology.* Nonetheless he was primarily a metaphysician – also a great mathematician of course, inventing Cartesian geometry – and also with an influential scientific metaphysics of space as a kind of aether with material corpuscles forming vortexes in the aether. This metaphysical scheme of local causation was replaced by Newton's particle mechanics, with gravity acting at a distance through empty space. But modern quantum theory of forces is in many respects like Descartes' vision of causation, rather than Newton's.

has replaced the traditional social metaphysics of previous centuries and civilizations, that sees humans living simultaneously in both a natural and spiritual world, as part of a natural whole, in an embodiment of conscious and moral existence within a more complex hierarchy of forms. It has replaced such traditional views with the arrogant modern view that humans own the world; that all meanings are human social projections, powered by the ego; that we humans can and should replace the messiness of nature by imposing our own tidy rationalised artificial reality on it. It also reflects the modern obsession with superficial emotional experience as the highest expression of consciousness; the transitory *psychological ego* and its drive for 'pleasure' replacing the sense of a deeper *spiritual identity* that previous societies had.

This vision replaces *laws of nature and natural morality* with *rules of society and legal rights of 'persons'* [114], and the result is an increasing modern dystopia. Rather than bringing wide-spread personal *freedom* (which our vastly expanded material production systems should have enabled), the mass of humanity is reduced to cogs within an industrial-legalistic system, slaves to the purposes of *other egos;* slaves to those who control the *meta-system,* those who *control the rules that define our existence in a framework of legal ownership.* Because it is such a delusionary metaphysics, it carries the seeds of its own destruction.

A 'rule' is a moral concept, and let us first go back to some basic moral philosophy. Educational psychology and moral philosophy identify distinct stages of conceptual development, which we pass through from childhood to adulthood. A well-known example are stages of abstract thinking about numbers and geometry, leading to mathematics. Children simply cannot master higher mathematical concepts until they have in place a 'scaffolding' of lower-level mathematical operations – a theme pioneered by Piaget, the famous mid-C20th educational psychologist. The child is not a *tabla rasa,* a blank slate, on which you write certain facts in a universal language that is already transparent in meaning to any rational person. Instead, learning involves the hierarchical construction of *concepts,* building upon each other, to create the language for comprehending mathematics itself. The progression of children through conceptual stages in mathematics is a well-established fact today.

A related example of conceptual stages is the development of *moral concepts.* Young children appear to learn morality initially through simple *behavioural rules.* Behaviourally, the rule states: *'In this situation, always do that'.* Or as a negative formulation, instinctive to authoritarians: *'Never do that! (Smack!)'* The behavioural rule is interpreted morally: *'Follow Mum/Dad's rule is **good**. Break Mum/Dad's rule is **bad**'.* Of course we want our children to take on our rules as their moral reality, as something they believe is *right,* not merely as behavioural programming. We want them to see and understand the *moral reasons*

---

[114] Corporates and other social institutions are now legally treated as 'persons'.

behind the rules. At some point, usually in teenage years, many (but not all) people develop a more mature level of moral concepts, where they realise that the *behavioural rules are not the moral principles in themselves.*

Behavioural rules are only *moral rules* insofar as they interpret deeper *moral principles.* It is much harder to interpret moral principles and concepts than just to follow specified rules. The army programs soldiers to blindly and unquestioningly follow rules: it doesn't want soldiers to wrestle with moral decisions, it wants them to follow the chain of command, to act like a machine. If you want to surrender yourself to this, decision making is easy. But *in real life* you have to make your own choices, and you have think through consequences, balance the values of outcomes, for yourself.

Anyone with moral insight realises that a simple set of rules cannot serve as a basis for moral behaviour *without doing the work of interpreting them.* For moral decisions and judgements – especially those that cause us difficulty, that we have disputes and disagreements about - require complex subjective calculations and intuitions, and any simple system of rules attempting to determine all human moral behaviour is inevitably incomplete, and comes into contradiction with itself in obvious ways. Often we must break one rule to satisfy a higher rule. Generally we must *infer or construct* moral decisions governing novel situations, from our deeper principles. We cannot find them in a pre-compiled list of behavioural rules, any more than we can find the right sentences to express ourselves in a list of phrases. Moral capacity, like language capacity, is open-ended, an ability to make creative constructions – constructions of meaningful sentences in the case of language, as famously emphasised by Chomsky in his theory of generative grammar (Chomsky 1965); constructions of meaningful judgements in morality.

This distinction is recognised in moral philosophy in two broad types of moral reasoning systems. One is called *consequentialism,* and holds that the principle for reasoning about morality lies in judging the *moral value of the expected <u>causal consequences</u>* of actions. E.g. you can break some 'rules' and do things that would normally be bad, if it is necessary to achieve a more important value. In consequentialism, rules are not absolute dictates. E.g. you may be justified in smashing a pharmacist's window in the middle of the night to take some medicine *if it is necessary to save someone's life.* Or more plausibly, you are justified in telling a lie if it is necessary to protect someone innocent from undue harm. You can break commercial confidentiality if you need to do so to expose corruption, or some greater threat. The focus is on achieving *outcomes of a larger process, seen holistically,* rather than a reductionist conformity to following micro-rules at every step – like a computer.

These are not uncontroversial examples though: and consequentialism is difficult in practice. Although most ordinary people would allow there are such situations where it is morally justified to break rules, to override laws, there are many people who would judge the opposite; for example, that you should *never steal medicine* in any circumstances, even to save a life. They can reason as follows. Corporate pharmaceuticals set the price of medicines to optimise profits, and this means they must be priced out of reach of many poor people (for without residual demand it is generally impossible to set prices to optimise profits); and people do indeed die and suffer needlessly in their thousands and millions because they cannot afford medicines that would help them. Now medicines are typically sold for 10 to 100 times their production costs, so that businesses and governments can make profits: otherwise there is no motive in the Capitalist state for producing medicines. So corporate executives can point in turn to a higher-level *meta-principle*: how could they make their profits and remain viable companies to produce any medicine at all, if poor people were allowed to *steal* medicine when they *needed it*? If you allow *one person* to steal some medicine to save a life, how do you justify stopping millions of others doing the same thing? The permanent value of sustaining the *capitalist system* by strictly enforcing rules of ownership, it is argued, outweighs the harm of allowing some people to needlessly die or suffer.

They are supported in this of course by Government bureaucracy: for how could the *system of power and control* be sustained if people were allowed to use their moral intuition to override laws, making exceptions as they please? Free-market capitalism as a rule-based system requires a large pool of poor people to serve the wealthy: this wealth differential is intrinsic to its very structure as a system. If impoverished people were morally permitted to steal medicines, or food and other resources to sustain their lives, how could the *system* possibly survive as a *system*? And how can the State maintain its own control of wealth through taxation, its underlying commercial imperative, unless businesses are able to maximise their profits?

Such people who perceive that *strict adherence to legal rules defines their own self-interest* in some domain will naturally believe that *these rules must be enforced without allowing exceptions on moral consequentialist grounds*, otherwise the system itself, the *meta-organisation* would be threatened. Maintaining the *system* itself is the highest and overriding 'moral' concept for most people in the establishment – that is, people whose self-interest the system serves. This mean enforcing its rules, even if they appear destructive, irrational and immoral in their immediate effects on real people.[115]

---

[115] For a real example, a few months ago a NZ helicopter pilot, Dave Armstrong, was convicted for flying without medical clearance, fined $5800 and lost his licence and probably any future career. An experienced rescue pilot, he flew to rescue an injured hunter, who lay dangling on the edge of a 50-metre drop with a broken leg, and with the weather closing in. The hunter would most probably have died if not rescued. Armstrong, 63, was the only pilot available. He flew for about 10-15 minutes. Now he had recently had a 'mini-stroke', and although recovered, he was not yet cleared again to fly with passengers. The NZ Civil Aviation Authority felt it was more important for Armstrong to follow their regulations than to act in an emergency to save a life. CAA lawyer Chris Macklin argued that Armstrong had demonstrated

Thus *moral consequentialism* based on individual values and judgements rarely suits those with power, and society runs another moral system in practice. This second type of moral system is called *deontological*, and holds that morality consists in *following the rules*. They are taken to be absolute rules that must be obeyed, defined in 'laws', 'regulations', 'commandments', dictating right behaviour. Actions are *right* if they conform to the rules, and *wrong* if they break the rules – just like Mummy says. This means *conforming to the letter of the law simply because it is the law*. (The newly popularised attitude of 'zero tolerance', i.e. absolute intolerance to all degrees of rule-breaking regardless of circumstances, is the new voodoo term for a return to this mentality that powered the Spanish Inquisition, Stalinism, McCarthyism, etc.) No calculation of moral effects, justice or value, is needed in this philosophy: the rules define moral actions.

Immanuel Kant, in the C18th, is the leading *deontologist* among the classic philosophers, and indeed practically the only major philosopher in the Western tradition to support this view. Most moral philosophers today are consequentialists, because there are too many contradictions in thinking that any simple set of rules can determine all moral behaviour. The underlying question is of course: who or what determines the rules in the first place? Insofar as we are talking about *morality*, rules are intended to *represent and interpret* something: moral principles. Let a bunch of second-rate bureaucrats write your rule book, and a bunch of second-rate lawyers will immediately find ways to circumvent their intended *moral* purpose. The laws of the land are our legal rules, but they are not morally right just because they are enacted by a government body: they are morally right only insofar as they reflect deeper principles of *justice, fairness, value, morality*. We have to *interpret* moral principles to behave properly. And we have to *use our moral judgement, make moral calculations, use moral intuitions, and judge larger moral consequences*, to do this. Most modern moral philosophers think about the logic of morality in terms of *moral values and larger moral principles*, rather than *systems of specific rules of behaviour*.

Of course, there are also people who reject any pretence that laws are about *morality* too – holding that it is merely the pragmatic survival of the system, or 'survival of the fittest' in the popular Evolutionist phrase, that matters. Capitalist economists, Corporate executives, Government managers and commercial lawyers are united in being *pragmatists* about their source of power and wealth. A ruthlessly legalistic interpretation of laws, disdaining

---

"contempt for aviation laws", and that a discharge without conviction would send a "very unhelpful message" to the industry. The judge, Tony Couch, agreed. Breaking the CAA regulation *threatened the health and safety rules*. Better to let a man die in agony than threaten the health and safety rules! Breaking the rule did not actually 'hurt' anyone, but it is treated as a meta-crime, outweighing *real* consequences. The man who was saved, incidentally, regards Armstrong as a hero. But *his* testimony was not considered relevant by the court.

consideration of 'morality', is central to their self-interest and privilege, and some are blunt enough to openly admit it. The ostensible principle in this world-view is that *people must obey the laws – whatever the laws may say – to make the system work*. The real principle is: *I can exploit the system of laws for my self-interest, and this is sufficient justification for me: it is simply a competitive struggle for domination*. Most sinisterly, this mentality takes control of the legal-political *meta-system* of modern democracies: *the meta-system for making the laws themselves, for controlling the rules of the game*. This meta-control is the primary role of the modern state-corporate bureaucracy, enabling it to override concepts of morality, the natural interests of democracy, and natural freedom and justice.

Although we are in a period where such legalistic mentality dominates our system of power, this is an essentially nihilistic (i.e. amoral) attitude, and is bound to fail for the population at large, because humans are intrinsically moral beings. We are driven to seek moral outcomes as part of our purpose; and the Neo-Darwinian vision of successful competition for 'survival of the fittest' as the only measure of morality serves only a small number, and justifies the vast disparities of power and wealth in our world, a primary cause of widespread misery and despair for real people. It creates a system that is intrinsically unstable, and it will come crashing down once again, as it has repeatedly in past civilisations, through its own excesses. Human society cannot live on rules alone, without being informed by an underlying sense of morality.

The need for *moral judgement and interpretation of rules* is also not a modern liberal invention, contradicting traditional ethics, such as religious ethics. It might be thought that the Ten Commandments in the Bible is a set of simple moral rules that we must follow absolutely (if we are Christian moralists), and that Christianity therefore has a rule-based morality, not a consequentialist morality. But that is not true. The Bible is full of stories of *moral interpretation*. The New Testament in particular represents Jesus as being confronted with challenges to make moral interpretations of Old Testament scripture, by enemies trying to trap him into heresies. He is accused of breaking certain Jewish commandments for instance, and challenged to justify himself against charges of heresy. Jesus is too astute to fall into these traps, and turns these confrontations on their head. He gives superior *interpretations* of morality. He interprets which of the Old Testament Commandments *are the most important*, for instance. He is not about simply blindly following rules, as we are meant to follow traffic regulations. He is about distinguishing 'righteousness', or a true inner goodness and morality, from the mere appearance of obeying laws. Stories in the Old Testament also typically represent issues of *interpreting* morality, not simply *legislating it*. And I think this goes for the scriptures of all the major religions – which is why they are vastly more interesting as moral texts than the tracts of modern legislation.

It is common of course for fundamentalist cults to take extreme literal in-terpretations of selected passages from the Bible, or other scriptures, and enforce absolutist interpretations of certain rules they read from it. But that is not the real meaning or intention of the Bible. In the Judeo-Christian tradition, Jesus and the other various prophets were not lawyers or bureaucrats or enforcement agencies. The Ten commandments is not a system of modern *legal regulations*. They are 'rules' of course, but they are primarily meant to express *moral principles*. Christians have the same problems as others in *interpreting specific moral behaviour* within their chosen set of broader moral principles. So do Muslims, Buddhists, Hindus, Confucians, and people of every genuine religious tradition. Only people living in narrow cults, dominated by dictatorial authority figures who want to *remove all capacity for choice and judgement from their followers*, think there have to be simplistic, absolute rules that apply universally to determine all moral human behaviour – and in proportion to their simplistic beliefs, their behaviour appears intolerant and irrational to outsiders. And these cults die with their autocrats.

The modern, legalistic interpretation of morality locks us into the cult of rules. Morality in the modern justice system has come to mean little more than obeying rules and regulations *for the sake of having a system*. Justice to the modern bureaucrat – the modern cultist of rules and regulations - means *strictly enforcing their own system of legal rules*. This corrupted vision of morality and justice has overtaken our own secular society. It has become the *metaphysical paradigm* for social organisation. Children are inculcated in it from the day they start school – the day they leave the informal organisation of the family, and start being *shaped* (the behaviourist term for training rats) to life in a public institution.

The expression of this is seen in vain attempts to write exhaustive legislation, interpreting every detail of every possible 'law' for every situation, to *exclude the need for any moral interpretation or human judgement by individuals*. Thus we have *thousands of pages – indeed, millions of pages - of legislation and regulation*, breaking down laws into infinitely finer sub-clauses, trying to cover every circumstance and every eventuality, and of course, specifying punishments for breaking each rule. This is where the social vultures we call commercial lawyers scavenge their huge incomes: finding loop-holes in the rules to manipulate outcomes to benefit those who can pay for their services - the wealthy keeping the greedy in business.

Drivers of the rule-based world view.

Yet it is hardly the lawyers who are the drivers of this rule-based vision of social reality.[116] I will mention two main groups who directly drive the expan-

---

[116] There are many different kinds of lawyers with different moral outlooks. Criminal lawyers, dealing with real people, generally have a real concern for morality that commercial lawyers, dealing with money and property, lack. Our sense of morality is deeply conditioned by our occupations.

sion of rules. They have different aims at heart, but often produce the same results.

One is a positive group of well-meaning people, 'social interventionalists' or 'social activists', who see genuine hurts, injustices and exploitations in the system, and press for legislative reforms, along with new regulations, new regimes of enforcement, Government control of domains, as part of campaigns to correct injustices. They typically have good justification – and many social activists have had to fight tooth and nail at great personal cost to have blatant travesties of justice recognised. Indeed, the rule-bound legalistic mentality is not their friend: activists are often fighting *against* unfair and immoral rules that are already in place – against travesties justified precisely by the failures of legalistic systems. It is good they press to reform flawed systems of rules and repeal bad laws, to institute checks and balances in the system to counter immoral practices by those who obtain their power from the manipulation of the law.

But it is seldom that purely legalistic measures work by themselves in any case. It is necessary to have support for reforms in law, but the deeper drive is to change people's thinking, to change their moral point of view: to change their *personal judgements*. This is what genuine social activists are usually after: campaigns to raise awareness of issues – of racism, sexism, slavery, violence, war, repression, abuses of power, economic corruption and injustices, environmental destruction, animal cruelty, entrenched poverty, systemic health issues, treatment of mental illness, suicide, drug or alcohol abuse; etc, etc. Reforms need legitimation in law: but astute activists want to change moral awareness, particularly by breaking down traditional prejudices. They want human judgement to come from inside. With such people I have every sympathy. Nonetheless, 'moral activism' invariably crosses a line into ideology, and the goal of enforcing principles in legislation typically slides into new forms of rule-based moralising, known as 'political correctness'.

However the obsessive rule-bound agenda is driven by a different set of people to the pioneering activists: the carpet-bagging ideologists who exploit the new territory they open up. Authoritarian government bureaucrats, pompous educationalists, corporate lobbyists, neurotic do-gooders, police and military totalitarians, sanctimonious moralisers and political reactionaries, seeking control for its own sake and their own ends – because they are 'tasked' by bureaucracies to expand their powers; because they have an exaggerated sense of moral righteousness; because they feel uneasy with uncontrolled activities and natural human freedoms; because they see opportunity for their own egos by manipulating the meta-system; or simply because they are fanatically convinced that they know best for everyone.

A typical expression now is seen in the intrusive regulation of 'risk activities' – covering play-ground and recreation and work-place – in the name of 'health and safety'. This is not addressing any moral issue: it is a bureaucratic drive to regulate human activities to a formal template of correctness. Its official justification is *cutting social and economic costs*, but it far outreaches any realistic practical goals in the name of *imposing rules*. It starts by identifying 'problems', which means people doing things 'the wrong way', and proceeds by introduced swathes of rules and regulations to force people to do things 'the right way'. 'The right way' is a standardised way defined for everyone. Bureaucracies today are so fixated on *bureaucratic regulation* as the only means to 'fix problems' that they cannot imagine anything else. They specifically lack a *scientific approach: careful identification of the cause and scale of 'problems'; creative development of solutions.* They are utterly fixated by a *political approach: how much more can we regulate, and will the public swallow it whole?* (Or do we need to soften them up first, with creeping regulation?)

Mention any 'problem', e.g. any form of accidental injury – children falling off swings for instance - and the bureaucrats and their political controllers will immediately think: *how can this sphere of activity be regulated?* "Impose mandatory safety belts on children's swings." What about children falling from trees? "Ban children climbing above 1 meter without safety harnesses". This is not a joke: it is real regulation. NZ schools typically ban children climbing above 1 meter, having innocent play fights, romping or wrestling in the playground, running on concrete paths, and so on. "Johnny might fall over and skin his knees. Terrible!" Well, this is a short-term hurt, and a universal experience of children over thousands of years of human evolution, from which children *learn long-term lessons,* including how to tolerate a little pain, and a *real lesson* in the need to be careful. But policy bureaucrats have a purely one-dimensional vision: *'hurt = bad = fix = redefine process = more rules required'.*

More controversially, children in many democracies are now forced to have vaccinations for multiple diseases, including traditional developmental diseases like measles, mumps and chicken-pox - despite many experts believing these diseases actually have long-term benefit. Many people also believe multiple bundled vaccinations can cause autism and other severe problems. But they are persecuted, even having their children removed from their care, if they do not comply with bureaucratic regulation to *force their perfectly healthy children to suffer complex immune system reactions.* The bureaucrats will claim they follow 'best scientific practice' – and this gives them the *right to force parents to treat their children* - but in fact they are ignorant of science themselves, and most of their 'expert science advisors' are incompetent puppets.

Adults are treated just like children too. It is now twenty years since every child and adult in NZ is banned from riding a pushbike without a safety helmet. A generation has grown up believing that cycling is dangerous, abnormal and immoral without the voodoo ritual of wearing a helmet.[117]

---

[117] The compulsory bike helmet law has failed to reduce bike accident statistics or serious injuries, originally claimed as its justification. It has lowered participation in cycling. It has probably cost the NZ public about half a billion dollars, and hundreds of millions of hours of time-wasting inconvenience. Most insidiously, it introduces a

What on earth can be done *to protect individuals from 'harm'*, authorities think, *except to impose more rules and regulations, along with regimes of surveillance and punishment to enforce them?* Protection from injury or risk-taking is only one example: there are rafts of legislation covering building, machinery, recreation, food, camping, tramping, biking, boating, driving, working, gardening, tree removal, pet care, child rearing, health – *any* activity you can think of is now swamped with regulation, in NZ and similar democracies. And regulations are still constantly expanding. For small businesses, bureaucratic 'compliance costs' are chronically out of control.

But although everyone complains about it, imposing blanket legal rules and regulations is the only option most people today can *imagine* to deal with 'problems', reflecting how deeply the rule-based mentality has become the *paradigm* of social organisation. A 'problem' is defined when anything goes wrong, any kind of accident or injury or misfortune happens, any 'hazard' arises, any 'bad practice' occurs that does not fit the official definition of 'good practice'. In fact, nothing has to *actually go wrong;* breaking a rule or regulation – *failing to fill out a permission form or pay the fee* – has become the main form of 'bad practice', a 'health and safety incident' in itself. People are even claimed to have a *right* to *risk-free environments.*[118]

---

new *neurosis:* making people *fearful and neurotic about a simple normal activity*, enforcing dependence on a pseudo-magical technology to make them feel 'safe'. It is a form of Voodoo: it allows the authorities pretend they have a solution, and gives cyclists a false sense of safety, without doing anything to *help prevent cyclists having collisions with cars.* The real cause of cycling injuries and deaths is *mixing cyclists with heavy motor vehicles.* Helmets do *nothing* to address this primary cause. In many high-risk sections of road, the only safe cycling route is to take the footpath – but of course *this* is against the rules, and punished as unsafe. Developing safe cycle-ways, making cycling *actually safer,* was too challenging or expensive for NZ transport planners – so they 'fixed the problem' by making a rule that means everyone pays to wear a piece of magic sticking plaster on their head whenever they go cycling.

[118] After a tragic incident in 2015, where two NZ Social Welfare workers were shot dead by a mentally disturbed client, 'Worksafe NZ' – a much-ridiculed bureaucracy with totalitarian powers over businesses – prosecuted the Ministry of Social Development for *failing to provide a safe work place.* All MSD offices now have multiple guards at every door. It is generally impossible to speak to anyone (even a receptionist) without first arranging a meeting via a centralised communication system. It takes MSD anywhere from thirty minutes to an hour to answer their phone ("we are currently experiencing high demand…"), and it will take a week or more to see someone. You will have to bring dozens of pages of documents with you, and if you overlook one, wait another week to continue the process. It is true they have a lot of angry clients – mainly because there is no recourse against their endless incompetence, errors and abuse. Government policy (called 'getting tough on beneficiaries') is to apply pressure and minimise payments of what are people's legal entitlements. Of course it is the poor stressed-out case workers who have to face the increasing anger of often desperate clients – the executives responsible for the fiasco of MSD are

These swathes of regulation are blunt and costly instruments of course. Their effect is to *criminalize and immoralise broad swathes of what used to be normal activity,* incrementally costing people increasing time, trouble and money, and intimidating people into conformist behaviour, with the obsessive intention of preventing a small number of 'bad practices'. What used to be common freedoms, activities left to individual or parental judgement and discretion, lessons learnt from experience about *taking care of yourself,* are now endlessly encroached and controlled by regulations. And most *dangerous incidents* are not prevented or detected anyway, because they are perpetrated by a small minority of constitutionally incompetent or reckless people.

Regulations serve as excuses for new power-and-profit centres, for new bureaucracies springing up to force new regulatory environments on people who are already doing activities very safely. In the meantime, there is a corresponding loss of perspective about *real, serious problems* in the swamp of ineffectual bureaucratisation of petty concerns. Real, large-scale economic and legal injustices that continue to severely degrade the lives of millions of

---

safely far far away, in another universe, having meetings in air-conditioned corporate offices. Have they thought that the flood of anger and desperation among their clients needs to be addressed in some way other than just by employing thousands of guards to satisfy Worksafe NZ?

Worksafe NZ itself is constantly trying to sensationalise dangers. Their web page opens with "Every year thousands of New Zealanders are killed or injured at work, or suffer from a work-related disease, resulting in huge personal, social and financial costs." Further down we find an average of 102 work related deaths and 378 serious non-fatal injuries, from 2.2 million workers, in 2008-2010. (*This is the latest statistics they have.* They claim there is a 'two year time lag' in getting statistics - from another government department - but it is now 2016, five years later!) The 'thousands of diseases' are dominated by asthma and 'musculo-skeletal diseases' (arthritis?). They are trying to sensationalise the stats for 'deaths' and 'serious injuries' to make their mission more important. They claim that the 'social and economic cost' is 3.5 billion dollars per year – partly by costing deaths at *millions of dollars each,* using the *life-time income of the deceased.* But they ignore that most workers will *spend most of their income on consumption – all tax revenue is consumed too.* They should count only the net part-life-time savings. And 30% or so of NZ citizens will leave the country. And what are the social and bureaucratic costs of Worksafe regulations? They have no idea. Their analysis is typical unreliable junk marketing stats.

I am all in favour of safe workplaces. I am against autocratic morons being in charge of regulating our lives. NZ historically has a poor safety record in some industries – what was 'Worksafe' doing for decades previously? *Oh that's in the past, lets forget that and move on.* What has changed in the last few years? *Hugely expanded rules and regulations, with totalitarian powers to punish, enabled on the back of information systems and a new wave of totalitarian political correctness.* What has *not* changed is the incompetence of management, the vacuum of scientific and analytic ability, and their inability to find *real solutions to work safety problems,* as opposed to imposing voodoo 'safety' rules.

people remain with no attention and no redress. Attention to real problems is blinded by petty bureaucratic rule-making, not enhanced by it.

For example, wide-spread *poverty and unemployment,* increasingly rampant in an expanding NZ underclass, is a *serious* social illness; but practically nothing is done about *this* phenomenon, that brings entrenched pain and hurt to generations of real people, because it cannot be dealt with by simplistic bureaucratic regulation. In fact, it is exacerbated by expensive bureaucratisation of goods and services, based on the model that affluent middle-class life is the norm. The incremental costs increasingly exclude the 'under-class' from participation in many activities. The bureaucrats want to stamp out this underclass – to their minds it is ugly and inconvenient - but increasing bureaucratic and financial pressures against the poor and low-educated and unfortunate actually just forces more people into poverty. The fact is that government departments and politicians will not tackle an issue like poverty that requires a *serious* social engagement, a *serious* revision of the system. They are stuck in the paradigm of bureaucratic tinkering with rule-making, and making pompous speeches about progress.

There is another major driver of the plague of bureaucratic regulation too, mentioned above: the *ideology* of neo-liberal State Corporatism, rampant among the managerial elite. This is the ideology that *money is the measure of all value,* with an ideal of measuring every 'atom' of human activity as a financial exchange.[119] Orthodox economic theory (although long discredited as having

---

[119] I fed the farm dogs for someone last weekend, as a favour so they could go away for a couple of days. Strictly this is illegal. I should charge them $20 for this, put it into my IRD return, claim back about $2.50 vehicle costs, pay about $6 extra income tax, charge, pay $3 GST if I am GST registered, pay ACC fees; and they should include it in their farm accounts, claim it against their income tax, claim back the GST, and fill in multiple Worksafe and ACC forms to register me as a worker, etc, etc – and have expensive discussions with their accountant about getting the legalities right. For feeding a few farm dogs! Legally, you are not allowed to do anyone an *economic favour* like this: it upsets the utopian perfection of the neo-liberal economic model. If the IRD found out I did some 'work' *for free* like this, they could prosecute me! Like a mafia protection racket, they have to take their cut of every transaction. The administration cost of tracking down and prosecuting this amount makes it impractical at the moment – but if it was a *days real work,* worth $US100, the farmer would *certainly* have to go through all these formalities – at real risk of prosecution by a number of agencies. Farm, small business, contractual and self-employed accounting and compliance is absurdly complex. Simply hiring a temporary worker for a day takes a large amount of paper work. What IRD wants of course is an *automated* system, where every tiny transaction, every human action, every fifteen minutes of your life, will be tracked and accounted for online. (Could we track people's locations remotely with GPS... and force them to declare all activities in legal diaries every 15 minutes?) NZ IRD has started a multi-billion dollar spend on computerisation towards this kind of goal. No cost to them of course: they will simply charge the NZ public for the cost of a universal surveillance system to monitor them work.

any *scientific reliability*) is that 'market forces' will automatically ensure 'optimal value choices' under ideal 'market conditions' where *every item of human activity is accounted for in a common system of monetary transactions.* In NZ, this model was introduced in the 1980's, under the government slogan of 'User Pays' – a free-market capitalist entrenchment introduced by the Labor government – remarkably, the traditional NZ *left-wing political party,* obsessed with 'Chicago-school' Friedmannian economics, initiated one of world's most trenchant State-Capitalist bureaucracies, with a program to privatise all traditional commons, and tax all social goods.

The fundamental paradigm here is that *the value of all human activities should be atomised and exchanged on a common scale of money.* This now apparently suits both the neo-left-wing and neo-right-wing agendas. (In the brave new world, *traditional liberals* and *traditional conservatives* no longer exist as political forces.) Regulating all activities as economic exchanges represents new opportunities for government to expand control over its citizens (the neo-left-wing agenda), and represents new opportunities for the government to act as a feudal profiteering corporate, ignoring wider social or moral responsibilities (the neo-right-wing agenda). Intellectually, it is driven by a *mechanistic scientific model of human behaviour reduced to monetary transactions.* The economic scientists behind this have projected a model onto reality: a simplistic mathematical model, originally derived from the atomistic reductionism of classical physics.

The informational overheads of putting this into practise (recording every single human transaction and interaction; making complex calculations of tax, GST, penalties, rewards, exceptions, etc) have been out of reach until recently, but modern information systems have now brought the prospect of universal data collection and universal surveillance. The next logical step is to remove physical money, and make all transactions electronic, and thus traceable; to remove all informal arrangements between people, breaking every human relationship down into electronically recorded contracts.

It is in this broader context of the bureaucratisation and state-sponsored commercialisation of all human activity that science now exists too. Science institutions are now a microcosm of the larger state-corporate institutions, in a scramble of self-interest in a rule-bound commercial environment.

Reflections of the rule-based world view in science.

This rule-based world view has a number of central expressions in the world of science proper, and it is an important theme in the context of the philosophy of science. I will mention a number of important connections.

The delusion of a rule-based scientific method.

The first connection is with the notion of a *rule-based scientific method.* We are constantly told by the scientistic ideologists that there is a 'scientific method' that guarantees the truth of scientific knowledge, and makes it superior to other forms of knowledge. But what is this method? Well, numerous philosophers have tried to define it in sets of rules. Modern versions almost always revolve around rationalised concepts of empiricism and experimentation as a method to determine truth, usually with inductive generalisation assumed as a rationalised method of theory discovery. But such proposals are *hopeless* as real rules for real science. They are just as hopeless as an 'artistic method' would be, if someone tried to spell out a rationalised set of rules for producing and judging beautiful works of art.

The fact is: *there is no remotely adequate scientific method defined by any explicit set of rules, either for testing truth, or for discovering scientific ideas and theories.* There are only more or less vague *principles*, that must be interpreted in context: for science is a method of enquiry with a large range of topics. These include principles like:

- the importance of *objectivity in evaluating evidence,*
- the importance of *achieving real explanations, with counterfactual validity,*
- the importance of *remaining open-minded,*
- the importance of *allowing sceptical attitudes and not clinging to dogmas,*
- the importance of *checking rationality of proofs and arguments.*

I will not argue at any length here for this: if you believe that there is a well-defined 'scientific method', coded in a set of procedural rules, that scientists learn in Science School, and that delineates 'good science' from 'bad science', and tells all scientists how to go about their jobs, then you are in the grip of a myth, a piece of propaganda, and either you have no experience of how science is really done, or you have not objectively compared your experience of science to the rationalised theories. If you think you know such a method, then try it out on some examples. How does it work for *physics and chemistry; for biology and evolution; for geology and ecology; for astronomy and cosmology; for geometry and logic; for statistics and model theory; for linguistics and information theory; for cognitive psychology and cultural anthropology; for social and political science; for business and management science; for economics and financial science; for explaining strange or rare events; for explaining psychic or spiritual phenomenon; for theories of time and space; or for your own theory of scientific method itself – is it scientific in its own terms?*[120]

The popularist notion that there is a defined cannon of rules of scientific method is a favourite delusion of the scientistic ideologists who would have

---

[120] See Appendix 5 for a proof against the positivist-empiricist theory of meaning.

us commit to a religious-like faith in science, and have us believe that they can strictly *demarcate* good science from bad science - demarcate 'nonsense' and 'pseudo-science' as the positivists and sceptics call it.[121] In reality, there is no such method, and most certainly it is not the *empirical experimental method.*[122]

The delusion that science is done by a *rule-based scientific method* is intrinsic to the drive to bureaucratise science, and to the model of science as an industrial production-line of technical specialisations.

Von Neumann and the delusion of inductive weather prediction.

A second example of a *rule-based mentality* in science is found in certain typical approaches to scientific modelling, with its classic expression in von Neumann. He was a great mathematician, who axiomatised quantum mechanics in the early 1930's (with serious flaws), and established game theory as a formal model of rational competitive behaviour (with serious flaws), and was instrumental in the development of modern computer software. In the latter connection, he pursued a program of *weather forecasting*, believing that the modern computer would allow us to develop precise long-term weather forecasting models. He assumed that by collecting more and more precise *initial conditions* about the weather, and using them as inputs to a scientific

---

[121] And it should be emphasised again that the modern 'sceptics' represent the antithesis of traditional *scepticism*, which means the questioning of theories or dogmas. They are about enforcing their own metaphysical belief in materialist reductionism and atheism, attacking alternative ideas, and claiming a privileged status for orthodox science. They are the enemies of heterodoxy and open science.

[122] To re-emphasise an earlier point, the enquiring mind has an active role in scientific reasoning and evaluation of evidence, and in some areas of knowledge, subjective observation is the primary source of information – e.g. in logic, linguistics, cognitive psychology, morality. We do not and cannot understand the subjective realms from 'pubic observations' or physical measurements. We can examine our feelings or emotions like pain, fear, pleasure, joy; cognitions like belief, rational inference, conceptual meaning; moral beliefs; aesthetic experiences. We understand a great deal about these – to an important degree, we understand what other people and animals are thinking and feeling. This gives us real knowledge and understanding. But it lies outside the definition of empirical method. The positivist-empiricist response was to deny reality to the subjective realm altogether – resulting in the absurdity of mid-C20[th] *behaviourism*, denying that people have conscious perceptions at all – to retain a dogmatic theory of scientific method. But 'mirror neurons' were discovered around 1998, which allow us to 'mirror' experiences of others in our own minds – to *simulate* subjective experiences we infer in others. This access to subjective knowledge does *not* conform to what empiricists think are 'rational inferences from empirical scientific data'. Yet it is still *empirical* experience.

model, forecasts could be made more and more accurate, and allow long-term forecasts without limit. By making inductive generalisations from vast quantities of computerised data, we could surely master weather prediction to whatever degree we want. This is the very epitome of the *empiricist-inductive* vision of scientific method. What could possibly be wrong with this eminently rational assumption?

As it turned out, everything. This rationalist paradigm was totally wrong. But it was not until the advent of *chaos theory* in the 1960s that this was widely recognised. Chaos theory is often explained precisely through the example of the 'butterfly effect': that the weather is a system so unstable with respect to initial conditions that a butterfly flapping its wings in North America one week can change local weather patterns in South America a couple of weeks later. Complex systems are generally *not* predictable in von Neumann's sense at all – not even in principle. This instability seems glaringly obvious to chaos theorists today – but von Neumann, an eminently 'rational' mathematical modeller and logician, failed to notice. His theory of games, which became modern *decision theory*, popular with neo-liberal economists, is equally a failure *in the real world*. As a model of real human behaviour and real decision making, it is so simplistic – representing social behaviour as a simple, first-order *competition* - that it is useless for predicting anything in practise, or for making real decisions about anything. It only applies to simple artificial games, defined by precise rules. Applied in neo-liberal economic theory, this idealised model is so simplistic and empirically inaccurate it is no more than a pseudo-science.

Von Neumann was also a fascist: a political philosophy characterised by an extreme rule-based mentality. His world view was extremely rationalistic: to him, it seems that every aspect of the world followed a model of *logical rules*. Fascism is a classic expression of this kind of world view in politics. The same mentality that led him to an intense interest in axiomatics (reducing logical thought to a systems of rules) and computer programming (ditto) and game theory (ditto) and weather forecasting (ditto) led him to an intensely rationalistic view of morality and social organisation as *a system of rationalised rules*. As a reductionist-materialist-evolutionist, it made perfect sense to him that morality reduced simply to the rule of competition: *whoever best survives, whoever wins the competition, is right*. Although a mathematician of the first rank, he appeared to have no philosophical insight, no dimension of social intelligence. He also has no discoveries in empirical science. He is the archetype for our age of the technocratic geek, capable of inventing the computer, but without the social and human understanding needed to use it for human values. This type of mentality is the reason ordinary people do not trust scientists and intellectuals to make political and moral decisions. This type of *technocratic geek mentality* is precisely what modern scientific education and institutions are designed to produce.

The delusion of AI and intelligent computer systems.

This brings us a third connection, with the 'science' of computer programming and dominance of the 'information industries'. There is a vast amount
of hype for the last 30 years and more about 'intelligent computer systems'
and the immanence of 'AI' (artificial intelligence) – hype that has dramatically
failed to materialise. In reality, even mundane business software development
is increasingly a disaster – *system design capability* breaks down as soon as
information sources become mildly complex, as soon as there are mildly
complex *data integration challenges*. Far from entering an era of advanced *intelligent* information systems, business and enterprise software systems are in a
war zone! It is analogous to the vast slums created when poor people crowded into urban environments nail boards and corrugated iron together to make
shanty dwellings, but *fail to create any integrated system of infrastructure for water,
sewage, toilets, power, transport, rubbish disposal, or maintenance*. The result is a huge
living mass of humanity – but living in an inextricable chaos, because there is
no organised *structure* to integrate the complex functions. Ditto with information systems in large organisations and businesses.

Computers are obviously useful in specific applications – but most of
these appeared 20 to 40 years ago.[123] Their more recent extension into 'enter-
prise-scale' information systems, built with gargantuan ambitions to control
and automate all the complex functions of a corporate bureaucracy, is a
disaster in practice. Frankenstein's monster has come to live in the Tower of
Babel. Failure is epidemic in the public and corporate sectors. The quality of
software system development has degraded over the last two decades, as programming has become locked into the fossilised software platforms spawned by
the giant multi-national software companies (Microsoft etc), software programming is overrun by dumbed-down technology graduates, large projects
are controlled by incompetent project management, and planning is made
captive to follies of bureaucratic ambition and mediocrity. This is something I
have followed closely through three decades or more, both through working
in the software-information industry, and through a dedicated research
interest in the scientific theory that lies behind information representation.

---

[123] Most large banking and bureaucracy core information systems still in operation
today were built from the 1970's (in programs like Cobol) to the early 1990's (simple
Oracle DBs) – they are 20-40 years old. Redevelopments of these in 'modern' software platforms are very expensive – frequently costing *hundreds of millions to billions of
dollars* – and invariably have serious failures, often disastrous ones. The failure rate
for corporate-scale IT projects is huge – not just in abandoned projects, but ones that
drastically fail their goals, that make systems *worse* than when they started, that end
with diminished performance, and leave businesses with huge maintenance costs and
constant breakdowns. If building contracts worked like this, cities would look like
war zones, littered with broken and abandoned shells of skyscrapers.

The fundamental connection is that computer programming is a domain of *rational rule-based models*. There is a gargantuan delusion now that software systems are able to encode rules without limit, to make 'smart systems' that duplicate the intelligence of human decision making. The standard jargon is that the programmer models the *business rules* of an organisation in 'data models', creating 'business intelligence' systems of vast power, able to exploit the power of 'big data', mining vast quantities of electronic data, etc.

The reality is that 'business intelligence' programming is frightfully poor. Systems are expensive, inflexible, vastly over-complex. *Hardware* quality has improved immensely of course – we are now in the era of 'big data' *physically*. But *software* quality has literally stalled at the stage equivalent to floppy disks.

The software industry utterly fails to live up the hype of 'information-driven decision making BI systems', and the main thing the 'big data' world is useful for so far is *mass marketing and mass surveillance*. The public mistakenly takes the success of IT technology in providing larger physical capacity, to store and transfer data – especially media data over the Internet – as success in *information technology*. But *information science* came to a grinding halt simultaneously with the commercial dominance of giant software corporates – Microsoft, Google, Apple, IBM, Oracle, SAP – a generation ago. It remains stuck in an era of unmitigated greed.

Methods of encoding 'business models' and 'business rules' – meaning logical rules to process *data* into meaningful *information* – remain primitive. There is a common delusionary image behind this, that systems of logical rules modelling 'intelligent behaviour' are flow diagrams with 'if-then' logic. You follow a series of steps, asking questions at each node in the diagram, and taking one turn or another depending on the answer, until you reach your decision. Like your IRD tax form. This is Boolean logic: the simple logic of transistor gates. But real logic is not like this *at all!* Flow diagrams of this kind cannot possibly give rise to any kind of flexible intelligence or learning. Coding 'if-then' rules is absolutely *not* the model used for programming complex systems. The 'flow diagram' logic for any complex process becomes outrageously complex if formulated in this kind of way.

Real logic requires, at very least, *higher-order recursion*, and *quantification*, and works by *generalising high-order functions*. It is not Boolean logic at all. A step to a more powerful logic is found in relational database systems (where SQL, the common DB programming language, is equivalent to a first-order logic), and in Object Oriented programming languages (with stronger recursive functionality than earlier procedural languages). But these are still just fragments of *real logic* – they do not even represent *second order logic*, discovered by Russell around 1900! Few information scientists even know what second order logic is, and are unaware that they program and think, at best, in first order logic.

At a higher theoretical level, the source of failure is illustrated again by Tichy's innovations in logic. A formal system of equivalent kind of power to *TIL* (with high-level recursion over constructions) is required to have any chance of programming a system capable of handling the logic of even a realistic fragment of natural language – the immediate language of our real thought and reasoning. Conventional computer languages and AI (although far beyond simple Boolean logic) still does not have anything approaching this kind of *representational power* yet. But almost no one has noticed this need. Modern AI has sidestepped the problem of programming natural language interpreters, because it is far too hard within their paradigm. Instead it has adopted limited tasks of developing *ad hoc* domains of 'pattern recognition' software.

The deeper theoretical failure is a rule-bound obsession with a *syntactic paradigm* in computer and information science, and ignorance of a *semantic paradigm*. Tichy's approach is called *formal semantics* – where a formal system (with syntactic rules, defining a *symbolic* logic) is used to model *semantic reality* – the world of *meaning* that symbolic expressions are used to communicate. Now if you talk to most computer scientists and formal logicians, you will find they have no idea what 'formal semantics' means. For them, it pro-gramming is all about systems for *manipulating symbols* – ultimately the 'o's and '1's in digital computer memories. They have a few basic paradigms for representing *information in symbols* – but these have barely progressed beyond the invention of the relational database paradigm, discovered in the 1970s.

The computer science and programming world is, as you might expect, as bastion of *rule-bound logic*. Now this paradigm – *syntactic or symbolic logic* - is inherent in the very definition of computers at one level. But what is missing is a higher understanding, that these logics need to *capture information through meaning*. In practice, ordinary programmers fail to understand that they are *building representational structures to capture the meaning of information*. Instead they are obsessed with the mechanisation of symbolic processing. This is analo-gous to the positivist's theory of scientific method, which is obsessed with the phenomenal layer of nature, and denies the need to discover an underly-ing *reality*, as captured in more abstract scientific theories.

But back in basement of human accomplishment, the simplistic delusion that programming means implementing 'Boolean logics', a vast network of decision trees representing *rules*, is widespread among low-level programmers, in-house system builders, and business managers who are put in charge of innovation projects. Incompetent programming by workers with basic train-ing in 'technology platforms', but no idea of functional programming tech-niques, dominates the software landscape in the large organisations that can afford to create their own systems. It leads to unmanageable, grotesquely overcomplicated systems, that are a dead end for the business – although it generates plenty of lucrative work for software consultancies – shamelessly exploiting both their business clients and their programming contractors. It is like teaching builders how to saw and nail planks to each other, and setting them to building houses. For a house is no more than a bunch of planks nailed together, isn't it? Information systems in most large organisations are no more than a bunch of planks nailed together.

Executive management is removed from the reality of what is going on in their information systems. Their *rule-based mentality* gives them the vision of enacting swathes of rules and regulations, enforced by collecting more and more information on clients, managed of course by computers. The fundamental new game-changer lies in these electronic information systems. All larger organisations are now completely dependant on computer systems to collect and manage their information, and to make processing and enforcement decisions. Government agencies seek to record activities of people in ever greater detail, collecting an expanding stream of quasi-legal data through regulatory forms, along with more secret surveillance data on clients and citizens, with dreams of omniscience. They want to match information from different sources to discover people's secrets. They think there is a high-tech industry able to provide them with technology slaves to create systems to implement their fantasies of social control.

But their IT systems generates endless errors. Poor structure makes it impossible to extract data of any complexity. Poor rule-modelling and error-ridden programming means they fail to handle numerous 'exceptions' to rules. Poor design means huge maintenance overheads, constant manual interventions to keep systems running, systems constantly breaking down. Ever tried arguing with a bureaucrat about some error in your files, or a mistake in their processes? "The computer can't handle that, the computer won't let me fix that, the computer won't let me enter that, I can't do anything about it". These failures let them conceal information, pervert rules for their own convenience; and victimise people they dislike. Individuals are helpless to respond against these abuses.

This context is both a symptom and a cause of the larger syndrome infecting science. It is a symptom of the intellectual failures of *computer science and logic,* degraded to a 'technology market'. It is equally a driver of the corruption of science to a domain of information collection and publishing, without conceptual thought or understanding. IT is the game-changing technology that empowers the vast bureaucratic take-over of the science industry, just as it empowers the vast bureaucratic take-over of human life.

I leave this theme here: the pathological role of computer systems is a key syndrome of our society's deepening mental illness, and requires an extended treatment in another place.

The delusion of economics and the science of value.

One more 'science' deserves special dishonourable mention here: economics. Not all economics, I should stress. There are many different branches of economics, and some areas of descriptive economics, political economics, social economics, macro economics, have very valuable material and insights. But the core of *foundational Capitalist economic theory – monetary theory, neo-liberal micro-economics, free-market economic theory* – is a paradigmatic scientific failure of our time. And it is a paradigmatic political failure. It is also the context for the social organisation of science.

Of course, this brand of economics has been criticised and ridiculed by thousands of researchers, and it is resented and despised by millions of people, perhaps more than any other 'science' in history. But this is water off a duck's back to the academic theorists and their acolytes in power. They appear immune to criticism. There has never been such a hugely influential science, responsible for national and global political ideology, that has been such an overt empirical and theoretical failure.

What is so wrong with it? It is not difficult to summarise. It is a *theoretical model of human economic behaviour, and the value created by that behaviour.* As such, it makes idealising assumptions – 'model assumptions'. From these it derives descriptions and predictions. Its idealising assumptions are wildly unrealistic, and its descriptions and predictions are wildly unrealistic. The main areas where its assumptions fail to match reality are:

- Its model of *value* is fundamentally wrong.
- Its assumptions about *market conditions* (such as 'perfect consumer knowledge', 'perfect producer knowledge', 'perfect rationality of players in the economic game', 'free choice of suppliers and customers', 'fair trading conditions', 'no subsidies or tariffs', 'numerous suppliers', 'numerous consumers') are fundamentally wrong.
- Its assumptions about *time and dynamics* are fundamentally wrong.
- Its assumptions about *the role of power* are fundamentally wrong.
- Its assumptions about *human behaviour and psychology* are fundamentally wrong.

Which is to say: at every point. Its entire theoretical edifice is wrong.

The core of its failure lies, again, in a peculiar kind of *rule-obsessed paradigm*. It takes a rule-bound vision of *human behaviour* as its starting point. It constructs a *rule-bound model* of an economic system as its theory, a model that runs by mechanical rules, without needing human interpretation or thought or judgement. It spawns an ideology of rule-bound *policy* as its outcome.

In the latter respect, it is very peculiar as a 'science', because most of its key idealising assumptions – the so-called 'free market assumptions' - *depend on the commercial environment that is legislated.* Since these conditions do not hold naturally, it spawns a political ideology intent on *creating the conditions to make its own theoretical assumptions true.* Thus it has a perfect excuse whenever its predictions fail: *the correct assumptions were not properly legislated to make the theory work as it should.* Nonetheless its acolytes claim it as a *scientific theory, representing natural laws of economics.* With this kind of starting point, it is little wonder they fail to take any evidence of its failure seriously.

Its failure is intrinsic to understanding the predicament of modern science too, because science has such a peculiar relation to *economic value* in the modern world. I will mention a few key points in this respect, as another exemplar of the obsessive rule-based metaphysics that dominates the scientific imagination, and subsequently the conditions of scientific practice.

Monetarist economics starts with a primitive theory of *value*, where value represented on a one-dimensional numeric continuum, called *money*. In theory, this is determined by the *relative exchange value of goods and services and labour*. The idea is that people will *exchange* their possessions or time (labour, expertise) for the possessions or time of others. The ratios of exchange represent relative values that we place on these things. The theory originated with Adam Smith in the C18th, and was refined into a *behaviourist theory of value* in the C19th by Jeremy Bentham and early capitalist philosophers (and notably a group of positivist empiricists in Britain, calling themselves the 'enlightened hedonists', including John Stuart Mill).

They did not believe in *value* as an objective concept. A product might have a great value to one person but little to another *("Jack Spratt could eat no fat and his wife could eat no lean...")* – and economic goods also have *diminishing value* as you get more – if you have a thousand apples, you will be happy to swap some for an orange. But they were faced with a problem of showing how a certain economic system (Capitalism) could *maximise value for everyone*, in spite of the subjective nature of *value*.

By taking *money* (a conventional legal token) as a conventional unit of exchange, they could first of all *objectify it as a universal token of value*. So far so good. Money is very useful like this. It is indeed a great technological invention. It lets people swap stuff they produce very conveniently, in a way that is impossible to do physically. It allows the creation of *capital*, where people can accumulate and loan money, and other people can borrow money and pay it back later. Thus it helps us *accumulate the value of our enterprise, and exchange it across time* as well. Money *per se* is not the problem with economics, and money does not in itself represent the demon of greed. The *system of economic rules* we subsequently place around it is the problem.

The second step in their theory of value is where the problems start. This says that *whatever we freely choose to swap for money – or swap our money for – defines its value*. If we swap $1 for an apple, then that is its value for us then. If we swap $2 for the same apple the next day, then that is its value for us then. What is its 'true value'? It has none: it only has the conventional value we will agree to *buy or sell it for on a particular occasion*. Thus value is reduced to a *rule of behaviour;* and value is measured on a *one-dimensional linear scale of $*.

At this point they start calling in other assumptions to make their larger theory work, to show that a *free-market monetary system will optimise value*. What happens ('in the perfect world'), they reason, is that producers will choose to produce goods that will sell for the highest profit. Since this produces the most money from their efforts, by definition it produces the most *value* – the value being had by the consumers. Thus a dynamic is set up: economic producers move to the highest-value domains they can, and when production and consumption – supply and demand – comes to a stable equilibrium, the net result ('our mathematical theorems prove') is the global optimisation of value. This just depends on people having *perfect knowledge of the market conditions* (so they know what will make them the most money), and having *free markets to sell in,* markets not 'distorted' by external factors (like subsidies or external redistribution of wealth), that distort the representation of value by monetary variables, and having fair rules (so players do not unduly influence the economic laws for their own benefit by political machinations that go outside the purely economic drivers).

On the basis of this simplistic theory, we have a huge ideological movement to set up a global system of *free market capitalism.* Oddly, the poor unemployed and working classes – whom it is supposed to benefit the most – are often opposed to it – ungrateful wretches - while the rich and wealthy classes, the owners of capital, the owners of the factories and production systems, who produce so much value for the masses, are wildly in favour of it.

We all know what the real outcome of a purely money-based culture of value is of course: *greed.* Greed is manipulative and obsessive behaviour to *acquire as much money as possible, at the expense of others.* Making money becomes a substitute for creating *value.* If the Capitalist player can make more money for himself by creating something he knows actually has *lower real value for the consumers,* he will do so, as his economic imperative.

But wait a minute – if *what consumers are prepared to pay for something is the only measure of value in the first place, how can this contradiction happen?* How can *making more money* result in *generating less value?* In the monetarist theory of value, this is logically impossible! Whatever people will pay the most for *is the most valuable by definition* – for the subjective judgement of value, reflected by money, *is* <u>true value</u> by definition!

And here we have a fundamental paradox that shows the theory is incoherent. For we must have an independent judgement of value for any theory about 'optimising value' to make any sense. If the theory is *true by definition,* if it optimises value *by the definition of value,* then it is not an empirical theory at all – for there is no way to objectively judge the *real value created.*

In reality, 'value' is multi-dimensional, and while it has a partly subjective element, *it is also substantially objective.* If we are sucked into buying some inferior product by slick advertising, then we *lose real value* we would have gained from buying a better quality product. For we buy products and services for *reasons, for goals, for purposes, for objective qualities.* We buy food for nutrition and taste – and while taste may be a subjective preference, nutritional value is essentially objective. We buy cars for performance and appearance – and performance is essentially objective. And so on. An economic system based on marketing and presentation and salesmanship and branding and manipulation of beliefs about objective value qualities distorts appreciation of real value.

This circular reasoning about *value* that forms the very foundation of Capitalist economic theory is pure sophistry. But it passes for conventional economic wisdom – because monetarist economists are too philosophically and scientifically ignorant to recognise or question their own evident delusions. They are in thrall to a *positivist-reductionist metaphysics*, the same one that deludes so many philosophically naïve scientists.

When we come to evaluate the success of an *economic system* or an *economic choice*, we must evaluate it *counterfactually:* what real value does it create, compared to another system or another choice? If it *actually* leaves millions of people starving, and a few people extremely rich, well, *that system is simply not optimising economic value for humanity*, no matter what any Capitalist economist tells us.

The term 'value' in economic contexts is synonymous with 'market value', i.e. 'price'. The value of my car or home or painting or book is *what I can sell it for.* But people who can only recognise monetary value are lacking a dimension in their souls. We also talk of many other dimensions of value: moral value, aesthetic value, scientific value, intellectual value, future value, sentimental value, personal value, public value, spiritual value. Monetary theorists insist we must reduce all such dimensions to one: *monetary exchange value.* This is intimately bound up with the paradigm of *ownership:* the ideal that *all things are privately owned by someone.* Value is then measured by *the amount of money the owners will exchange their possessions for.* If we object that all things are not owned – the types of values mentioned above are not generally things owned, and they cannot be traded between people – the economist has a ready reply: *we must define the world so all things are privately owned!*

Conventional economists also take the stance that, although there are other dimensions of 'value', we should *define* economics as being only about the domain of *monetary value.* This is exactly analogous to the positivist-empiricist fallacy of *defining* science as being only about *empirical observation.* In fact we find that many sciences – domains of systematic knowledge developed about the natural world - are *not defined empirically* – and we find equally that many economic activities are *not mediated by money.* Economics is about the production and exchange of goods and services. There are many pre-monetary societies – e.g. where physical goods are traded directly – do they not have economic activities? Even in our money-obsessed society, people still frequently engage in productive activities, that fall in a natural continuum with production of goods and services – from unpaid housework to favours to volunteer services to art projects - that are not mediated by monetary exchanges. Do we simply define them as *non-economic activities?*

"Well yes", the monetary economists say: "We must use *money to demarcate economics.* But where there are productive activities have not been reduced to financial exchanges, *they should have a monetary model imposed on them.*" Their ideology is thus to *bring all activities into the economic fold.* If there are activities that this does not work for – then *they do not really have value, and the economic system should force them out of existence.*

Here we find a close connection between scientistic positivism and economic positivism.[124] Both want to *demarcate their domains by an arbitrary theoretical standard,* that subsequently drives an ideological obsession. Both want to impose a simplistic *rule-based model and methodology* on the complex activities in their domains. The scientistic positivists impose a model based on a mechanistic theory of scientific method, then counting 'peer reviewed publications' as the measure of scientific value – a *rule-bound system dispensing with human judgements.* The economists impose a model based on a mechanistic theory of economic productivity, where *human beings are units of labour,* atomic inputs to the economic machine. They maintain that *economics will work perfectly, as a mechanical system, simply by imposing a set of simple rules on economic behaviour, and ignoring the need for any human judgements of value, ignoring human intuition and morality – as long as we set up the system properly!* If we set up the 'free market' with the correct rules, and let it run as a natural process, and *Hey Presto!* the 'invisible hand' of Adam Smith will make everything work!

I have criticised the economists' theory of *value,* as the deeper metaphysical foundation of their fantasy world: but in fact *all* the idealising assumptions of their model enumerated above are wrong. They are so widely criticised over the last fifty years that there is hardly any need to repeat the details here; but I will note some of the vital points of failure.

The economist's theory of *time* and *dynamics* is especially delusionary. There is no 'equilibrium state' in the real world of modern economics – our economic systems are inherently unstable, on macro and micro scales. John Maynard Keynes, the great macro-economist, showed this in the 1920s-30's, arguing that pure Capitalist-based economics would crash and burn (as it did in the 1929 depression) without stabilising intervention and control. Keynesian economics was widely adopted by governments in the post-depression and post-WW2 eras, up to about the 1970's, when a new wave of radical monetarist ideology took over again, spawned first by the academics, and then eagerly taken up by the Capitalist elite and their political stooges.

Today we see huge instability in the dramatically widening split in wealth between rich and poor, especially in the rich industrialised countries. The richest 1% now owns some 50% of all wealth – increasing from around 35% in the post-WW2 period. The split has accelerated over the last two decades,

---

[124] Ayn Rand, a popular right-wing American philosopher and novelist of the mid-C20[th], once a cult hero among the conservative American political establishment (in the 1960's she attacked hippies, peaceniks, socialists and free thinkers) is a perfect exemplar of this connection. A fervent philosophical positivist, she funded her right-wing Capitalist ideology from this bankrupt foundation.

and now increases by about 1% every two years. This reduces large sectors of the population to inescapable economic slavery. Equally, we see violent instability in financial markets. The political instability resulting from this will eventually overrun the Capitalist system; just as the desperate conditions of the French proletariat overran the Aristocratic ideology, some 200 years ago.

On the micro-scale, the idealising assumption that *production will systematically shift from low-value to high-value goods* in response to demand and innovation, and thus steadily increase total value, is delusional. It takes huge resources and long periods for industries and national economies to gear themselves up to new industrial production – large companies crash, whole national industries become unemployed, entire vocations disappear. Without extensive social welfare and subsidy systems in the capitalist countries, large sectors of the population would simply starve to death between bouts of unemployment.

Time is the real Achilles' heel in the economist's model of *value*. Their value theory applies only to *consumable goods in the immediate present*. It does not take future value into account at all. It models human beings as stimulus-response mechanisms, simply reacting in the present. But human activity *intrinsically* revolves around acting and planning for the *future*. How on earth do we calculate the *future value* of innovations, scientific discoveries, education, health services and infrastructure programs? Capitalist theory gives nothing here.

The idealising assumptions about *market conditions* – perfect consumer knowledge, perfect competition, etc – are equally delusional. Markets are created by *marketing* far more than by quality. Marketing is aimed at precisely the opposite of consumer knowledge: it is aimed at creating subjective and psychological illusions about product qualities. The 'free market assumptions' of the model are simply absurd. They were modelled on the image of an C18th *fish market*. The image of *village produce markets* is still the predominant example used today to teach the 'free market' concept in Capitalist economics classes. The global economic environment has absolutely no resemblance to a village produce market.

This relates to another major part of the Capitalist propaganda imagery, which is to conflate the concept of *free enterprise* (where numerous *individuals and small businesses* are *free* to undertake *enterprising activities* in *fair competition* with each other) with its diametric antithesis: *free market economics*, which means a system where huge monopolistic or oligarchic corporates are *free to control and dominate markets and labour conditions, and use the power of their wealth to destroy competition and optimise their operating environment.* Industrial markets today are overwhelmingly dominated by huge corporates – who wield huge power over the *meta-system*, the system of rules and regulations. There are said to be some 50,000 *full-time professional industry lobbyists* in Washington, all working to shape legislation for their economic benefit to write the rules for their own game. The payoffs and advantages of this *meta-system influence* are *huge*.

The legalistic rule-bound environment of the 'free market', controlled by government bureaucracy, is really the *power base* for these players: for they are the only ones who can afford the expensive services of lawyers, spin-doctors, marketers, lobbyists and media, required to control mass perceptions and political perceptions.

And finally I will observe that *there is no scientific or intellectual input into government calculation of economic policy to speak of – certainly not in* NZ. Economic policy is governed by political ideology, conceived by management drones, not by *any* scientific intelligence or analysis. It is autocratic and paper-thin. In NZ, a handful of political executives – the Minister of Finance and Prime Minister and a few politically-appointed senior drones in Treasury and Reserve Bank – chosen as Free Market Capitalism ideologues, and ignorant of science - interpret and control the economic destiny of four million people. Other countries surely have some smarter executives; but they are equally controlled by tiny political elites, captive to political ideology.

As Lopez Corredoira observes: *Capitalist economics and its culture of greed is the real god of the modern world, and it has become the god of science now too.* This threatens to override the independence that *science* once had, as an autonomous domain, operating with its own values. In conclusion, conventional economics is a rule-driven domain of political control, deeply ideological, and working in opposition to scientific values.

Capitalist economists and their bureaucratic stooges are the last people we should allow to tell us how to organise science.

Wasting the time of our lives.

Time has been my central preoccupation in philosophy. My interest in physics has centred on the nature of physical time, with different aspects implied in the theories of relativity, quantum mechanics, thermodynamics and cosmology. Certainly this is a scientific subject that fascinates me. But my interest stems from the fact that time is the precious commodity of our lives. Time is fleeting, and our time will soon be over. The fleetingness of time is at the heart of all human philosophy and its search for meaning, because it is at the heart of our existence. The *human condition* is defined by our mortality: the comprehension that our incarnation in these human egos will *inevitably* be over. This is about the nature of our existence in time.

We may wish and dream that we have a more permanent identity, a spiritual embodiment transcending time. Well, that is as it is. Studying it will not change its reality, whatever it is. But how we understand time and mortality has a very real effect on our attitudes to life and value and meaning, and consequently what we do in our lives. Personally, I think we do have a more lasting spiritual identity, but what it is I cannot say. But whatever the case may be beyond this life-time, our *personal embodiment as the humans we are* is what we have *now*, and it will soon be over. We will one day all be gone from this world, just as surely as we were absent in the past. We do not take the lives, the psychological egos, that we have *here* with us. Whatever value there is in *this* human existence can only be realised in this short space of time that we have to live here on Earth. And *this* is the miracle of existence for us as human beings.

My preoccupation with understanding time and the nature of existence is why I particularly resent the shallow modern philosophy that tells us modern physics has proved the flow of time is unreal. The flow of time is the reality of change, the fact that we live in a world of *temporal existence*, with a past, a present and a future. This is what we must come to terms with in real philosophy, in the philosophy of life. The nonsense peddled by the scientific philosophers in the subject, that *time is space*, that we live in an *eternal unchanging bloc universe*, with its implications of determinism and nihilism, trivialises any attempt to understand our real meaning, and undermines real philosophy. It is a waste of our time as philosophers, and as scientists, and as human beings. It is prompted I think by the academic's sublimated fear of death, which makes them afraid of life.

The recognition of the inexorable passage of time threatens us with a catastrophe of meaning: the terrifying prospect that the meaning of our lives is destroyed by the inevitability of our death. But this is only a catastrophe if life has no intrinsic meaning in its own terms, which I believe it has. Recognition of our mortality is difficult, the most difficult thing of all perhaps; but it is our truth, and it is what makes life urgent and important. Given the shortness of life, it is urgent *not to waste our time*, for if we waste it, we waste our own meaning. We waste our own unique miracle – for to appear here and now, out of the vast stretches of cosmic time, out of the vast improbability of the Big Bang and the chaos of billions of years of swirling dust, where any single tiny microscopic event over billions of years would have prevented our birth, to appear in this blazing world of light and life, as intelligent and feeling creatures, with language and art and science, seemingly out of nowhere, is truly a miracle.[125]

But what should we use our time on Earth for? Well that is the question for everyone to work out for themselves. The process of living is our process of working that out. But there is at least a partial answer everyone can appreciate. Our nature is to have certain intrinsic drives, for higher-order goods, over and above the basic drives to eat and drink and procreate that we share as animals. These are creative and social and what you may call 'metaphysical' drives, including philosophical and artistic and intellectual and scientific and moral and spiritual and religious drives. We all have these inside of us in

---

[125] It is a miracle in scientific terms, because there is no scientific explanation of this: *why do I exist?* (Everyone must take 'I' to refer to themselves).

different ways. They are expressed in some degree by a need for contemplation and understanding, sometimes for religious devotion or spiritual experience. More universally still, they are expressed in our need to *work*, in purposeful and constructive or creative activities.

The happiest people are those who are happy in their work. Or more than happiness: those who live through their work. This is evident in those who succeed in work that fulfils their natural sense of vocation – notably in successful artists and craftspeople, musicians and writers, social and health workers, philosophers and teachers, activists and evangelists; and in many scientific vocations, extending to entrepreneurs and inventors and engineers and architects and many other too. People who are truly engaged in their work are filled with purpose. They feel that their life would lack its natural purpose if they were *not* allowed the opportunity to engage in their vocation. For them, the miracle of existence is grounded in this positive foundation.

But this is relatively rare. Most people go to work as a chore, as drudgery. Most people today are *forced to work* for money. (In the past often pressed by the whip: now more invidious pressures have been perfected.) Meaningless drudgery in our work is what wastes the quality of the time of our lives above all else. And for many people with true vocations, degraded conditions of their work come to destroy their purpose in it: for it is often turned into drudgery, perverted into rule-bound bureaucracy, forced into greed and competition for power, by the systems of *employment and management* that control our conditions of work. So it has become with the vocations of science. In the context of modern economics, the sciences have become just another mode of *employment*.

We should value the miracle of our own existence above all the money in the world. The excessive demands of our bureaucrats and executives are there to waste our time. They are there to pervert our intrinsic purposes to their purposes. The rules of economics are there to make us to trade our time for money. The rules of the labour market are there to turn our precious living time into dead time, performing drudgery in servitude to others. The bureaucrats and capitalists want to take *the time of our lives* – the only thing we really and truly have - for themselves. They want to take our work to satisfy their own egocentric delusions.

The bureaucrats break us slowly on their wheel, forcing on us more and more rules and regulations, to shape us to their purposes. Today they force on us more and more compliance requests, forms to fill, permissions, processes, information demands: their tools of control. They demand our submission in servile and demeaning relationships, where we must ask their permission for everything and anything we want to do, and we must do it *their way*. The 'compliance tasks' they demand of us cost them nothing, for we must do them *in our own time*. They may seem little things – another form here, another process there – but increment by increment, they encroach our lives, locking us into the mental prisons of their rules. Where once slaves wore the chains of the masters on their legs, we now wear their chains in our minds. We have been programmed.

Already, the demands of work consume as much of our time as the system can force on us. Outside the factory or office, our grid-locked city transport systems mean we spend hours of our own time travelling to and from work, stuck in traffic, waiting in queues. Inside the office, the business processes defined by management force us to spend increasing amounts of time in pointless tasks, in meetings listening to childish management clichés, herded like sheep in pens. "But you are getting *paid for it*, so don't complain!" people will say. But this is the question: how much money is worth the sacrifice of the time of our lives? It is a question not just about our individual choices to work – we may have no realistic personal choice - but about the larger culture of work we are caught up in, the system our society forces on us collectively, in its worship of economics and money.

Our work-time is micro-managed, broken down into 15 minute intervals, budgeted and charged out. We must be constantly busy, at our computers, service counters or machines; there is no time to think about what we are doing. When we get home after work, exhausted and stressed after ten or twelve hours of travelling and working, we have an hour or two left to see our children or partner – and this precious time is further encroached by the bureaucratic and financial organisational tasks that demand our evenings. As the economic web around our lives expands, we pay the cost of endless financial contractual relationships that lock us into our jobs, rent, mortgage, power, phone, media, water, insurance, banking, finances, accountants, tax, lawyers, cars, transport, schools, doctors, purchases, appliances...

Our time and income is in a circular churn largely spent sustaining ourselves to keep going to work. Most of us are left with little quality time *for ourselves*. When we do get time to ourselves, we have run out of energy to *think*. We end up blobbing in front of a TV screen, or anaesthetising ourselves with alcohol or drugs as stress relief, to shut out the flood of the world that overwhelms us.

And yet most people are addicted to this lifestyle of constant *busyness* and economic grind. They see it as normal. They see it as *moral*: suffering the pain of pointless work, the loss of the time of our lives, is a sacrifice to be embraced – rather than the squandering of our meaning, to be regretted. Unemployment, or living outside of the capitalist economy, without a 'real job', i.e. a job defined in servitude to others, is despised as lazy and parasitic.

Most people are locked into this frenetic mode of existence as an *economic imperative*. People are very afraid to lose their jobs, not because they bring any satisfaction or fulfilment, but because they need the money to sustain the materialistic churn of their life-style. People fear destitution and homelessness if they become unemployed. This is an ever-present fear in modern society. People are locked into numerous predatory financial relationships they cannot escape, short of walking out on their lives. Modern civilisation is locked into an *economic churn* as never before. Politicians and bureaucrats preach economics constantly, it is the public concern more than any other single issue, they constantly promise to deliver us economic security (which never comes), and constantly threaten us with looming economic disaster. *Making more money by a constant churn of work* is a permanent necessity, and comes to be seen as the natural motivation for human activity.

I pause here to observe that this is one of the great paradoxes of modern life: we have powerful economic technology, that makes us vastly more productive, letting us produce far more with less time; and yet people in employment today work longer and have more fragile economic security than preceding generations. We have more consumer goods, yet quality of life has diminished in real terms for large sectors of the population. For millions of unemployed and impoverished people in the wealthy capitalist nations, there is no hope of making a living income at all. Millions of lower-skilled workers are effectively trapped in a subsistence lifestyle, with people sacrificing all their time to labour in a constant struggle to survive.

In pre-technology times, people had to sacrifice their time in grinding manual labour to subsist, just to produce food and shelter - and to work for the tribal estate or the land owner or tax collector. Large numbers were trapped in serfdom or slavery – which in one form or another has dominated most forms of civilization throughout recorded history. Only a few privileged souls had sufficient wealth or power to own their own time.

The promise of modern economics was to free people from this drudgery for survival, from the uncertainties of survival and servitude to masters, and *give the time of our lives back to us.* Technology has created vast productive power – allowing us to produce food and clothing and housing and transport in abundance; and taking away our need to spend the time of our lives in hard labour for someone else's purpose. Technology has made us about *100 times more productive* in traditional labour-intensive tasks. A single bulldozer can do the work of a thousand men with picks and shovels. And yet this has still not freed the majority from the intense demands that work makes on our time; and it has not brought the majority wealth or security.

Instead, wealth and power has become enormously stratified. In Western capitalist countries (typically the US) it goes roughly like this.

- The top 1% of people are extremely wealthy, owning some 50% of all wealth – which generates equally vast incomes for them without work.
- The next top 25% (or 24%) are comfortably wealthy – middle-class professionals, executives, lawyers, managers etc – and they take 50% of the remaining 50% of total wealth (25% of total wealth.)
- The next 25% are skilled workers with good incomes, enough to make some economic surplus after paying the bills – and they take the next

50% of remaining wealth, or 12.5% of the total – leaving 12.5% in total for the remaining 50% of poorer people.

- The next 25% are workers with incomes they can survive on, but not by much, struggling to pay their bills week by week. They take two thirds of the remainder, or 8% of total wealth.

- And the last 25% have the remaining 4% of wealth to divide. They are low-paid workers in drudgery, the poor, the unemployed, and the homeless, living on the poverty line. They are constantly in threat of losing the necessities of life – housing, power, food and clothing.

The essential shape of the wealth curve is shown in this graph, with the vertical axis representing income (starting at the richest 1%). The numbers here are only illustrative, but reflect approximately the real scale of annual income (in industrialised countries). The shape of the curve is the essential thing. It is

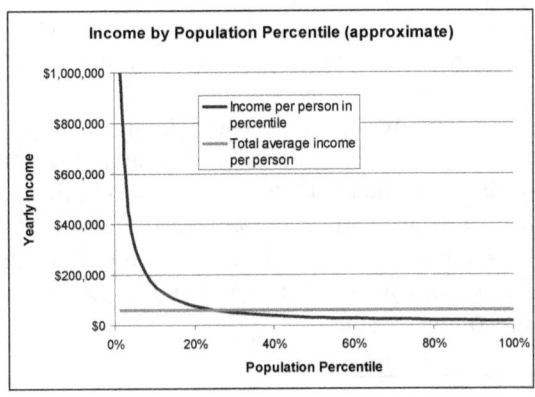

exponential. It shows how capitalist economic wealth is *really structured*.

The vast amount of wealth is owned by a tiny percentage – and it sky-rockets at the rich end. The curve flattens out at the poor end, with the poorest people shown here getting about $10,000 per year – just enough to subsist with no surplus or savings – no way out of poverty. The *area under the curve* (its *integral*) represents the cumulative total wealth or income. 50% of it is concentrated in the first 1% of population. This graph indicates the richest 1% averaging about a million dollars per year - but this does not show that the curve rises exponentially for the tiny number of *super-rich* (this would dwarf the scale of the graph too much to be able make out the curve). The very richest people today grow constantly richer, with incomes on the scale of *a million times more* that lowly workers.

The instability of the monetary system is reflected by the fact that the curve is getting significantly steeper every year. The split is now 1% owning 50% of wealth –at present rates, in ten years this will be 1% owning 55% - in 20 years, 1% owning 60%… This means the remaining 99% are left with 40% - an impoverishment by 20% from their 50% share of wealth today.

Is this sustainable as an economic system? Of course not. It is a social time bomb waiting to explode!

Is it fair and equable and just? In any economic or moral terms: absolutely not.

In economic terms, the very rich as a group contribute nothing of *value* to society in proportion to their wealth and income. The rich make their wealth not from work but from their legal ownership of capital, through the labour of the masses. They cannot possibly *create that amount of wealth from their own work*. In fact, they contribute very little *real value to the economic system* by their own work. They have been able to exploit the *meta-rules of the system*, the legal-property-political framework.

This is seen by considering the counterfactual: *what would happen if all the 1% of richest people suddenly disappeared off the face of the earth. How much would this change the real productivity of goods and services?* Production would not even diminish by 1% let alone 50%! The very rich are in figure-head executive power roles: they can easily be replaced by others, and it would make little difference to the *productivity of the system below them*.

Alternatively, what would happen if 10% of the lowest-paid factory workers, garbage collectors, labourers, fruit-pickers, lower-skilled service workers disappeared? Productivity would drop *dramatically* – by far more than 10%. For all kinds of *essential and valuable real work* would not be done. Yet consumption would decrease by only a small amount - their share of the social spoils for their labour is tiny. Society would suffer a large, very real loss of economic value.

It is not quite so straightforward to interpret the total effect of the wealth stratification, because the very rich 1% do not necessarily *consume* all their wealth for themselves. To the extent that it is invested in productive capital – factories and infrastructure etc – it can generate productive goods, and it is to some extent an abstract matter who owns it. But the extreme stratification certainly means that the poor are disenfranchised, and the rich have an exorbitant control over the use of wealth. This is evident in the chronic lack of economic control and security that the poorest 50% of people have. It is evident in the kinds of projects that private wealth is used for. It is used for the purposes of the very rich: primarily to make more wealth for themselves, in obscenely excessive consumption, to make monuments to their self-importance, and to disperse among a tiny number in their personal circles.

Creative intellectual activities like the arts and humanities and philosophy and science, as well as public infrastructure, are left out in the cold by the Capitalist rich. If these do not suit their personal fancies, and they rarely do, they are not going to *give wealth away to enable them*. Usually at best they may give some token wealth away for their own agendas. This is the economic condition of private wealth for scientists and intellectuals.

As a result of course there is huge pressure for public redistribution of wealth, meaning forced acquisition by the government bureaucracy, to provide social infrastructure of transport, education, health, justice, defence, emergency welfare – and science. Without this redistribution, of course the poor would quickly become totally destitute in a Capitalist system – and the wealthy would soon be destroyed. This redistribution of economic surplus through bureaucracies now forms the context of science: essentially *only 'science projects' controlled through bureaucracies are funded.* And even when private wealth is donated to science, it is channelled through the same bureaucratic processes – not to *scientists* to use according to their judgement, but to institutional managers according to *their judgement.* The public system is locked into the social visions of public bureaucrats. It is perverted to their ideas of *control and acquisition.* They become the state-owned corporates. Herein lies the institutional death of science.

In any case, there is a vicious cycle at work here in term of conventional monetary economics. Automated factories can produce far more with less labour. But this only increases the wealth of the owners if the products can be sold. It can only be sold to the masses. What do the masses have of value they can exchange? At the end of the day, only the *time of their lives.* Money fundamentally buys people's *time,* so that the rich can command the services of the poor. But if there is so much produce available to the masses, then the imperative to exchange the time of their lives to labour in servitude vanishes too. This is a fundamental paradox for a monetary economy that becomes efficient enough to allow everyone to live comfortably on a tiny amount of labour. Money threatens to become meaningless – so artificial means are invented to force people to remain in servitude, to force them to sell their time.

The vast wealth disparities are only half the reason for the stressed economic state of the masses. Sure, this is a critical form of *rationing wealth,* to keep the masses poor, and locked in time-servitude. But there is another reason equally important: an economic *churn.*

Given that we have increased *productivity* in primary productive industries by about 100 times in the last few hundred years – since the Middle Ages, say – and a proportion of people still made enough to live comfortably in the Middle Ages – you would think that surely there should be a *huge* economic surplus now, sufficient for *everyone* to enjoy at least a good basic living standard, at least in provision of the basic essentials of life – food and water, clothing, housing, health - the essentials required to enjoy *the time of our lives.* And yet the majority of people are trapped in a desperate economic churn – in a subsistence economy, poorer in real terms than the well-off 10% or so of Medieval people, who lived in a society with a *tiny* economic production.

While *primary physical* productive power has expanded by perhaps 100 times today, the real increase in *productive value* created is probably only *one tenth* of that (and then only a fraction of that trickles down to the masses; the rest being taken by the government and the very wealthy. Where did all the potential productivity go, to lose a whole scale of *magnitude?*

It is lost through escalating costs of *layers of infrastructure underpinning industrial, financial and social systems.* This includes both physical and social infrastructure.

First we have layers of physical infrastructure required to support increased production systems – roads, transport networks, vehicles, ships, power stations, machinery factories, construction industries, computer factories, chemical and medical factories, warehouses, supermarkets, etc, etc. These are all needed as the *industrial base* to support the factories and farms that ultimately produce the actual goods we want. For example, a modern farm is probably 100 times more productive than a medieval one *per unit of farm labour* – but it is supported by layers of physical infrastructure: tractors and farm machinery, energy and power systems, roads and fences, water and irrigation, fertilisers, herbicides, pesticides, with special vehicles, aeroplanes and helicopters to apply them, animal medicines and vaccines, seed technologies and planting machinery, computers and accounting systems, trucks and vehicles to transport produce. (And farming is a relatively *low-tech industry!*) For all this technology to be available requires layers of high-tech industries behind them. And of course, the produce gets to the end-consumer through a network of processors, packagers, middle-merchants, product marketers and supermarkets.

As a result, the farmer makes on the scale of 10% of the end-sale price of their produce as *income* (depending on sale prices, which fluctuate wildly in the 'free market economy' – so whole farming industries would go out of business in most countries without stabilising subsidies from governments). The remaining 90% goes to production and distribution costs. (And more is siphoned off in taxes and mortgages.) Thus the primary produce of one farmer's labour (the stuff that is *actually useful as a product – all the intermediate farming products are only useful to produce this)* is supporting the equivalent of 5 – 10 other jobs in the supply and distribution chains.

Second, we have multiple layers of *social infrastructure.* And this is where things get really expensive, and really artificial. This includes intellectual and organisational and financial and legal infrastructure: systems for educational, financial, bureaucratic, political, planning, legal, technological, communication and scientific services. This is where bureaucratic, financial, legal and managerial domains come into the picture of course. These are *meta-activities, multiple layers of meta-activities,* used to control and manage the primary activities of people.

The organisation of a government department is a good image of this. On the ground floor there will be workers performing the *real* function of the department. Processing and issuing passports perhaps. Interviewing clients and authorising benefit payments. Etc. Above them, there are team managers, controlling small teams of workers, looking after their relationships with the organisation – allocating tasks, making sure they come to work, get paid, etc. So far, so good: this is the typical *small business structure.*

Above the team manager is a second-tier manager. They oversee a number of team managers. They control budgets, managerial performance reviews, work processes, etc. They seldom deal directly with team workers, but have some awareness of the details of the work. In a small-medium business, this is the function of the *executive-financial manager*, usually the business owner. (And that is where it stops in SMEs.)

Above these are third-tier managers. They are politically and policy oriented. They command the second tier managers, dictating high-level *policies* they want implemented. They review high-level reports, sign off on budget requests, and so on. They are primarily intent on implementing the *bureaucratic culture*. They rarely speak to workers or team leaders, and have faint contact with the work-place. The 'customers' themselves are beneath their contempt: mere tokens in a process.

They report in turn to their master, fourth-tier managers, who control a whole *business group* – the Brigadier General in an army. These elevated creatures make only high-level decisions, tasking the managers below them to implement their fancies. They spend their time mainly in meetings and social occasional with others of their rank and VIPs. They never meet the workers or team managers or first-tier managers, except to make speeches on formal occasions, and they have no real idea of the jobs people do, the realities on the shop floor. They talk almost entirely in political, policy and business jargon. They sponsor business initiatives. They report to the CE, the figurehead of the organisation.

The CE, at a fifth level removed from the business, is the conduit between the organisation and the Minister in charge of the department, and CEs and other VIPs from similar organisations. They control *relationships* and high-level signoffs and so on. The Minister and their political party cronies dish out budgets and high-level executive power, legislating for the legal powers of the Department.

These layers above the first and second level management are essentially a bureaucratic churn, deigned to maintain hierarchical power and control, to force the soldiers to conform to the executive command chain. It is obsessed with the world that obsesses the Minister: *public appearance, image marketing, power relationships, political policy, the exercise of power and wealth.* Hierarchically it is structured like the army (except the army is far more competent: its generals were once front-line officers.)

The overwhelming urge of the bureaucratic executive is to control all aspects of life within its domain: the lives of the soldiers. The big difference with the army is that the sphere of control of the army hierarchy (in peacetime anyway) *ends with the army.* The government hierarchy sees itself as controlling the *civilian population.* It exercises its powers over the citizens of the country. The civilians often do not realise this: they think the department is providing them *services*, and expect to be treated with the same kind of respect as commercial businesses display. But the government department is about *controlling the public*, intent on dictating their behaviour in as much microscopic detail as they possibly can. This is natural to them: they exist in a hierarchical organisation, where their bureaucratic powers extend across the organisation. It is natural that they force their bureaucratic churn on the population. This is the social model – the social metaphysics – they have internalised from years in their power-cliques. This level of micro-control is what governments seek to expand today, beyond any historical precedent, through the medium of computer technology.

Thus we have an exemplar of a *social churn*. The government department does not merely provide services: it provides layers of control and bureaucracy over the top of its service department. Corporates are essentially similar: their large size allow layers of executive hierarchy. These high-level executive layers suck up major income – they produce the 50%-wealth-to-1%-of-people scenario.

The *social churn* of the modern economy extends to the expensive roles of *business professionals* - lawyers and accountants and bankers and financiers. These are not engaged directly in meaningful production– nobody values legal documents as commodities in their own right. You can't eat them or drink them or read them. They're too shiny even to wipe your butt with. Legal and financial products have no *intrinsic use* in themselves. They are only there to keep the legal, property and contractual labour systems churning over. Banks and financiers similarly extract huge amounts of money for controlling the *meta-system* of business.

The costs of physical infrastructure are necessary to keep modern production systems going – although they may be more complex and expensive than justified. But the further costs of *artificial social infrastructure*, steadily expanding today through swathes of regulatory legislation, are the real killer. They create millions of jobs, you might say. But they are *intrinsically meaningless jobs*.

The fundamental problem is that we have no idea of how the products of modern industry could be distributed *except through a traditional monetary-employment system*. For we can only conceive distributing goods to people *by taking their time in exchange for money*. The drive of the governments of wealthy nations in recent decades is not at all to provide solutions to the structural paradox of over-production and unemployment and poverty – it is the diametric opposite: to privatise ownership further, to *charge commercialised rates for government services themselves*, to increase the churn of economics, to increase the wealth divide, to keep the masses in economic stress and make them amenable to the conditions of forced labour.[126]

---

[126] In NZ there is presently an outcry over *zero-hour contracts*. Employers can force workers to sign contracts that *require them to be available for work when demanded, at short notice*, but do not guarantee them any work. So they have no idea if or when they will work in the next week. Conditions for workers have eroded over the last two dec-

These are fundamental reasons why the vast increase in productive power enabled by technology filters down to such a modest increase in real economic benefit for the majority of people. The administrative overheads of the State-Corporate economy, the vast hierarchy of non-productive activities used to control social and business processes, and the manipulative behaviour of professional, financial and legal services used to suck money out of the productive system without creating any value, swamps the economy in a vast churn of frenetic activity devoted to power and greed.

This frenetic mode of activity appeals especially to *business people*, who are constantly busy 'multi-tasking': organising processes and attending meetings, constantly taking calls, juggling the complexities of business decisions in their minds, churning over the bureaucratic and legal tasks required to satisfy business processes. They spend their working lives *calculating how to make money and manipulating power through business relationships*. To them, this is the natural mode of work, and they think everyone should work like this too. But they can only manage this by having a *shallow intellectual life*.

People like scientists and artists and writers and intellectuals generally require deep *concentration* on single complex tasks – not shallow concentration, flittering between multiple tasks. Imagine a musician answering phone calls in the process of creating a song – or an artist being constantly interrupted in the process of creating a painting. This simply destroys the creative process, wherein a large complex conception needs to be visualised holistically. Similarly with scientists. Periods of intense concentration are required when you are trying to visualise complex theories, work out explanations, and internalise complex ideas. Time for contemplation and thought is needed. I will often stare blankly into space for an hour, trying to mentally visualise structures or complex mathematical functions – before scribbling down an equation, or starting to build a new program element. It has to be done *until it is finished* – if you are interrupted in the middle, you lose it. It is like mediation, which

---

ades, with the replacement of *employment* by *short-term contracts*. Thus workers contract the supply of their labour just like other suppliers of physical goods. Contracts are then mediated by *employment consultancies* – who take 10%-20% of worker wages for a few hours of pseudo-legal work. My contracting jobs in the last few years have been through multiple layers of *consultancies sub-contracting my services to larger consultancies contracting me to bureaucracies* – each taking 15% of the final fee – with contracts extended a month at a time, at the employer's convenience. The costs and uncertainties are unsustainable, and I have ceased contracting work. This is forced by government regulation too: bureaucracies are only allowed to contract services through a small number of selected corporate consultancies; smaller niche consultancies must sub-contract through them. Bureaucracies and corporates dominate all contracting work, because costs are so inflated and performance so mediocre they are only organisations rich enough to employ specialist contractors.

requires deep calm concentration. (Descartes once spent some days in an oven...) A typical chunk of theoretical work takes two to four *weeks* of such continuous concentration; filling your mind with the complex connections required. The distraction of jumping around to other mental tasks, even if they are relatively minor, undermines the process. It is like stopping in a marathon to mow someone's lawn here, do someone's dishes there. This is why scientific work is often so mentally draining and exhausting, in a way most people cannot understand.

The take-over of science by business culture brings precisely this clash of understanding. Bureaucratisation places the research scientist in a management hierarchy, choked with business processes – grant applications, permissions, budgets, approvals, sign-offs, progress reports, time-lines, project plans, presentations, etc. *"Well, most of these don't take too long, what's the big problem? I do my routine paper-work before lunch so I have my afternoons free..."* replies the business manager. But does the business manger do *science* after lunch?

The levels of bureaucratisation are intensely distracting and time-wasting for scientists. They make the working environment fraught with *uncertainty*, they dramatically slow down processes, they sap significant mental effort that you are *already trying to push to the extremes for your real work*, they multiply relationships and dependencies; and they place the scientist in a degraded power relationship. The processes that make sense to the business project manager do not fit a scientific research process at all. You can't necessarily *make* a project time-line because you don't *know* what you are going to find or even what you are going to do. You can't *define* all the steps you will go through because a *scientific method doesn't exist like that*. You can't *explain why your results will be important or predict their profitability* because you *don't know what the results are yet*. You can't justify their scientific value to business executives because they won't understand the background of the science anyway. Project applications and grant applications, etc, are all formally structured by *business managers like business processes*. Scientific research isn't like a business process.

"Oh but we have to administrate and monitor you! We can't just give *money* without careful oversight of what you are doing with it! What if you *waste* it? (Money is very *important* you know... that's why they pay me so much to look after it carefully...)" Well, what if you waste my time? And with it my talent in my real vocation? Is that important? In the end, you are stuck with trusting the judgement of the scientist in any case, and merely covering it with a paper trail.

Many people in our society are so brain-washed into following processes defined for them, so dependant on others to define their purposes and meaning for them, that they have no idea what to do by themselves. They can only operate in the pre-defined context of a 'job'. Those of us who do have a genuine sense of our vocation and do know what we want to do and how to do it and are driven by our internal sense of purpose should be given the freedom to use our initiative, without answering to bureaucrats and managers. Or else we should take our freedom, and leave the institutions and bureaucrats behind.

In the end, this is what is at stake: the time of our lives. It is up to each of us to interpret what to use the miracle of our lives for, in our own way. Modern economics provides *vast resources* to enable those of us who want to use the time of our lives for constructive purposes in science or other intellectual work to do so. There is no economic imperative to work in drudgery or economic slavery for others. That is merely a myth sustained by those who wish to spend *their lives* obtaining money and power. Well, that is their choice: but we should not sacrifice our lives to them. But this is the economic and bureaucratic context of modern science: to be forced into a position of servitude and slavery.

The context of modern institutions and the world view behind them have been sufficiently emphasized, and I now turn conclusions of this essay, and our possible responses.

PART 4. RESPONSES TO THE CRISIS OF SCIENCE

What should or could be our response be to the decline of science? I will briefly comment on three aspects. First is at the micro-level of individual behaviour or culture within the scientific establishment. A second is the prospect for reform of the larger bureaucracy of science and the larger role of science in bureaucracies. A third is the personal response of independent scientists or intellectuals, wishing to work outside the establishment as free and autonomous agents.

López Corredoira has a number of suggestions to reform aspects of the systems of research science, to remove certain injustices, make resources more fairly available, redress prejudices against heterodox intellectuals, etc. He hardly sounds confident that any such reforms will be adopted or be effective, and then suggests that if we do not like the suggestion of reforming the system by such means, and choose to let it continue in its present way, we should mete out punishments for failure and corruption. We should punish individuals who make bad decisions that result in injustices to others:

"We could apply a system of compensations and penalties according to the biases of the system. We would let the system run as it is now and, when we realized that some decision was unfair, we could correct it by punishing the people responsible for the bad decision and rewarding the scientist who suffered the consequences of that decision. For instance, let us suppose that a researcher is unable to publish a paper in a high profile journal, so he/she has to publish in minor journals or not at all. ... They might even have to leave research and dedicate their time to other things. Let us suppose that after many years or decades, the results of this researcher are discovered to be of high importance in their field, innovative, and a revolutionary landmark in that subject. This is not very usual, but it might happen.... In that case, if it is not too late, that is, if the researcher is still alive, they should be given a prize, but an honorary prize is not enough, they should also be compensated for the money that they could not earn with many years of research. On the other hand, all those people involved in the negative evaluation of that work should be announced publicly if anonymous and punished; it is fair, in order to compensate the damage in the career of the scientist who was rejected. The referees of major journals who rejected their papers should be punished economically or even fired if their unfairness is demonstrated.

He thinks this punishment should apply to institutions too.

"The institutions which refused to give the researcher a position could be punished financially. Some institutions should be even closed if it is shown than in recent years they have invested their money in following the wrong lines of research, damage the innovative research. For instance, imagine that within some decades it is shown that ideas about dark matter and dark energy in present-day cosmology are totally wrong while some other heterodox idea whose researchers suffered obstruction was correct. In such a case, the many tens of research institutes which are now dedicating most of their funds to standard cosmology and the search of dark matter or similar topics, while closing their doors to scientists with other, alternative views, should be closed. It is fair. If the lives of scientists with better ideas were shipwrecked in favour of a cast of mediocre scientists, it is fair that the working lives of those mediocre scientists be now cut short and the better scientists elevated to a higher position. If a scientist feel that he has the right to cut short the scientific life of a person by judging the quality of their science, it is fair that the system cuts short the scientific life of that judge if he has made an unfair decision."

My initial reaction to this is sceptical, because I do not intuitively like the idea of *punishment* as a primary means of redress for problems. Or more accurately, I do not like the idea of all the rules and regulations and bureaucracy that would have to be set up around a system of punishment. Who knows where it stops, who makes the rules, how it will be abused? Will it come to be used against the freedom of scientists to express forthright opinions?

But on reflection, I think López Corredoira is right. It is not a question of punishing scientists for being *wrong* or even incompetent. But should we hold scientists, researchers, academics to account for incompetent judgements that destroy the careers of better scientists – especially when they reflect deliberately manipulative anti-competitive behaviour? Yes: it is a serious problem in science, and they should be punished, in the same way they unjustly punish others. As López Corredoira says: it is fair. It is the only option consistent with the present organisation of science, which currently deals out rewards and punishments based on these people's judgements, and which they have learnt to exploit for their own benefit. If you are prepared to live by the sword, you must be prepared to die by the sword. People should lose positions of privilege when they abuse those privileges. This behaviour destroys the careers of real scientists, who have become deeply fearful of upsetting the wrong people by expressing heterodox ideas.

People in other occupations are punished for failures of incompetence and self-serving conspiracies. Why are these occupations – senior academics and scientists, along with executives and bureaucrats - considered different? Simply because they are gate-keepers of a feudal power hierarchy? The people who obtain these positions come to believe they have an entitlement to them. But they do not have an entitlement. They are paid from public coffers. They come to believe that since they have won power by achieving their positions, this gives them a right to destroy their enemies. But they do not have the right to destroy the careers of others. They are in public and official roles to be responsible, not as a permission to be vicious. These people now act like the aristocrats of a bygone era. We should not support a new aristocracy.

But while addressing important injustices, this is not going to solve the larger crisis in science. The problem is widespread incompetence and mediocrity, the systematic degradation of scientific culture. It is a general problem for academia. The *research* competence of university academics is generally very poor. Academics are traditionally paid about half their salaries for teaching and half to do research, but most are not competent to do meaningful research. Far more are passable teachers than are competent researchers.[127] Perhaps half might pass muster as competent teachers (depending on the subject; teaching in mathematics is much more competent than in 'business science'). So they have a real value as teachers. But *very few* are competent to do research, to direct the research of others, or to judge the research of others, which they do as peer-reviewers and gate-keepers. What right do they have to control the fate of other researchers?

Few academics today have a genuine *vocation* for their roles: rather it is a career path they embark on. Very ordinary scholars can achieve positions by passing conventional hurdles – getting good grades, flattering the right professors, obtaining fatuous publications. Once they have achieved an *academic position,* an *office* in the academic system, they have no need to perform. To retain their positions, they only need to turn up to work; they don't have to do any. Those who choose, simply go through the motion of teaching, collecting good salaries in a pampered environment, where they will remain for decades until they retire. They undermine the system and destroy its value, discouraging and misdirecting the brightest students, and blocking career opportunities for real intellectuals, especially those motivated to do real *research work*. There are strictly limited positions available in any field. These people, like incompetent bureaucrats, should be removed from the system. Similarly, incompetent peer reviewers for major journals should be removed from those roles, and their failures acknowledged.

But would such reforms of the system, or other reforms such as López Corredoira suggests, make any real difference? The pragmatic answer is: *No.* It is not practical or possible to reform these systems now in place. It is

---

[127] The meaning of *research* has degenerated in the modern age of course. 'Research' nowadays typically means *any kind of self-education activity,* keeping abreast of the state of play in a field, with a common popular meaning of *collecting information from the internet.* It rarely means *advancing new knowledge and re-evaluating concepts,* i.e. scientific research in its original meaning. Academics certainly do need time to keep abreast of their fields.

simply not realistic to think this could be done. Modern *institutions* of science and academia are systemic failures because of a deep structural failure in their very mode of organisation. In this they are exactly like public bureaucracies: they are structural failures. It is proven by the fact that *exactly the same kind of failure-ridden performance is evident in all of them.* It is not that some bureaucracies do well and others do poorly. Their incompetence is universal. There are always isolated teams within any bureaucracy and any university that do perform well. But these islands of quality are drowned in the futility of the larger whole.

The level of incompetence is exacerbated by neo-liberal corporate-style business models applied to organisations that are not businesses. But it is deeper than that: it is because *the fundamental challenges before them, presented by the increasing complexities of the modern world including modern science, have overtaken their natural level of competence.* No reformation of "business structure" will ever make these institutions viable to do their jobs effectively. No plausible changes within bureaucracies will bring them scientific competence. We should stop worrying about this aspect that we cannot change, and focus on our own roles: on how we choose to spend the time of our own lives.

What should we do as individuals trying to live in such a deeply dysfunctional society? Our response will differ according to our position, including our age. The best and most sincere people within institutions will want to keep trying to 'fix' them, eventually get exhausted with it all, and if they can see an alternative they will exit – the *stratification feedback loop.* Certainly there are gross failures and injustices that should be addressed. But the larger problem of reforming existing institutions to be *competent at their intended function* is a lost cause. At some point, we have to decide as individuals not to waste our time and energy on lost causes.

I write now from the point of view of an independent intellectual committed to working *outside the system:* someone with no viable place in our system, and no desire to get back into it. What should people like us do? People who are creative intellectuals by vocation and personality, and need to find a meaningful place and mode of life, outside the system of capitalist economic production.

Instead of trying to bend over backwards to compromise ourselves, we should remain independent. I think the time has come when we should give up the idea of working for the system, and any thought of 'fixing' the conventional institutions. They have failed over decades. They are in the condition of general paralysis of the insane, and will keep failing. We should forget them. We should withdraw support and participation, stop worrying about them, stop taking them seriously, and pursue our own goals. We should stop trying to participate in the cultist rites of journal peer reviews, publications, academic prestige and promotion as the goal of our work.

The role of the independent intellectual must revolve around valuing ideas. For this is what intellectuals are: *people who value ideas.* Intellectuals must value their ideas, and they have a responsibility to try to nurture them properly; to make sure they are properly developed, and are given a proper chance of success. I think we should regard them like our children. We want to give them their best chance and opportunity to achieve their potential and live their lives, and have their own children (or not as the case may be).

This is a point of view about the value of the diversity of ideas: every intellectual has his or her own personal ideas, and wants to see them succeed. It is necessary, to do justice to the ideas, to give them a proper chance. But feeding your own children does not mean taking bread from the mouths of other's. Raising your own children well does not mean depriving other children of sustenance. On the contrary, you should be benevolent to other people's children as well as your own. But of course, you have a special responsibility towards your own, and you must naturally treat them specially. You place a special value on them. We can call this the Family Model of ideas. What makes this metaphor appealing is that we can freely adopt other people's ideas, and treat them like our own; as we do.

The Competitive Model (underlying the industrialization of the intellect in the Capitalist framework) sees ideas differently: your *original* ideas are *your personal property*, and are *part of your intellectual ego*. They are in *competition* with other people's ideas. You try to demolish alternative ideas to your own, and give your own ideas a monopoly. The greater prestige they gain, the greater your personal success. Ideas compete in a Darwinian struggle for survival and popularity and success; by definition, the winners are the best, that is, the most suitable to survive.

Seen in the light of the Family Model, this Competitive Model makes the mistake of interpreting our children's successes in terms of our own egos, and of living off our children and through them; of exploiting them. It also makes the mistake of sanctioning genocide against baby ideas of other people. You can kill a baby idea, as one large corporation will kill a small company, or one gangster another, for the sake of giving your own enterprise a better chance of domination. This is what some kinds of snakes do with other snakes' babies. It is what ruthless businessmen do, and what criminal gangs and political tyrannies do. But however valued this competitive behaviour may be in Wall Street or Microsoft or NASA, it is not the right behaviour in the family environment.

The modern academic environment is not a family environment; it is not a fully Capitalist environment either; it is a bastardised commercial environment, crossed with a Public Service ethos of hierarchy and complacency. It has its own gangs and gang leaders, in disguise as scholars and intellectuals. It has its own honest workers, and its own elite performers, but many academics have to belong to intellectual gangs of one kind or another to survive. Their territories are well-enough carved up that they have formed a static peace between tribes; but within a territory, they wage wars on interlopers, and they demand trials of initiation before allowing entry to the hierarchies of power. Because loyalty is strongly demanded, the tribes value the homogenei-ty of their members, and are suspicious of those who have other parts to their lives than their gang membership.

The fundamental power of the independent intellectual is freedom. This translates to freedom to choose what to work on. The most subversive thing an intellectual can do is to insist on working on ideas they consider intrinsi-cally valuable, and refusing to work on ideas that are not. The forces of Capitalism and Bureaucracy and Academia believe they can buy intellectuals to work on *their* ideas – which are all too often misguided, reactionary prod-ucts of group-think.

But there are two basic problems scientists or intellectuals who choose a path of independence need to face.

- How do we find resources to sustain ourselves in our vocations?
- How do we find outlets for the results of our creative activity?

These are similar problems whether your natural vocation is as a scientist, or philosopher, or writer, artist, social critic, historian, anthropologist, or any-thing else. Public resources for these activities are the monopoly of the bureaucratic institutions: they will not be shared with outsiders. As far as resources go, we simply have to survive as best we can, where we can. This has always been the case for intellectuals in the past, and it is the same today. There is no ready-made place for us. But there are more ways to survive in modern society than for intellectuals in the past. It is still a precarious exist-ence, you will not have the financial security of the middle classes: but that is the price of being independent. Living in poverty may be the price of *buying back the time of your life* from the economic servitude the system wants you to sacrifice yourself to.

One mode of life is to remain *semi-autonomous* by contracting your services out to institutional projects in applied science, as a researcher, statistician, programmer, or whatever role you are good at, and try to save enough energy to pursue your own research in your own time. There is a lot of benefit in this, especially while you are younger. You get to see inside the institutional world, and that is interesting. You are exposed to a variety of different sci-ences and research domains, and that is valuable. However, this has its limits, and it is exhausting over a long period. You end up wasting a large amount of your time and energy on pointless activities. In the business environment in NZ, and I am sure other countries are the same, this mode of work is in-creasingly unsustainable too, because you are forced to spend so much time and energy on bureaucratic relationships serving the government-enforced infrastructure.

It also does little to solve the problem of specific scientific resources for your own projects. E.g. López Corredoira uses the example of getting *telescope time* for an astrophysicist, and many sciences require some significant resourcing to do even basic experimental laboratory work. This is a basic limitation on the kinds of science independents can pursue. He is right that the time of 'amateurs' in primary experimental fields is effectively over.

One way around it however is to obtain experimental data from others – there are vast amounts of data from empirical studies in every domain that have never been competently *analysed* yet. Social research data collected by government departments is a prime case in point: they rarely do any meaningful analysis, and if you can get data from such projects, there is plenty of very interesting science to be discovered. The same goes in physics and other subjects too. the production of data has overrun the resources of analysis.

Primarily, you have to be adaptive: seek out areas of science that are cheap and accessible. There is actually a large range of topics in applied sciences and social sciences that are still in primitive states of development, typically with large data collections available if you can get access to them, but with poor theoretical development, poor explanations and explanatory frameworks, and stifled by desultory work from institutions. This is where independent scientists can make their real contribution: *for institutionalised science cannot honestly understand what its research means.* It is too busy presenting its results as propaganda. Propaganda for its next round of research grants, its ideological agenda, its economic imperative.

But there is one resource most essential to intellectuals who have survived and succeeded in the past, and that is finding a community of like-minded thinkers. The French Impressionists provide a classic exemplar: iconised by starting the *salon des refuse* for the refuse, the garbage, of art the Academy spurned. These are the glorious paintings by Monet, Renoir, Manet, Sisley, Van Gogh, Cézanne, Seurat, and so on, that now sell for millions of dollars, and make fortunes for the art dealers. The impressionists revolutionised art by establishing a strong *independent art movement,* mutually supporting each other. And the same for many other writer groups, and scientific groups in the past too. The fact is, you can survive hunger and poverty, but it is hard to maintain yourself in a long struggle with ideas and research without some supportive collaborators. So the *most* important thing is to connect with some independent community of like-minded independent intellectuals. The major pressure you have to face as an intellectual in our society is psychological.

In this respect, the Internet is the game-changing resource of our time, for you can find like-minded individuals at a distance. I know of no one in NZ with a research interest in the philosophy of physics and time – it is a small specialised field even internationally - but there are nonetheless a few thousand individuals, scattered around the world, that I can potentially communicate with. In this respect, it is interesting to consider the demographics of specialist topics.

There is probably one person in a thousand with a serious bent for physics; and perhaps one in a thousand with a similar serious bent for realist philosophy. The two are quite independent: physicists are rarely interested in philosophy research, and philosophers are rarely interested in physics research. So there is perhaps on the scale of only about one person in a million with a serious interest in both – e.g. in the *philosophy of physics and time*. Nonetheless, in a world of about 5,000,000,000 adults, this still means there should be somewhere on the scale of 5,000 persons potentially interested in this niche. Perhaps it is only 1,000, perhaps it is 10,000, but it is a significant number in any case. The Internet allows such individuals to find each other and establish small communities and micro-networks.

By this reckoning, there would be about 5 such people from NZ. I don't know who the other 4 are (they probably left the country). There are no academics in this field in NZ, so you cannot find a community through an academic center here. For intellectuals living in a country like NZ, there was little prospect of finding any intellectual community before the Internet. But now I can find others through the Internet. And I can find their work, without having to depend on the journal stacks in a university library. It is true there is a lot of garbage to sift through on the Internet, but nonetheless. In this respect it is more viable to work as an independent intellectual today than ever in the past.

Nonetheless, there is also no substitute for human contact, for most people anyway. Electronic communication only goes so far. It meets a certain need: the formal communication of ideas, the reassurance that there are others in the world who think the same ideas are important as you do. But human beings do also need personal contact, to communicate living feelings and emotions. In the end, intellectuals migrate from the provinces to the centres of ideas, where they can find a living community of like minds and souls to share a world and a common purpose with. From this point of view, independent intellectuals have a certain freedom of movement that those bound to jobs in provincial universities or unsuitable institutions lack.[128] Independent intellectuals who do not migrate from provincial states like NZ (or provincial state towns in Australia or the US, which are the cultural equivalent) generally die slow deaths of the spirit. If you have some gift to contribute, is important to leave such places when you are young, and find out where others of your own kind live. The idea that you can contribute intellectual value in the wrong environment is also a mistake.[129]

---

[128] Interesting too that Popper, who was born in Austria, lived in NZ for a few years, as a political refugee in the period (1937-1946), working at Canterbury University in Christchurch. He wrote *The Open Society and its Enemies* there. But he found almost no one to talk to, and he was glad to be able to escape to London, and the LSE. Had he remained in NZ his fate would likely have been the same as Tichy's.

[129] But science is also not my primary interest in life. I saw its trivialisation that Corredoiara talks of when I was young, and became more interested in philosophy and society and human nature. You can only find out about these by living in the

The second problem, finding an outlet for your work, is problematic. The doors of orthodox science are closed to heterodox ideas even when they are proposed by academics with strong academic positions and connections, let alone for independents. One of my own theories, the unified theory of the *"geometric universe"* is a good example. It has no prospect of publication in any orthodox journal of physics. I haven't even bothered to try placing it on the physics *arXiv* preprint site. I know typical peer reviewers or mediators will reject it without reading a word because it proposes an alternative foundational theory of physics. But what heterodox scientists must ask ourselves in this respect is: *what outcomes do we want?* What represents *success?*

There is a wide-spread assumption, that comes from starting in the orthodox world (as most heterodox scientists usually do, starting at university, often as star students, and then retreating or being excluded), that success ultimately lies in being accepted by the orthodox world. People think that if you discover a new scientific result, the measure of success is to have it accepted by the mainstream academics, by the 'scientific authorities', to have it become the *new orthodoxy*, published in text-books, and so on.

But in the modern scenario of the death of science, I think *this* way of thinking is what should be rejected. It comes from the image that there is an official center to the scientific world, a kind of corporate monolith that represents scientific truth and accomplishment. It goes along with the idea that there is single scientific truth, rather than a genuine diversity ideas (as accepted in art and music and philosophy). It conforms to the larger cultural myth that success is a matter of public recognition, fame, prizes, offices, etc. The ultimate accolade for most scientists, Nobel prizes are like Academy Awards in the movie industry: the measure of 'ultimate success'.

At a deeper level, this goes back to a belief that you achieve 'immortality' by public recognition. You conquer death through fame. Your ego lives on through your name. This is certainly a delusion. Your ego does not live on. Once you are dead, your ego is gone. Your soul or your spirit may survive – depending on what the deeper truth of personal identity is. But your ego is gone. Your personality, your experience, your history, all the things you loved and clung to in the human world – these vanish when your life ends. Your *name* in the future does not refer to anything real of your ego left in the human world. A few people may remember you for a few years – or rather, remember their mental image of you. But once the clock ticks out a few more years, everyone who has ever known you personally will be gone too: and there will be nothing left of you, except an empty name. It makes absolutely no difference to *you now* whether anyone in the future knows your name or not. Unlike López Corredoira, who is a materialist, I do think our personal identity survives physical death in another form, and I think our actions in

---

real, dysfunctional society of the majority, not by living in the artificial, privileged society of academic institutions.

life do carry implications over to our continued existence in another form. But I completely agree with him that this is not the *human ego we have in this life*. This delusion of placing our own ego at the center of our world is wrong. And it leads us to make the wrong choices in life.

Success for creative intellectuals cannot mean public success, at the start or at the end.

At the start, it means *success in your own private terms*. Genuine creative scientists and natural philosophers, all those who make real discoveries about the natural world, are driven by personality and spirit to pursue what they do. They are driven to achieve a kind of *understanding*, like Beethoven and Mozart were driven to create music. Success lies primarily in fulfilling these personal creative drives. Of course, some are also greedy for other things – for things of the ego – for fame and fortune and recognition and so on – but these are add-ons. Newton is somewhat criticised for an egocentric feature of his personality. But I think Newton would have done exactly what he did in science *whatever his personal success turned out to be*. Indeed, he delayed publication of his greatest book for some decades, primarily because he struggled to complete a single proof he wanted to perfect his theory of gravity. (He never found the proof: it made no difference to the reception of his theory). He was vastly more concerned about the intrinsic quality his work than anything else. He did not do science for his ego: he did it from a primal creative drive. He just happened to have certain egotistical drives as well, and, like thousands of other people of the most mediocre achievement, he exploited the opportunity to satisfy these when he got the chance.

And at the end, insofar as you are creating a *product,* something that endures through time beyond yourself, you do science or philosophy (or art or writing) *for the future*. The future audience for works of real quality is much greater than the present audience; and the impact of revolutionary ideas is primarily something for the future, not the present. Any scientist who recognises a new revolutionary discovery (their own or someone else's) realises that it is a new beginning, and will have a greater value as time goes on. Creative intellectuals, like everyone else, have a social need to contribute some value to society, and this means to the *future society*. Why, if our egos are no longer there? Well, we don't need to answer that question: *we just do have that need.* Everyone who values human life has it. And it is right.

I am not saying we should sacrifice all the time of our lives to the future: we must live in the present, and have some fun and enjoyment and happiness and fulfilment here. That is the heart of *human* existence. As the Buddhists emphasise, *living in the now* is very important. Always living in the future is a recipe for neurosis. Always sacrificing *the time of our lives now*, to collect money to be used in the future, is a delusion our commercial society tries to brainwash us into too. It robs us of life. But nonetheless making some sacrifices, of our time and immediate pleasure, for the future, and for future generations we will never know, is also necessary for human beings. If our ancestors did not do this, we would not be here ourselves, we would not have this civilisation that others have built for us in the past.

What does all this mean? It means that creative heterodox scientists should not be consumed by the idea of success as public success, or as success through the orthodox institutions of the present. The more revolutionary and more important your theories, the less chance you have of recognition in the present, the more powerful enemies you will make. Today, popular success is only achieved with propaganda. The power cliques have the propaganda channels for science sown up. Instead of trying or expecting to have new discoveries accepted by the 'central powers of science', heterodox scientists must be satisfied to have their ideas and discoveries communicated among just a few, just enough to carry them forward into the future, beyond their own time, into a new time when they will be appreciated. Your ego will not be there to benefit. But you can feel good about it now.

This is why I think the innovation of official free publication of work on preprint servers, rejecting the *peer-review* system, and consequently decentralising the academic journal industry, is perhaps the most important practical step we can take at this point. This means explicitly *redistributing power away from the institutional centres*. It challenges the central power hierarchy of science, the notion that there is a single, authoritative *center and truth* of science. It means that numerous smaller, heterodox movements may flourish. It means that 'scientific truth' becomes diverse, that a diversity of views may flourish. I do not mean to contradict López Corredoira's view that *there is an objective truth about reality and nature*. That is called 'realism'. But I do mean to contradict the view that *modern science is capable of representing that truth in a single monolithic system of information*. That is not realism, it is arrogance. And it is a bad idea, like the notion of a Single World Government ruling all nations according to a common set of rules and regulations. That is a very bad idea of the same kind.

I think that creative heterodox scientists and philosophers today should try to work independently of institutional systems, and build their own peer-networks of communication and their own communities. Even if you do discover the next great idea in your field, you do *not* have to publish it in *Nature* or *Science,* or think that everyone in your time should be convinced of it. You do not have to have it recognised at the *center of the official science world* at all. All you really need is a small group of people with a serious interest in it – enough to keep it alive. A dozen is enough. Tichy, with one of the indisputably great theories of modern logical semantics, probably has no more than a dozen supporters who understand his work. But they are a very astute and determined dozen, and they will eventually bring it to wider attention, as the ramifications of his theory are developed, and as other theories continue to fail. To get a hundred people to appreciate a new idea or theory is plenty to my way of thinking. A thousand is far more than enough. Intellectual ideas in science are not best-selling novels. They are seeds for the future, they will sprout when the time is ripe.

Of course, many seeds die: they are often rediscovered anew, not nurtured from their first discovery. But people then also look back, and find the seeds of the new in earlier ages. And this is a gift to the future too: to show that people in the past also had the same ideas, the same intellectual gifts that we have, and show us how fragile the seeds are. We are interested in our history partly to reclaim a sense of affirmation from it, to reaffirm that the thugs and barbarians who dominate the political and economic history of the past were not our *only* ancestors. To know that the positive spirit of human accomplishment and genius has always been there. It is important that the future can look back at our modern science and philosophy, and affirm for themselves that it was not all dross and corruption, that an intellectual spirit lived in our time too.

There is an important practical distinction between *heterodox and orthodox* science here too I think. I will send research papers in some areas of orthodox science to niche orthodox journals that I know have genuinely interested followers, because they are interesting to others. But I would no longer send *heterodox papers* to orthodox journals. And I would no longer send any material to a philosophy of science journal. If such ideas cannot be published as part of real science, they must remain as heterodox ideas, outside the boundaries of the science industry. Heterodox ideas are what independent scientists specialise in, but these ideas do not have to go into the mainstream science journals at all. Today, they belong on preprint servers, or in independent monographs, for the community of independent scientists to pick up.

That is my view of the attitudes independent scientists or philosophers should take. We should not be trying to 'help the system', we should be trying to satisfy ourselves. What of the larger prospects for institutionalised science? López Corredoira thinks it will fade away into irrelevance, not immediately, but slowly, over the next decades. It is in its twilight, because its meaning has lost relevance for our civilization.

"The death of our science, of our whole culture indeed, is coming slowly. We are at the beginning of a transitional period which will transform our present culture into a new one. Within some centuries, humankind will contemplate the ruins of our civilization in the same way that romantic poets contemplated the ruins of past cultures with the sweet allure of decay, death and destruction. Science will prompt an emotional response again and will be revived. P.194.

My instinct is that we are in for a more violent and dramatic future. Past civilizations have risen and fallen, and tipping points are occasioned by new technologies, new conditions that destabilise old orders. Old power structures and old institutions prove inadequate to manage new conditions. Are our present institutions permanent or robust to see us through the future? No. We are in a period of unprecedented *global history*. Globalisation, new technologies, ingrained ideological conflicts, instability of core institutions of finance and economics, the rapidly changing demographics of age and poverty, the increasing gap between the wealthy and the poor, the degradation of natural resources and environment, looming crises in energy and food production, the fragility of over-complex infrastructures, prospects of regional wars, prospects of global climate shifts, all combine to present us with a frightening scenario for the future. The thing is: we have reached the edge of our planet's resources for continued expansion. We have nowhere else to go. And we have no solutions in sight.

We will not survive the coming crises with our present institutions of power intact, our government and corporate institutions, driven by their corporate business models and megalithic rule systems. They are too incompetent and too self-centred to provide the planning and organisation required for our global civilisation to survive its crisis. They are not capable of providing solutions through creative scientific or intellectual or philosophical thinking and planning, because creative intellectuals are excluded from them. Intelligent thinking is excluded by ideological thinking.

There is a superstition that scientific technology will save us: but it will not save us. Technology solutions in the first instance are used to *enslave* us. All the prospects are that our civilisation will crash. As in all such apocalyptic scenarios, brute force will come into play. The vast accumulation of weaponry and military power, held in check only by fear of mutual self-destruction over the last half century, will be unleashed again. It is inevitable. That will be the end of science for our age, it will become a luxury as we will turn back to survival. This end for our present mode of civilisation is the prospect we face within our own life-times.

But that will not be the end for the human species, or for science. As López Corredoira says, in some centuries hence, sooner or later, a new civilization will contemplate the ruins of *our* civilization, and with this new mode of civilisation, a new science – or *natural philosophy* - will be resurrected. The gigantic egos and childish theories of our present time will not be celebrated then. The scientists and science of the present decades will not be seen as the heroes and the high points, but as the final period of decadence. Our present generation will be seen as the corrupters and destroyers of the golden age of modern science. The future will talk about our era as *the time when science died.*

PART 5. APPENDICES

Appendix 1. T.D. Lee. Symmetries, Asymmetries, and the World of Particles.

Tsung-Dao Lee (T.D.Lee) is one of the youngest ever Nobel laureates, awarded for physics in 1957 at age 30 for his work with Chen-Ning Franklin Yang, showing parity violation in K°-meson decay (or the weak nuclear interactions). This means failure of *space reflection symmetry*. It was subsequently extended to a violation of *time reversal symmetry*, at the level of fundamental particle interactions. This is a profound result, and a contradiction of popular claims that fundamental physics is time symmetric.[130]

The book section reproduced here from (Lee, 1988) contain some of his mature reflections on the progress of particle physics, with a number of simple but penetrating observations in the philosophy of science – backed up by *empirical sociological data*. I have never seen any reference to this in philosophy of science. Lee is a rare case of a *real philosophical scientist,* not a technocratic academic physicist, nor an armchair "philosopher of science".[131]

Discoveries and the Laws of Physicists

When each new accelerator is proposed, theorists are employed like high priests to justify and to bless such costly ventures. Therefore it pays to look at the track record of theorists in the past, to see how good their predictions were before experimental results.

Figure 14 lists almost all the major discoveries made in particle physics for more than three decades. It is of interest to note that, with the exception of the antinucleon ($\bar{p}$ and $\bar{n}$) and the intermediate bosons ($W^{\pm}$ and Z^{0}), *none* of these

p. 38.

---

[130] It is often noted as a 'minor exception' to the general time symmetry of physics – but there is no such thing as a 'minor exception'. That is like saying: "there is no empirical measurement to determine a universal frame of reference for space (with the minor exception of the Cosmic Microwave Background Radiation field)."

[131] And for direct contemporary relevance, his comment that *"none of these landmark discoveries was the original reason given for the construction of the relevant accelerator"* may be contrasted with the situation of the vastly more expensive LHC, which appears to have discovered nothing new, beyond (probably) confirming the Higgs particle it was constructed to find.

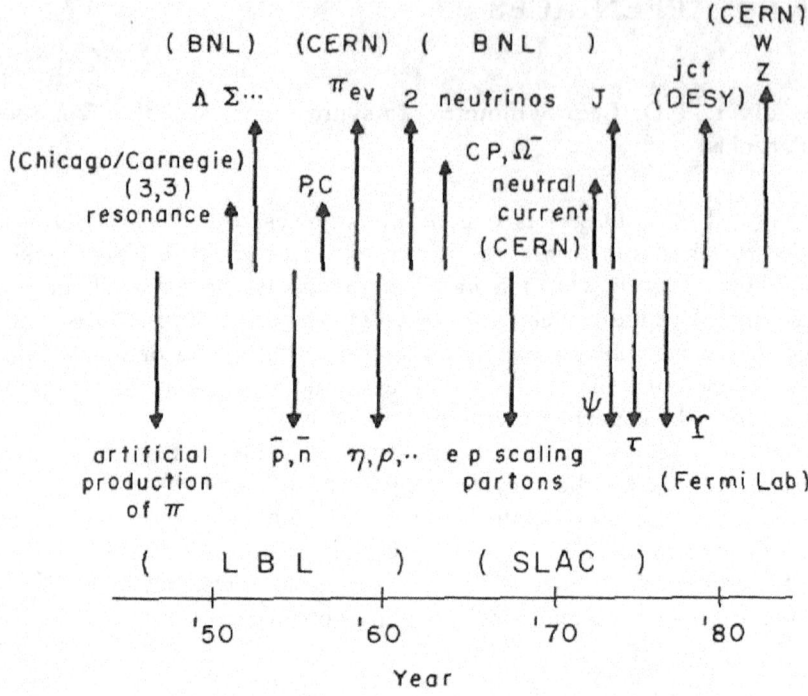

Figure 14. Major discoveries in particle physics.

Each line indicates a major achievement; its position gives the approximate time. Each arrow points to the name of a discovery and the laboratory responsible. Starting from the left, we have the artificial production of π made at Lawrence Berkeley Laboratory (LBL) in 1947–48, the discovery of the (3,3) resonance in 1953 at the University of Chicago and the Carnegie Institute of Technology (now Carnegie-Mellon University), the investigation of the dynamics of Λ, Σ, and other particles at BNL in 1953–54, the discovery of the antiproton and antineutron in 1955 at LBL, the discovery of P and C violations in 1956–57 (at Columbia and other institutions), the discovery of π → ev at CERN in 1958, and so on.

p.39

landmark discoveries was the original reason given for the construction of the relevant accelerator.

Let us start with the first discovery on the chart. When Ernest Lawrence built his 184-inch cyclotron, the energy was thought to be below pion production. Therefore, after the cyclotron was turned on, even though pions were produced abundantly, for a long time nobody noticed them. They were finally discovered accidentally and turned out to be the most important contribution made by that cyclotron.

The progress of particle physics is closely tied to the discovery of resonances,* which started with the (3,3) level first produced at the Chicago cyclotron (the second item in Figure 14). Yet even the great Enrico Fermi, when he proposed the machine, did not envisage this at all. After its unexpected discovery, for almost a year Fermi expressed doubts whether it was a genuine resonance. A similar story can be told about the next landmark discovery. When the Cosmotron was constructed at Brookhaven, some of the leading theorists thought that the most important high-energy problem was to understand the angular distribution of proton-proton collisions, which remains mysteriously flat even at a few hundred MeV, although at that energy the dynamics of the collision is quite complicated. Many different levels (s, p, d, f, g, . . .) are all involved. Why should they conspire to make a flat angular distribution? But, as it turned out, when the energy increases the angular distribution of proton-proton collisions no longer remains flat and becomes quite uninteresting. Instead, it was the production and decay dynamics of the strange particles, Σ, Λ, . . . that put the Cosmotron on the map.

---

*A resonance in particle physics means simply a fixed energy level. The word was carried over from the tuning fork, which vibrates (resonates) only at certain frequencies, similar to the radiation emitted during the transmutation of particles from one energy level to another.

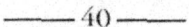
—— 40 ——

p.40

We could go on and on, and the same pattern would repeat itself throughout this list. This leads to my *first law of physicists:**

Without experimentalists, theorists tend to drift.

There is no reason for us to believe that it will change, nor should we expect too much from our present theorists for the prediction of the future.

Look at Figure 14. You will notice that the density of great discoveries per unit time is quite uniform and averages out to about one in two years. Let us hope that this long-standing record of constant rate of discovery can be maintained. In order to achieve that, we must have good experiments.

We now come to my *second law of physicists:*

Without theorists, experimentalists tend to falter.

An example is the search for "neutral currents" in weak interactions. When high-energy neutrino experiments were proposed in 1960, it was suggested that perhaps this could be a tool to uncover neutral currents.† After two neutrinos were found in 1962, intensive experimental efforts were made to search for such a current. At that time, however, there was no theoretical guidance as to its magnitude. A year later, in 1963,

---

*To understand the laws of physicists, it is not at all necessary to know any law of physics.

†In a typical weak interaction, a neutron may beta-decay to become a proton (giving off an electron and an antineutrino). Conversely, a neutrino can convert a neutron into a proton plus an electron. Because there is a change of electric charge between neutron and proton, such reactions are called charge-current processes. Neutral current refers to a new kind of weak interaction in which a neutrino simply collides with a neutron without converting it into a proton.

p.41

an upper limit was set for the ratio of neutrino events due to neutral vs. charged currents (Proceedings of the Siena Conference):

$$\frac{\text{neutral current events}}{\text{charged current events}} < 3 \times 10^{-2}. \quad (1963)$$

A decade later, however, in 1974, after the theoretical progress made by S. Weinberg and others, new experiments were performed. The same ratio turned out to be much larger:

$$\frac{\text{neutral current events}}{\text{charged current events}} = .42 \pm .08 \quad (1974)$$

in good agreement with the theoretical model. The reason for the large discrepancy between these two results was never explained.

Another good example is the history of the "Michel parameter" in μ-decay. The momentum of the final e in μ-decay

$$\mu \to e + \nu_\mu + \bar{\nu}_e$$

varies from 0 to its maximum value. The electron distribution can be plotted against

$$x = (\text{momentum of } e) / (\text{its maximum value}),$$

and is characterized by the well-known Michel parameter, ρ, which can be any real number between zero and one. The

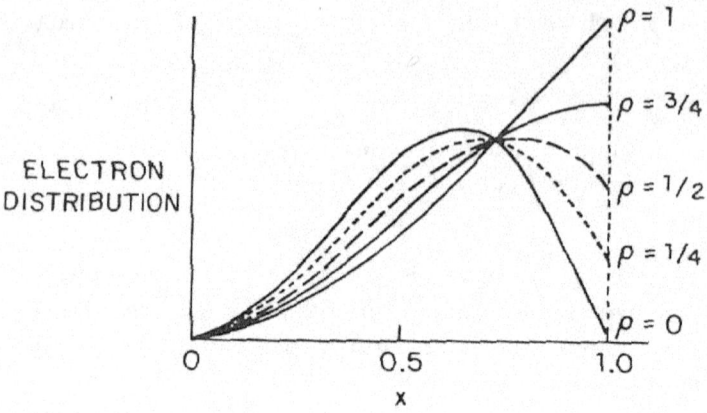

Figure 15. The distribution of electron energy in μ-decay.

Michel parameter measures the height of the endpoint of the electron distribution at the maximum electron momentum x = 1, as is shown in Figure 15. We see that different ρ-values give quite different electron distribution curves.

The Michel parameter has been under intensive experimental investigation since 1949. It is instructive to plot the experimental value of ρ against the year when the measurement was made. As shown in Figure 16, historically it was first found that ρ was near 0. Subsequent experiments, however, yielded different values; ρ slowly drifted upward. Only after parity nonconservation in 1957, when the theorists were able to make a precise prediction, ρ = 3/4, did the experimental value also begin to converge, finally reaching excellent agreement with theory in the sixties. When one looks at Figure 16, one is struck by the remarkable fact that at no time did the "new" experimental value lie outside the error bars of the preceding one.

———43———

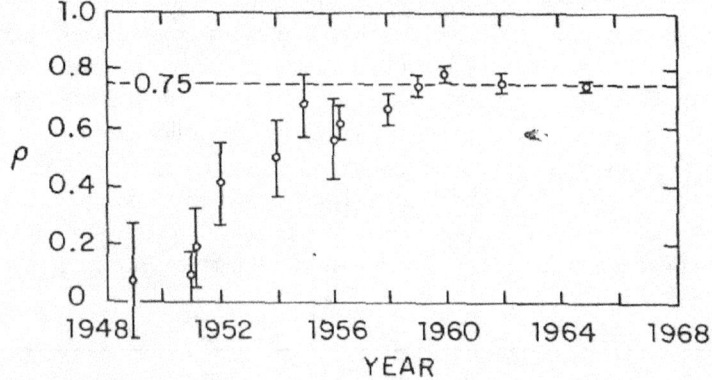

Figure 16. The change of the Michel parameter ρ from year to year.

I hope these examples explain the two laws of physicists and illustrate the interdependence of theory and experiment.

Present Status

The close collaborations between theorists and experimentalists over the past three decades have led to our present status, which is summarized in Table 1. There are three general classes of interactions: strong, electroweak, and gravitational. The strong interaction describes the forces that form protons and neutrons and combine them into various nuclei; the basic building blocks are the quarks. We think there are three families of quarks, with each family made up of two members. They are

> up (*u*), down (*d*),
> charm (*c*), strange (*s*), and
> top (*t*), bottom (*b*).

 44

Appendix 2. Einstein's view of the Aether.

Einstein is famous for rejecting the 'luminiferous aether' in his earliest work on STR, and the concept of the aether is now often ridiculed by modern physicists as 'metaphysical'.[132] But in fact Einstein revised his early views after developing GTR, and far from ridiculing the concept, he discussed the aether very seriously, and even asserted its existence! The following extract is from the end of an extended discussion, some 3,600 words, given in 1920. By this time he had matured from his early STR phase, when he was prone to naïve positivist reasoning, into a careful realist. He is now careful not to use the careless epistemic arguments of his youth: the fallacy of inferring *reality* from *observability*.

Aether and the Theory of Relativity.
Albert Einstein, PhD.
Address delivered on May 5th, 1920, at the University of Leyden, Germany.

"... The Special Theory of Relativity forbids us to assume the Aether to consist of particles observable through time, but the hypothesis of Aether in itself is not in conflict with the Special Theory of Relativity. Only we must be on our guard against ascribing a state of motion to the Aether.

Certainly, from the standpoint of the Special Theory of Relativity, the Aether hypothesis appears at first to be an empty hypothesis. In the equations of the electromagnetic field there occur, in addition to the densities of the electric charge, *only* the intensities of the field. The path of electromagnetic processes *in vacuo* appears to be completely determined by these equations, uninfluenced by other physical quantities. The elec-

---

[132] Lorentz famously persisted in his belief in the aether long after the advent of STR, in fact all his life; and he is often ridiculed for this, by physicists who repeat a myth that Lorentz could not understand STR. Nothing could be further from the truth. Einstein himself admired Lorentz more than any other scientist of his time.

"It is as a representative of the German-speaking academic world and in particular the Prussian Academy of Sciences, but above all as a pupil and affectionate admirer that I stand at the grave of the greatest and noblest man of our time." (From *"Address at the grave of H.A. Lorentz"*, 1928. (Einstein, 1954)).

"At the turn of the century the theoretical physicists of all nations considered H.A. Lorentz as the leading mind among them, and rightly so. The physicists of our time are mostly not fully aware of the decisive part which H.A. Lorentz played in shaping the fundamental ideas in theoretical physics." (From *"H.A. Lorentz, creator and personality."* Address delivered in 1953. (Eistein, 1954)).

tromagnetic fields appear as ultimate, irreducible realities, and at first it seems superfluous to postulate a homogeneous, isotropic Aether-medium, and to envisage electromagnetic fields as states of this medium.

But on the other hand there is a weighty argument to be adduced in favour of the Aether hypothesis. To deny the Aether is ultimately to assume that empty space has no physical qualities whatever. The fundamental facts of mechanics do not harmonize with this view. For the mechanical behaviour of a corporeal system hovering freely in empty space depends not only on relative positions (distances) and relative velocities, but also on its state of rotation, which physically may be taken as a characteristic not appertaining to the system in itself. In order to be able to look upon the rotation of the system, at least formally, as something real, Newton objectivizes space. Since he classes his absolute space together with real things, for him rotation relative to an absolute space is also something real. Newton might no less well have called his absolute space "Aether"; what is essential is merely that besides observable objects, another thing, which is not perceptible, must be looked upon as real, to enable acceleration or rotation to be looked upon as something real.

What is fundamentally new in the Aether of the General Theory of Relativity as opposed to the Aether of Lorentz consists in this, that the state of the former is at every place determined by connections with the Matter and the state of the Aether in neighbouring places, which are amenable to law in the form of differential equations; whereas the state of the Lorentzian Aether in the absence of electromagnetic fields is conditioned by nothing outside itself, and is everywhere the same. The Aether of the General Theory of Relativity is transmuted conceptually into the Aether of Lorentz if we substitute constants for the functions of space which describe the former, disregarding the causes which condition its state. Thus we may also say, I think, that the Aether of the General Theory of Relativity is the outcome of the Lorentzian Aether, through 'relativation'.

As to the part which the new Aether is to play in the physics of the future we are not yet clear. We know that it determines the metrical relations in the Spacetime continuum, e.g. the configurative possibilities of solid bodies as well as the gravitational fields; but we do not know whether it has an essential share in the structure of the electrical elementary particles constituting Matter. Nor do we know whether it is only in the proximity of ponderable masses that its structure differs essentially from that of the Lorentzian Aether; whether the geometry of spaces of cosmic extent is approximately Euclidean. But we can assert by reason of the relativistic equations of gravitation that there must be a departure from Euclidean relations, with spaces of cosmic order of magnitude, if there exists a positive mean density, no matter how small, of the Matter in the universe. In this case the universe must of necessity be spatially unbounded and of finite magnitude, its magnitude being determined by the value of that mean density.

If we consider the gravitational field and the electromagnetic field from the standpoint of the Aether hypothesis, we find a remarkable difference between the two. There can be no space nor any part of space without gravitational potentials; for these confer upon space its metrical qualities, without which it cannot be imagined at all. The existence of the gravitational field is inseparably bound up with the existence of space. On the other hand a part of space may very well be imagined without an electromagnetic field; thus in contrast with the gravitational field, the electromagnetic field seems to be only secondarily linked to the Aether, the formal nature of the electromagnetic field being as yet in no way determined by that of gravitational Aether. From the present state of theory it looks as if the electromagnetic field, as opposed to the gravitational field, rests upon an entirely new formal *motif*, as though nature might just as well have endowed the gravitational Aether with fields of quite another type, for example, with fields of a scalar potential, instead of fields of the electromagnetic type.

Since according to our present conceptions the elementary particles of Matter are also, in their essence, nothing else than condensations of the electromagnetic field, our present view of the universe presents two realities which are completely separated from each other conceptually, although connected causally, namely, gravitational Aether and electromagnetic field, or as they might also be called space and Matter.

Of course it would be a great advance if we could succeed in comprehending the gravitational field and the electromagnetic field together as one unified conformation. Then for the first time the epoch of theoretical physics founded by Faraday and Maxwell would reach a satisfactory conclusion. The contrast between Aether and Matter would fade away, and, through the General Theory of Relativity, the whole of physics would become a complete system of thought, like geometry, kinematics, and the theory of gravitation. An exceedingly ingenious attempt in this direction has been made by the mathematician H. Weyl; but I do not believe that his theory will hold its ground in relation to reality. Further, in contemplating the immediate future of theoretical physics we ought not unconditionally to reject the possibility that the facts comprised in the quantum theory may set bounds to the field theory beyond which it cannot pass.

Recapitulating, we may say that according to the General Theory of Relativity space is endowed with physical qualities; in this sense, therefore, there exists an Aether. According to the General Theory of Relativity space without Aether is unthinkable; for in such space there not only would be no propagation of light, but also no possibility of existence for standards of space and time (measuring-rods and clocks), nor therefore

any space-time intervals in the physical sense. But this Aether may not be thought of as endowed with the quality characteristic of ponderable media, as consisting of parts which may be tracked through time. The idea of motion may not be applied to it."
(Albert Einstein, 1920.)

Appendix 3. The time reversal operator in quantum mechanics.

Without trying to repeat the arguments in my paper (2003 (m)), I briefly note some background for readers wanting to make sense of the dispute. It is about the choice of time reversal operator in quantum mechanics. Everyone admits there is something peculiar about the usual choice. On the literal reading of *time reversal* as the transformation T mapping: $t \rightarrow -t$, time reversal would render QM *irreversible*, or *time asymmetric*. So this is rejected. Instead, a complex transformation, called T^*, is adopted in QM (due to Wigner) with precisely the purpose of rendering QM *reversible*, or *time symmetric*. This is a celebrated result, underpinning popular claims that *fundamental physics is time symmetric and does not distinguish the two directions of time*. On this basis, a profound metaphysical implication for the nature of time is taken from fundamental physics: viz. that there is no physical basis for 'time flow'. This orthodox analysis is taught in all advanced textbooks on QM, and widely quoted in the philosophy of physics and naturalistic philosophy of time.

However, the *conceptual foundations* of QM time reversal are very murky, and have been seriously questioned a number of times, notably in the 1950's by Watanabe (1955, 1965, 1966, 1972), Costa de Beauregard in the 1970's, Healey (1981), and also in my own PhD thesis (1990). The debate was suppressed almost out of existence in the 1990's, but the same questions rose to the surface yet again in the 2000's. The critical problem is whether this choice of time reversal operator is *objective*, or just a *conventional choice* cobbled up to make QM appear reversible. The orthodox result is invariably justified by the fact that *the orthodox reversibility symmetry (T*) <u>actually holds</u> of QM*[133], *while time reversal symmetry taken literally (T) <u>does not hold</u> of QM*. The orthodox symmetry is therefore of real use in the mathematics of the theory – but that doesn't show that it represents time symmetry. My paper (2003 (m)) is a clear representation of the problem, and shows some of the reasons for the *conceptual* failure of the conventional analysis. Among a few specialists in the last 15 years, this has become an open question again. David Albert, a physicist and philosopher, managed to raise the problem to prominence in his 2000 book - to a chorus of scorn from many orthodox physicists. It is highly embarrassing to admit that such a basic question has gone unresolved within physics for so long. But in fact the whole topic of *time reversal and time flow in modern physics* has been a conceptual mess for decades.

---

[133] But with the exception noted previously, that Lee and other showed experimentally in the 1950s-60s that mesons have not only *parity asymmetry* but a small but distinct *time asymmetry* too, as noted in Appendix 1.

Appendix 4. A realist interpretation of time in Special Relativity.

I illustrate here the simple concept of a *realist interpretation,* illustrated with the issue of interpreting *time flow* in special relativity theory. First I say a few words about the concept of a *realist analysis.* If we consider a sentence or equation, a *realist analysis* of it means an interpretation that takes it literally, as if it were stating a proposition, about the things it mentions. Giving a *realist analysis* does not commit us to its *truth* – i.e. to realism about the entities it *assumes.* That is something we can only investigate after we understand what it is intended to mean. Thus statements about all kinds of things that we may or may not believe are real (e.g. phlogiston, celestial spheres, Santa Claus, classical space, moral values, God, ufos, esp, etc) are *interpreted realistically* in the first step; and evaluated for their truth subsequently.

Why do we need to interpret statements anyway? Or why do we have to make a dedicated exercise of it? Because very often their meaning is unclear – or they have multiple possible interpretations – and they often contain implicit assumptions and preconceptions that, with our fuzzy human intellects, we are not really aware of until we examine them closely. There are many famous examples in science that are still unresolved.

E.g. Newton interpreted *space* as 'absolute', meaning that individual points of space are real (they are entities with identities through time; there are absolute facts about motions of objects w.r.t. points of space). But Liebniz interpreted space as 'relational', meaning only distances between objects in space are real (there are no absolute facts about motion w.r.t. space, only relative motions of bodies w.r.t. each other). This is a famous dispute, still unresolved in modern philosophy of physics. More recently, it is controversial what the quantum wave function, usually symbolised as Ψ, is supposed to refer to – a real physical object, or a mathematical variable, or a concept of probability, or a code for some observational procedures? In social science, concepts like *culture, ideology, value, institution, race, law, democracy* and so on may mean very different things to different people. Trying to define concepts precisely is one of the core functions of Western philosophy ever since Socrates and Plato, and it is not merely a 'trivial semantic' exercise (as many scientists often scoff): it is very difficult indeed!

The idea of a *realist interpretation* may seem like simple common sense to most people, and essentially it is. We simply consider all terms in an expression and *try to identify exactly what they are literally meant to refer to.* But mainstream philosophy throughout the C20th has been dominated by *anti-realist meta-semantic theories,* which *deny* the possibility of realism! Versions of 'linguistic philosophy', or 'natural language philosophy', particularly dominant in Britain and its ex-colonies in the mid-late C20th century, denies that philosophy is about anything except *'linguistic behaviour'* ('language games' in the Wittgensteinian phrase). It produced nothing but stultifying tomes of observations about 'grammar' and futile meta-philosophical ramblings about method.

The idea that philosophers should be trying to evaluate the *literal meanings* of claims and beliefs, including metaphysical beliefs, was rejected by these academics for decades. (This is one reason scientists often disparage philosophy.) But just as virulently, *anti-realist scientistic* movements (Logical Positivism, Behaviourism, Pragmatism, Instrumentalism, Operationalism, Deflationism, etc) try to give *anti-realist reductionist accounts of meaning,* on the same pattern that chemists give *reductionist accounts of matter.* Chemists can say: "this substance you call *water* is really H_2O *molecules* ... its phenomenal properties, (continuity, transparency, temperature, etc) are *appearances:* in some degree, subjective perceptions of the observer, not fundamental properties that define the real nature of water." Well, that is *a reasonable scientific theory about the atomic structure of matter.*

The scientistic anti-realists intuitively adopted a similar reductionist model for *meanings,* saying: "the meanings of sentences (or linguistic terms) are *not* the propositions (or entities like Platonic *Universals*) that they naively appear to reference; such things are abstract entities that do not exist; rather meanings must be reduced to observables like *linguistic behaviours,* or *operational rules for confirming 'truth'* or *pragmatic conditions for agreeing on the use of terms* – things completely *unlike* what meanings seem to be like; just as H_2O molecules are completely unlike what water seems to be like. We reduce 'meanings' to more fundamental 'atoms' – based on empirical observables. 'Meanings' taken literally would be *unobservable.*"

These anti-realist 'sematic philosophies' are intellectual junk – and they have poisoned modern philosophy for much of a century. The fact that *water is really made of H_2O molecules* does not imply that *the term 'water' means 'H_2O'.* (Or no one could have known what they meant by 'water' for thousands of years). The fact that we judge something is 'water' by looking at certain cluster of properties does not imply that *the term 'water' means that cluster of properties.* The term 'water' *means water.* It cannot possibly mean anything else.

Propositions stated in natural language as well as in scientific language, *are meaningful in their own terms,* not by having their meanings 'reduced' to *some other theoretical terms that are utterly unlike meanings.* But instead of wasting our time arguing against anti-realist meta-ideologies of academic semanticists, let us simply illustrate how a realist analysis works with a real example.

Consider Pythagoras' theorem, applied to describe physical space:

$$r = \sqrt{(x^2 + y^2 + z^2)}$$

The *realist interpretation* of this identifies a *reference for all the empirical terms in the equation.* On the left, we have the term r, which we interpret as a *distance between two points in space.* On the right hand side, we have x, y, z, which we interpret as three separate distances, in mutually orthogonal directions. The equation represents a *generalised proposition,* relating these entities – physical distances – to each other. It is implicitly generalised as a fact about the distance between *any pairs of points.* As a theorem of Euclidean geometry, the purely mathematical proposition is analytically true, but interpreted as describing empirical space, it may be true or it may be false, depending on whether *there are real properties of distance between points of physical space,* and whether *they really conform to this relationship.*

Note that in this form, the equation is 'perspicuous', or 'transparent': it relates two clearly identified and interpreted quantities, on the left and right hand sides respectively. We can rearrange it, and mix the terms up if we want. E.g. we could write: $\sqrt{(r^2 - x^2)} = \sqrt{(y^2 + z^2)}$. Mathematically this is the same proposition – its truth or falsity is *exactly equivalent to the first equation* - but it is much less perspicuous, it is more difficult to *interpret its physical meaning.* The two different forms relate two different *constructions* on the left and right hand sides. In physics, it is not just the truth (extension) of a statement that is important: the *construction it represents* is also important to our understanding of it.

Note the difference between this *realist interpretation* and *non-realist interpretations,* as proposed in positivist or instrumentalist theories. The latter tells us that terms of physics do not refer to *elements of the real world* at all. Instead they merely refer to *measurements and measurement procedures.* Thus the equation does not tell us that *there are real distance between points of space, related in this way,* but that *if we use a certain measurement procedure to obtain numerical values for r, x, y, z,* (e.g. using a standard meter ruler to assign numbers for distances), then *the numbers obtained through this measurement procedure will conform to this rule.* This circumvents any direct reference to real-world properties of things, and turns everything into a calculation of *measurements.* However this is nowadays widely recognised as a complete failure as a semantic theory.

The reality is the opposite: we make measurements *to infer properties of real entities.* We make measurements *of something.* This is explicitly recognised in many other sciences that have had to deal more seriously with measurement theory than physics. E.g. in educational research, it is explicitly stated that *test scores are measurements from which we (try to) infer underlying traits of individuals.* (We are trying to use test scores to infer traits, so that we can say that a certain test is more or less *accurate* as an assessment tool). Models of such traits are explicitly stated as the basis for interpreting measurement theory. In bird count surveys, a measurement technique in bird ecology, researchers explicitly state that *bird counts are a measure of the conspicuousness of the bird population present in a survey plot.* We are interested in inferring the real feature of the bird population – *abundance* - from the measured counts.

But physicists persist with their *instrumentalist semantics* – and resort to it in situations when they need to justify some conceptual difficulty, automatically referring back to instrumentalism as if it is an authoritative theory of semantics. In most of their work, physicists are intuitively realists: they take their terms to refer to properties of real things. The major exception is in quantum theory, where they cannot find anything 'concrete' to identify realistically with the *quantum wave function,* Ψ. This is because it assigns complex numbers to points across space, and physicists have difficulty imagining that something physically real is described by a complex number. This is the subject of a vast amount of unresolved debate stretching over many decades. The notable feature of the debate is that the dominant orthodox view in physics simply *appeals an instrumentalist or positivist theory of interpretation as if it is the final and definitive answer* – whereas it is well accepted for decades by logicians and semanticists to be incoherent.

The example I wish to go on to now is the interpretation of the *space-time metric of special relativity.* This is the core empirical prediction of the theory when shorn of all its metaphysical baggage, and normally written like this:

$$c\Delta\tau = \sqrt{(c^2\Delta t^2 - \Delta r^2)}$$

This is similar in a general sense to the Euclidean metric, except it now contains *terms referring to time and space mixed up together on the RHS.* (See Holster (2003, B) for a more detailed realist explanation of this equation). The interpretation is that $\Delta\tau$ is a quantity of *process time or clock time,* Δt is a duration of *real time,* and Δr is a quantity of *spatial distance.* However, we cannot literally add *durations of time* to *spatial distances.* This makes no sense: they are two different types of quantities. Instead we convert the quantities of *time* into quantities of *distance* by multiplying by a speed, *c,* the speed of light. Thus on the left hand side we have the construction of a quantity of *spatial distance: c*$\Delta\tau$. And on the right hand side we have the construction of a quantity of *spatial distance, c²∆t² – ∆r².* The latter is called the *space-time interval,* and since it is literally a *spatial distance,* we see that the *space-time manifold* is a spatial entity.

Now the standard interpretation of relativity theory is that the left hand side, *c*$\Delta\tau$, is an absolute or invariant quantity, because it is the same however we choose to measure it. We essentially *count discrete physical events (such as rotations of a clock hand, vibrations of a particle, or counts of radioactive decayed atoms)* to determine the quantity of $\Delta\tau$. On the right hand side, however, there are various choices of ways to measure the quantities of Δt and Δr that leave this equation true. They are measured by assigning *coordinate frames.* Basically, we can increase the amount of Δt (real time between two events) if we also increase the amount of Δr (distance between two events). If we do this appropriately, we will get the same result for $\Delta\tau$, keeping the proposition true. The rules for assigning alternative *coordinate systems* (i.e. measurement systems) are called the *Lorentz transformations.*

The famous implication taken from this is *the duration of real time is not objective or absolute* – because it is only sensible when combined in a construction with space, to give the correct physical invariant or objective quantity, $c\Delta\tau$. Now this interpretation means that, for two events that are relatively close in time, Δt, but separated by a significant distance, Δr (called *space-like separated*), the temporal order appears different in different measurement or coordinate frames. Thus, by choosing one frame they will appear simultaneous, in another frame, one event is past when the other is present, and in another frame again, this order is reversed.

Einstein and Minkowski famously interpreted this to mean that *time flow is subjective*, and in particular, that there are no objective facts about the content of *the present moment*. However this is a very peculiar scientific theory, and a troubling inference to make. For the *equations of Special Relativity themselves* do not rule out that there might be some property of simultaneity – they only imply that *without adding something more to the physical interpretation than the bare (flat) space-time manifold*, simultaneity relations are not present and not measurable. There have been lots of challenges by philosophers to the inference that Special Relativity shows time flow is *impossible*. Indeed, it is widely accepted by more astute physicists that time flow (i.e. an absolute frame of simultaneity) is not a logical contradiction of STR, but rather, at worst, a *redundant hypothesis, because if STR is the whole truth, there is no way to physically measure what the correct frame of simultaneity is.*

This can be challenged too though: for STR is certainly not the whole truth. The discovery of the cosmic microwave background radiation actually *does* give a unique and universal frame of simultaneity across the known universe. And the 'instantaneous' (faster than light) transmission of information in quantum wave function collapse is difficult to comprehend without an *absolute causal order among distant events*.

But in any case, when I began to perceive multiple inconsistencies and errors in the standard treatment of this subject, along with the subject of time symmetry and reversibility, in the course of my PhD studies, it occurred to me that the thing to do is not simply to keep searching for arguments to deny time flow, but rather to *add the assumption of time flow back into relativistic physics, and see if it has interesting consequences* – or if it just leads to inconsistencies. In the latter case, this would be a kind of *reduction ad absurdum*. In the former case, it might give an interesting alternative theory – as happened with the invention of non-Euclidean geometries, in an analogous situation, when no one could prove the 'parallel postulate' from the other axioms. So what happens if we do this? Well, if time really is absolute, the natural thing to do is to rearrange the *metric equation* above, to the form:

$$c\Delta t = \sqrt{(c^2\Delta\tau^2 + \Delta r^2)}$$

This is mathematically equivalent to the original, but it shows a different type of *construction*. We now have a term containing *time* on the left hand side, and a term constructing a *spatial distance* on the right hand side. Indeed, this construction can be rearranged more simply as a '*speed postulate*':

$$c = \sqrt{(c^2 \Delta \tau^2 + \Delta r^2)} / \Delta t$$

This simply says that: *all particles travel at a universal speed, c, through a spatial manifold that has a distance metric:* $\sqrt{(c^2 \Delta \tau^2 + \Delta r^2)}$. Orthodox physicists can hardly complain that this is empirically wrong or 'nonsense' or makes no sense – since it is simply a mathematic rearrangement of their own equation!

The final step in a realist analysis is then to ask this question:

- *If* $\sqrt{(c^2 \Delta \tau^2 + \Delta r^2)}$ *is the distance function for a spatial manifold, then what could be the physical interpretation of this manifold?*

Well, this looks like the metric for a Euclidean physical space. If so, then the quantity: $c\Delta\tau$ must be a *distance through some spatial dimension*. But the term *r* already contains *the three dimensions of ordinary space*. Thus, we would have to postulate *extra dimensions of space*. And we would have to postulate that the underlying 'periodic events' that let us measure clock time, $\Delta\tau$, ultimately reduce to periodic motions in these new dimensions. And because the new dimensions are not macroscopically evident – we do not see things dispersing in more than three dimensions – we would have to postulate that they are very small or 'curled up' dimensions.

From this we immediately get a new type of fundamental theory of physics, simply by questioning and reinterpreting a well-known equation. I eventually guessed a simple topological solution for the new dimensions – a *torus* - that works nicely for modelling the two primary mass particles of our universe, electrons and protons. The 'curled-up space' model immediately give their *quantum properties* – relativistic quantum wave functions, with quantum spin as real angular momentum. It also gives a theory of gravity, similar to the GTR, but with a slight modification, just on the limit of present empirical data to distinguish. These coincidences seemed too good to be accidental, and I have developed a unified theory with a striking cosmology that shows how well this model works.

You might think this is the kind of theoretical exploration physicists are supposed to follow up: explore new interpretations, try alternative explanations. But alas! It is not so. Suggesting such an idea in physics today is considered shameful and *blasphemous*! For the orthodox *interpretation* of space-time – a metaphysical theory - has come to be considered an inviolable principle, an Immaculate Conception, fundamental to the Faith of Physics.

Appendix 5. A logical contradiction in positivist semantics.

The following proof is an example of an important negative result in the philosophy of science, based on a simple but rigorous argument, using formal logic. It is a quite well known proof, which I have seen in different forms, but I have lost the source, and I cannot recall who it was originally due to. I think it appeared in the 1940s.

A positivist-empiricist principle of semantics trumpeted in the 1920's is that *any meaningful true proposition must be verifiable or knowable at least in principle.* That is, if P is a true proposition, it must be *logically possible for P to be verified or to be known.* There are lots of true propositions that are actually unknown of course. That is why science is needed: to try to determine unknown truths. But the positivists were concerned with the demarcation of what they called 'nonsense' or 'metaphysics' from meaningful scientific propositions. And they thought that propositions that are impossible to verify in principle are meaningless and non-scientific.

A classic example for some is that *God exists.* Some argue that this is un-knowable – that there are no scientific observations that could prove the existence of God. Rather than conclude that *God does not exist* – this is un-knowable too – they want to resolve the matter by concluding that *the apparent proposition that 'God exists' is unknowable, and therefore meaningless.*

For a more scientific example, the 'luminiferous ether' of Lorentz – en-tailing facts about absolute position in space - is *impossible to detect* on the assumptions of Einstein's Special Theory of Relativity. So many positivist physicists hold that the postulate of an ether is *meaningless.* Metaphysicians about space (or Isaac Newton) might postulate that absolute motion w.r.t. an ether still exists, despite being undetectable in principle. But positivists see this as meaningless. Their principle, above, lets us draw this conclusion.

But their principle suffers a logical paradox, shown in the following proof. Let us suppose that *P is an unknown but true proposition.* There must be such propositions in their view. This is stated in symbolic logic as:

(1)   P and ~K(P)

Here '~' means 'it is not the case that ...' and K means 'it is known that...'

The positivist semantic principle means that, for any proposition, call it Q (to avoid confusion with P):

(2)   Q entails that ◊K(Q)

Here the diamond symbol, '◊', means *'it is possible that ...'.* This symbol is seen throughout model logic, the logic of possibility and necessity.

Now proposition (1) above is itself a *true proposition*. We can therefore take it as an example of Q in (2). Substituting (1) for Q, we conclude that:

(3) $\Diamond K(P \text{ and } \sim K(P))$

This says that it is *possible to know that (P is true and it is not known that P is true)*. This sounds like a contradiction, and it is. How can we *know P but also know that it is not known that P?* But logic can be tricky, so let us derive the contradiction carefully and explicitly, to be sure.

To derive a contradiction more precisely, we also need the explicit principle that: *if: K(A and B) then: K(A) and K(B)*. I.e. if it is known that *(A and B)* is true, then it is known that *A* is true *and* it is known that *B* is true. Thus: $K(P \text{ and } \sim K(P))$ entails: $(K(P) \text{ and } K(\sim K(P)))$. We also need the explicit principle that: *if it is possible that A is true,* and: *A entails B,* then: *it is possible that B is true.*

Then from (3) we can conclude that:

(4) $\Diamond (K(P) \text{ and } K(\sim K(P)))$

Now: $K(P)$ also means that *P is true.* For we use K to define here the concept of knowing where *only true propositions are known.* Thus: $K(\sim K(P))$ *entails that:* $\sim K(P)$. Hence (4) entails that:

(5) $\Diamond (K(P) \text{ and } \sim K(P))$

But this says that: *it is possible that (P is known and P is unknown).* But ther inner clause is an explicit logical contradiction. It is never possible that: *A and $\sim A$,* no matter what *A* is. So (5) is a logical fallacy.

Yet (5) follows simply from the assumptions that (i) *there is some true but unknown proposition* (which is true), and (ii) the positivist principle (2); and (iii) some simple and normally uncontroversial logical principles.

Hence, we conclude that the positivist principle (2) must be false. This is an example of *reduction ad absurdum.*

You may wonder if it is possible to reformulate the positivist's principle to avoid such a contradiction. But no one has been able to, despite extensive efforts. Positivist literature is full of attempts at *reductionist theories of meaning,* principles like: "*the meaning of a proposition is the conditions under which it can be verified*". But modern semanticists realised that this kind of reductions is simply a delusional idea. It is a fundamentally *wrong kind of way to analyse meaning.* You just cannot reduce *meaning* to *espistemology.* There are still many philosophers still trying to do this – under programs like 'pragmatism' – but they do not know any modern logic or modern formal semantics.

What about the *ether* example? The claim that the *ether* is physically undetectable is still widely used in physics to dismiss it as *meaningless.* But this is very confused thinking. First, even the positivists only hold the weak claim that it should be *logically possible* for a proposition to be known – not the strong claim that it must be *physically possible*. The ether is said to be impossible to detect *physically*, but this is only on a particular theory of physics, i.e. STR. We must empirically verify this theory. It is logically possible that STR is wrong – in fact it is known to be wrong. *Very wrong* when compared with GTR. It is not *logically impossible* for an ether to be detectible - there are other viable empirical theories that might make an absolute spatial frame detectable after all. In fact the microwave background radiation determines a unique global frame at every point of the known universe, and any modern ether theory can appeal to this.

That is one confusion: the assumption that modern physics, in the form of STR, has *decisively proven any ether is physically undetectable*. It has done nothing of the sort. But that is still not the primary conceptual fallacy. The conceptual fallacy is that *absolute position must be detectible in principle for the ether to be considered real. Otherwise it is metaphysical.* This is a positivistic fallacy that has become widely popular among physicists. When we try to state its rationale as a general principle, it turns out to be incoherent – as in the proof of inconsistency given above.

But more generally, physics theories are full of theoretical assertions that are not *factually observable or directly verifiable*. Rather, they provide *simple ontologies, or systems of entities, that have a powerful explanatory value*. E.g. space is modelled as a *continuum*, like the real numbers. This means it is infinitely 'dense' with points infinitesimally close together. But it is impossible to observe any such feature directly or indirectly. It is adopted because it supports differential properties, giving a nice mathematical model for trajectories. Physicists rarely object to this. In general, it is perfectly legitimate to infer the existence of *unobservable entities*, as long as they play a real role in a larger explanatory framework. Thus the legitimate reason for rejecting the ether is not that it is unobservable and thus a meaningless concept – but that *in the context of STR the ether ceases to have any further legitimate explanatory role*. (Note that it is not at all *inconsistent with the empirical predictions of STR*.) In many other possible theories, including theories consistent with the empirical predictions of STR or GTR, an ether may (and does) have a legitimate explanatory role.

The field where this positivist principle really caused endless trouble however is quantum mechanics, where reality is denied to the quantum wave function because of its 'unobservability', and yet the wave function is postulated as central to the theory! Here physicists have tied themselves in knots for decades supporting *anti-realist philosophies*, to avoid having to deal with the problem of interpreting their theory. As a result, realists have been driven from the field, and major conceptual difficulties hidden under the carpet. Physics remains firmly in the grip of this philosophical arrogance.

Appendix 6. Possible world metaphysics.

Here I illustrate some key concepts and questions of metaphysics using *possible world diagrams*. Physicists have special diagrams, like *space-time diagrams* and *phase diagrams*, to illustrate how physical systems are structured and behave over time. They are an aid to interpreting equations or concepts of physics. Semantic logicians have similar diagrams, called *possible world diagrams*, to illustrate theories of how *worlds as a whole* are structured. These are an aid to interpreting semantics of logical concepts like: *actuality (@), present time (Now), propositions (P), logical possibility (◊), logical necessity (□), laws of nature (**N**), determinism (**W**), counterfactuals (If), self (I), place (Here), probability (Prob)*.

Points in a *possible world* diagram are individual *worlds*. The frame of the diagram represents *possibility*, and is called the *logical space*. What is a *world*? It is something that determines the truth of propositions. The initial idea (the intuitive beginning of metaphysics) is that a *single actual world* exists, and all factual propositions are actually true or false of that world. But our knowledge of the world is incomplete, and there are many propositions that we do not know, and many *logical possibilities* that are not actual. Possibilities are represented by *worlds* in the larger space that are *not actual*. In the first diagram below, we start with the simplest possible theory of worlds: a single point in a frame representing only *the actual world*. This is not a plausible logical theory, because it denies any real content to *possibility*, but it illustrates the first logical step in constructing a theory (and it is also what many philosophers such as nominalists believe).

This section is meant to illustrate two critical points. First is that logic is not merely a theory of *'logical syntax'*, i.e. a formal symbolic game, as most introductions to logic present it. Logic is about the content of certain concepts: *logical concepts*. These are concepts (like *truth, propositions, actuality, worlds, necessity, etc*) that are foundational to the domain of *reasoning about propositions, truth, inference, etc*. In this sense, logic reflects a domain of metaphysics. Analogously, mathematics is not a study of the *syntax* of mathematical terms (as formalists would have us believe): it is the study of *mathematical concepts*. Capturing theories of such concepts in a syntactic calculus, in either maths or logic, is merely a tool to express theories of the concepts. Exactly the same can be said of physics too: an axiomatisation of a theory of physics is merely a tool to help express the concepts, not the *subject* of physics. This contradicts the *formalist* philosophy of mathematics, logic and semantics, a popular Positivist doctrine in the C20th.

The second point is that a central goal of logical semantics is to discover the *construction of the logical space;* but this is not pre-determined by logic or rational intuition. There are various open possibilities. The classic mistake of modern philosophers is to give circular logical arguments, based on making assumptions about the construction of the logical space, and then deriving the consequences as truths. E.g. there are numerous attempts in the C20th, starting with McTaggart, to prove that *time flow or change is logically impossible*. But these all implicitly or explicitly start with an assumption about the logical space: viz. that *there is a single actual world that determines all past, present and future truths; and the content of the logical space is itself unchanging*. If we assume this, of course we can prove that there is no time flow, i.e. no change in the world – for time flow requires that some propositions change their truth values. (Some propositions about the *present state of the world* change from being false to true, and vice-verse). But such proofs are circular, for the conclusion is smuggled into *a priori* assumptions about the logical space to start with.

The real question is about *the construction of the logical space itself*: is the *logical space* static and unchanging (the Bloc Universe view), or is it temporal (the Time Flow view, as illustrated in Theory 6 below)? This is a real metaphysical question. It is not answered by analysing language, or syntax, or making *a priori* judgements from logical intuition. It can only be answered by its coherence with our broader system of knowledge, including empirical knowledge, phenomenology and conscious experience, and theoretical physics. The connection with physics becomes clear when we consider how logical spaces may be constructed for non-deterministic theories.

1. The actual world, @.

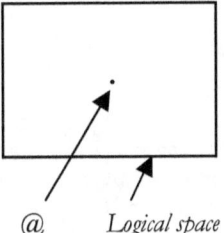

@   *Logical space*

> **Theory 1.** There is a unique actual world, @, and that is all. @ is the complete *world history*, containing all facts at all moments about the actual world. The *logical space* contains no other possible worlds. There is no fact about *the present time*. This reflects *nominalism*. It also means all empirical truths are logically necessary – for the only logical possibility is the actual world. (It is not a serious theory.)

In this simplest theory, *a single changeless world-history* is the only possibility. This appeals to pragmatists, who do not want to deal with the complexity of language, logical reasoning, and the metaphysical assumptions implicit in them. What exists are simply the concrete facts we observe in actuality. What we call 'possibilities' are not real, merely a 'mode of speech', or a 'convention of reasoning'. This view goes hand in hand with the *bloc universe* theory of existence popular with relativity theorists, which says that past, present and future are merely subjective categories, and all 'real facts' are timeless propositions. But this theory, although apparently eschewing 'metaphysics', instead represents a very simplistic metaphysics.

2. The actual world, @, in a Logical Space of possible worlds.

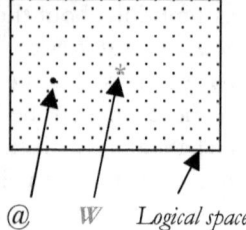

@   W   *Logical space*

Theory 2. There is a unique actual world, @, with many other possible worlds, like *W*, which are not actual. Worlds are *world histories*. There are still no facts about *the present time*. The *logical space* contains all possible worlds. They may be related to the actual world as variations of what is actual.

Allowing many *logically possible worlds* allows us to interpret false propositions as still being about something. Distinct worlds can contain alternative facts about the same entities or individuals as in our world – e.g. a world similar to ours, but where John Lennon was *not* murdered. Anti-realists dismiss *non-actual worlds* as non-existent, and so meaningless to refer to. But in reality we do refer to non-existent facts, and reason about how things might have been different. To have a theory of meaning that can describe propositions adequately, we must have some way to allow for false propositions, propositions about non-existent possibilities.

---

3. Factual Propositions.

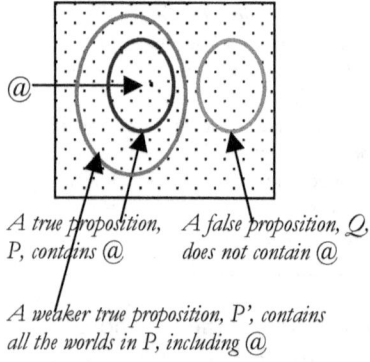

@

A true proposition, P, contains @

A false proposition, Q, does not contain @

A weaker true proposition, P', contains all the worlds in P, including @

Theory 3. Propositions are represented by classes of possible worlds. Any *true proposition* contains the actual world. Any *false proposition* does not contain the actual world. Logical conjunction of two propositions (P *and* Q) gives a new proposition as the intersection of P and Q. Logical disjunction (P *or* Q) gives their union. Negation is the set compliment.

This develops a theory of *propositions* in the world-based metaphysics of *Theory 2*. Intuitively, a proposition picks out a set of ways things *might be*, explicitly represented by *worlds*. We say that *propositions are true at worlds*, *W*. Affirming a proposition P is to say that this is the way things *actually are*, thus *P is true at @*. Logical implication is a subset relation: *P entails P' iff P is a subset of P'*. (Because whenever P is true at a possible world *W*, *P'* is true too, since *W* is in *P'*.) The only *tautology* is the class of all possible worlds. The only *contradic-*

tion is the empty class of worlds. There are still no facts about *the present time* however.

4. Multiple similar actual worlds.

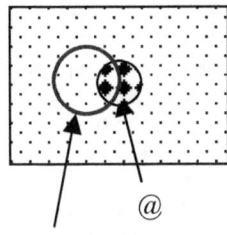

A partially true proposition contains part of @?

Theory 4. There is not a single unique actual world, but a class of actual worlds, @. I.e. 'actuality' is multiple or 'indeterminate' to some extent. If all actual worlds have the same individuals but with slight differences in properties, then the actual world has some 'fuzzy' or indeterminate object properties – sometimes quantum mechanics is modelled rather like this.

Here is a first challenge to the simplest possible world theory: there could be more than one actual world. Why assume that @ must be *unique*? Perhaps there is a 'cluster' of actual worlds that are all very similar, so most propositions are definitely true or false, but some may be 'partially true' or 'indeterminate'? (What would we say about a proposition that contains only a part of @?)

---

5. Multiple independent actual worlds.

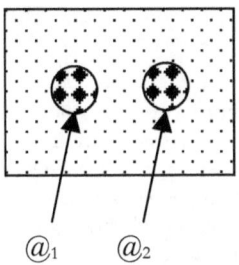

@₁   @₂

Theory 5. There are many disjoint parts to the class of actual worlds, @₁, @₂, etc, containing different individuals, logically independent of each other. This is like a theory of multiple parallel realities. But how could we know of the existence of other 'realities'? Do we assume that we are in one, but others may be in another?

Here we imagine the more radical possibility that multiple 'realities' go on in parallel, possibly with different individuals existing in different *actual worlds*. This is a theme of many sci-fi stories – and some interpretations of QM. These last two variations illustrate that it is at least possible to doubt the assumption of the *uniqueness of the actual world*. We can depict an alternative metaphysics. Thus our metaphysical assumption has some content. The fact that this is seen as a real *scientific theory* by some shows that it may be empirical. But usually the empirical versions hold that distinct 'actual worlds' are causally connected through their pasts, and may interact in the future – requiring the role of *time* to be acknowledged.

6. The present moment: world-times and world-histories.

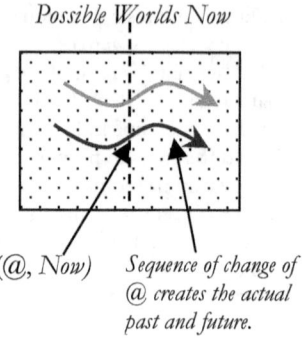

Possible Worlds Now

(@, Now)   *Sequence of change of @ creates the actual past and future.*

> **Theory 6.** There is a single actual world existing *at a moment in time*, the present moment: *Now.* The *world-now* is singular, but it changes. There is a unique momentary actuality: *(@,Now)*, and many other possibilities, e.g. *(@, t)* the actual world *at another moment,* and: *(W, Now)*, another possible world *Now,* and: *(W,t)*, another world *W* at another moment, *t.* The logical space is now structured by *time.*

Here we have returned to a single actual world, but take it as a *changing world.* Points in logical space are now *worlds at moments of time*, or *world-times.* Worlds have identity through time, and *world-histories* are sequences of *world-times,* represented as lines through the logical space. The motivation to include *time* as a fundamental part of the structure of the logical space itself reflects our experience that *what is actual is changing,* and that *propositions generally apply to the world at moments of time, and change their truth values with time.* Thus propositions like 'John Lennon is dead' are true at some times and false at other times. This supports propositions about the actual world at past and future times.

7. Temporal Indeterminism.

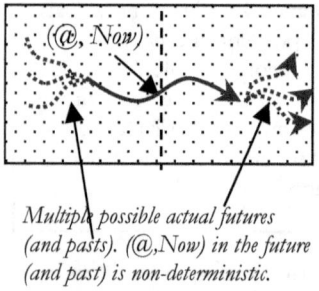

(@, Now)

Multiple possible actual futures (and pasts). (@,Now) in the future (and past) is non-deterministic.

> **Theory 7.** The actual world has a non-deterministic and *non-unique* future (and past). Future and past may be *determined* by *(@,Now)*, the *actual world now,* for a certain period, but eventually bifurcates. This means that multiple *world-times* are possible in the future (and past) – and there really is *no unique proposition about the future (or past).*

The dynamism of this model means that when @ gets to what is *Now* the future, some of the present future possibilities (dotted lines) disappear, and only one becomes actual. This reflects a popular ontology of *indeterminism* for quantum mechanics: future bifurcations are undetermined outcomes of 'wave collapse' events. Everett's many worlds QM alternatively takes *all bifurcations to remain real,* giving multiple actual world-histories.

8. Indeterminism with many actual worlds.

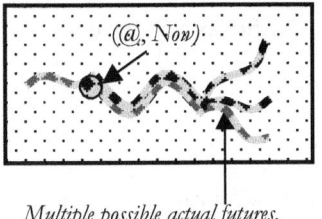

Multiple possible actual futures. (@,t) in the future is non-deterministic Now.

Theory 8. Combine multiple actual worlds with multiple possible futures but a unique present. When the actual future is 'realised' (e.g. the blue outcome), its past-Now is determined too, as a sub-set of possible worlds (when the past was the actual present) that are consistent with what the future actually became (e.g. the blue *Now*).

This kind of ontology is similar to some interpretations of QM, where it is proposed that multiple possibilities in the future reflect multiple actual worlds in the present, but worlds so similar they have not yet been distinguished. How do we distinguish which 'actual state' a QM system is in? We perform a future 'measurement': one outcome is actualised, and this determines the *past* more closely. This connects *dynamism* with *indeterminism*.

9. Particle-State Propositions and Atomic Reductionism.

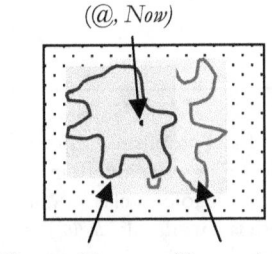

(@, Now)

The complete proposition about particle A.   *The complete proposition about particle B.*

Theory 9. Assume complete propositions about *fundamental particles* are the only fundamental propositions. All other propositions are logical combinations of these. Thus particle states determine the logical space completely. Each fundamental proposition captures one particle state completely, but leave states of other particles only partially determined. The complete states of all particles determines the *actual world* exactly. This is the assumption of *atomistic reductionism*.

Reductionism claims the set of complete propositions about 'fundamental particles' uniquely determines the actual world. In examining 'particles', we are now *'looking inside worlds'*, considering their internal construction. The construction of our world in terms of *atomic particles and their properties* appears to us as *empirical knowledge within the world*, but it reflects the *structure of the logical space*. We can take a fundamental theory of physics proposing a specific range of particles with properties and laws, and construct a logical space *to represent this physics as the theory of the world*. (However quantum entanglement means classical atomism is false: the complete state of two QM particles is not the intersection of their two states taken separately).

10. Laws of Nature: Symmetries.

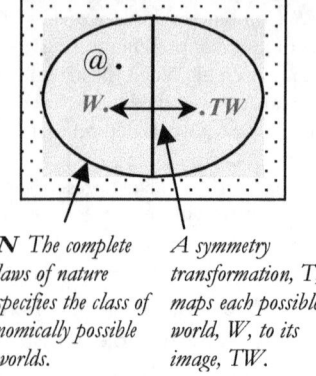

N *The complete laws of nature specifies the class of nomically possible worlds.*

A symmetry transformation, T, maps each possible world, W, to its image, TW.

Theory 10. After specifying the *logical space* for a fundamental theory, we also specify the *laws of nature* as a proposition, **N**, within that space. There are *logically* possible worlds that are not *nomically* possible – e.g. it is *logically* possible for particles to move about randomly, but the *laws of physics* tell us they move only in certain ways. Theories of *symmetry transformations* mean that each world, *W*, in **N** has its symmetric image, *TW*, also in **N**. If *T* is a natural symmetry, then: $T\mathbf{N} = \mathbf{N}$.

Possible symmetry transformations include *space reflection (P), time reversal (T), charge reversal (C), space translation, rotation, velocity boost, energy and momentum.* But fundamental physics has now lost key fundamental symmetries it thought it had established in the 1950s. Either the theories (QM) are not fundamental, or nature is not constructed according to symmetries as expected.

---

11. Laws of Nature: World Dependence.

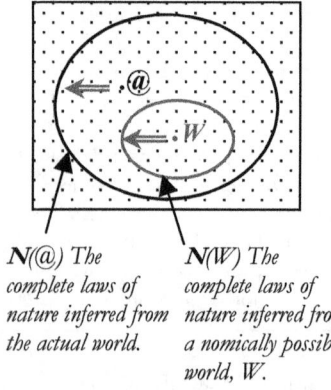

N(@) *The complete laws of nature inferred from the actual world.*

N(W) *The complete laws of nature inferred from a nomically possible world, W.*

Theory 11. *Nomic necessity* is a *modal concept*, not a fixed proposition. From our actual world now, we might infer the laws of nature to be a class: **N**(@). Among these is a world, *W*. **N**(@) implies it is possible that this might have been the actual world – or the actual world may evolve to *W*. But the laws of nature inferred from *W*, i.e. **N**(W), may be much simpler than **N**(@). They may not even contain @!

The *laws of nature*, **N**, is a world-dependant concept too. We think there are the *actual* laws of nature. But we suppose there are possible worlds, with laws possibly different to our own. We must suppose that *alternative worlds come with their own laws of nature.* (Otherwise the laws of nature would be logically determined, not empirical). Thus **N** is not a simple proposition, but a function from *worlds* to their *laws*. **N**(W) maps from a *world*, *W*, to the *laws of nature for that world.* Since the actual world is considered *contingent*, how do we know whether what we 'see' as the laws of nature in our world, at our time, are the full or complete laws? We may be in a simple world-state where the laws appear much simpler than they are – in fact this seems almost unavoidable for many possible worlds. There is generally no unique logical mapping from *W*'s to *N*'s. If so, *the laws of nature* is not an objective concept.

Perhaps most important though is that to write the laws of nature, we first have to *find the right logical space to represent them in*. *N* is not just a simple set of 'empirical generalisations', or law-like propositions, in some universal language of 'observables' or 'measurements'. It requires a theoretical ontology: a *logical space*. Little scientific work has been done on this concept. Rather than investigating the possible construction of logical spaces, physicists assume *the logical space of nature must suit their preconceptions of physics.*

---

12. Temporary Existence of Particles and Changing Logical Space.

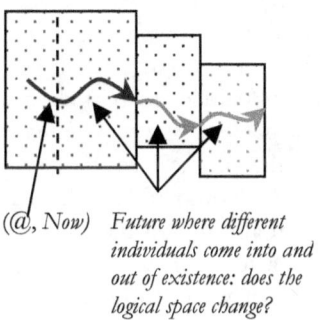

(@, Now)   *Future where different individuals come into and out of existence: does the logical space change?*

> **Theory 12.** Particles appear and disappear from the actual world through time. If particles are the fundamental *individuals* defining the logical space, does this not cause the logical space itself to change? How do 'new individuals' get 'into' the world? If they are destroyed, what facts remain about their past? Is there a single stable *logical space* that lasts eternally?

Physicists intuitively think about fundamental particles as the basic individual entities whose states define the actual world, and the 'fundamental particles' we know of are being constantly created and destroyed. But then what happens to the *logical possibilities* for those particles in the past or future? Can temporary particles really be logically fundamental entities, on which the logical space is constructed? Does that permit any stable, universal logical space? Or does 'reality' change radically over time, turning into different 'kinds' of reality; an evolving logical space, with evolving laws of nature?

13. Subjective existence: the I.

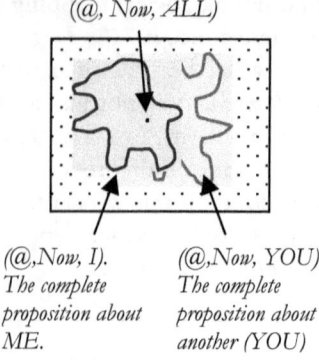

(@, Now, ALL)

(@,Now, I).
The complete
proposition about
ME.

(@,Now, YOU).
The complete
proposition about
another (YOU)

Theory 13. There is a specific *actual world Now*, along with a set of *subjective propositions* in the logical space. Each is an 'I', representing a 'self'. These do not exist 'inside' the actual world (like physical particles): they exist as higher-order propositions, in a frame of logical worlds. They cannot exist without *the frame of logically possible worlds existing*, because they are differentiated by the LPW's, not by particle states in @.

Representing *subjective existence* (the *I*) has long baffled both science and logic. Many philosophers have wondered if reality itself is *objective*, or if it has an intrinsically *subjective* aspect. In this possible world theory, multiple *I's* or subjective individuals are represented. They look similar to particle propositions: they represent *the complete proposition about the state of a subjective individual*. But they are taken to exist independently of the physical contents within a world. For the subjective individual, all the worlds compatible with their state appear equally real. This requires strong realism about possible worlds. What exists is *the space of possible worlds as a whole*, with a set of *subjective entities* (inducing propositions) as real *second-order* entities. What we call the 'actual world', @, is a common world contained 'in' everyone. It models our interaction with others: we interact through sharing a common 'objective world'. This is a more radical metaphysics than those previously. It is a form of *strong metaphysical realism*. If it seems strange that existence includes *many possible worlds*, it may be observed that *subjective existence also seems strange*.

14. Subjective entities as eternal.

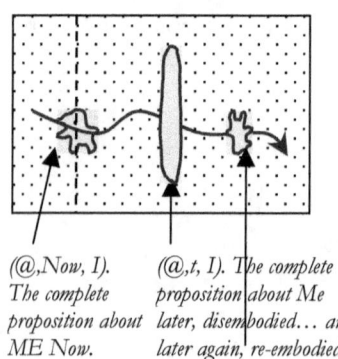

(@,Now, I).
The complete
proposition about
ME Now.

(@,t, I). The complete
proposition about Me
later, disembodied... and
later again, re-embodied.

Theory 14. *Subjective individuals* in the logical space are eternal, changing content but not identity. Each represents a personal identity. When we are embodied in a physical body, the content of a mind is causally connected to the state of a physical body, and becomes very detailed. When disembodied, on physical death, the content becomes very diffuse. It may become re-embodied again.

This ontology provides an interpretation of personal identity as existing independent of the physical state of your body. Personal identity is carried by a real entity – a 'boundary' around worlds or container of worlds – that is logically like a proposition, not a physical object. Death is not the annihilation of the individual (i.e. personal identity), but the dissolution of the detailed *content* of its proposition. Alive you have an extremely detailed *propositional content*. After death, this turns into a very fuzzy proposition. Logically, the individual can be re-embodied. The mystery is how these 'free floating' propositions in the logical space become entangled with the 'particle propositions' represented by our physical states within a single actual world.

15. Subjective entities as eternal.

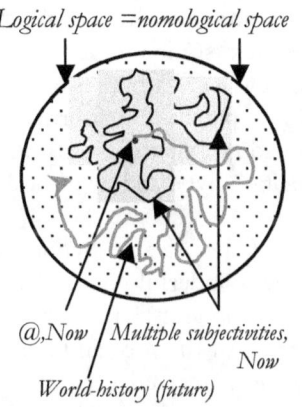

Logical space =nomological space

@,Now / *Multiple subjectivities, Now*

World-history (future)

> **Theory 15.** The logical space is the nomological space. There are multiple subjective propositions (individuals). The actual world traces a 'space-filling path' through possible worlds. Subjective individuals influence or determine the future path. The actual world is the intersection of all subjective individuals. Physical particles 'within worlds' are equivalent to determinate propositions on the logical space.

We finish with the most radical possibility so far considered. The *nomological space is identified with the logical space*. Thus there are no 'laws of nature' *except* the logical space. All logical possibilities are law-like. Or: if we discovered the correct *'logical form' of reality*, this would tell us all the *laws of nature* too. Worlds can evolve with time, into different states. If the actual world-history is a 'space-filling curve', or is cyclic in the logical space, then there is little *contingency* about the world: the only real contingency is *the time we are at now*. Moreover the *actual world* is not a special world, but rather is picked out in its present state by the intersection of all subjective propositions.

The purpose of these examples is to illustrate that *there are real metaphysical choices, they can be articulated precisely, and they are sometimes reflected by scientific theories*. Such choices are not decided directly by science, or by experience, or by rationality alone: they are *precursors* of science, experience and rationality. To investigate this subject properly requires formal theories of *possible world semantics* to help study how logical spaces are constructed. This is a subject which has hardly been explored.

Appendix 7. The Alvarez Controversy.

The scientific controversy over the 'Alvarez Hypothesis', that the dinosaurs were killed by a meteor (or asteroid) strike some 65 million years ago, is iconic of our time. It raged through the 1980's, and was notorious for its virulence and personal spite. It typifies the real nature of modern scientific controversies, and propaganda wars within science. The following extract is from Michael Benton (1997) in *Scientific American*, who reviews two books, one by Louis Alvarez, the other by Officer and Page, two of Alvarez's bitterest enemies.

"Most geologists and paleontologists initially objected to the Alvarez theory. They argued, first, that dinosaurs and other groups died out gradually (over at least a million years) rather than instantaneously; second, that the iridium layer was a local feature that represented some minor peculiarity in the sediments; third, that Alvarez and his crew were a bunch of physicists and chemists who should stick to their own patch; and fourth, that the whole notion was pitched at the press and had no scientific basis. From the start, the debate mixed science and personalities, hype and hypotheses.

"Officer and Page, who began as skeptics of the Alvarez hypothesis and never wavered, thread their book with some wonderful gossip, backbiting and accounts of scurrilous deeds by impact proponents. They report extensive evidence for bias in reporting and funding of the pro- and anti-impact viewpoints. They are uncomfortable with the perceived pecking order in science: math and physics good, chemistry a form of physics and so acceptable, biology and geology pretty dodgy, psychology and geography beneath contempt. The authors characterize Luis Alvarez as little more than the devil incarnate, one of the most extraordinary character assassinations of a recently deceased person that I have ever read.

A fascinating early review, when the controversy was still running hot (and shortly after the death of Louis Alvarez), is found in Fellman (1988). Extracts from this article are below. Fellman acknowledges and even justifies the existence of virulent and spiteful controversies in science – unlike much scientistic propaganda of today, which makes 'science' out to be a saintly, rational process.

From Bruce Fellman. 1988. "Shootout at the K/T boundary".

"The acrimony of the whole debate has left many scientists reeling. 'I like to think of science as pristine, pure, unsullied—with scientists doing their work for the same reasons that Mozart composed and Rembrandt painted,' says McLean, whose work has been roundly condemned by the impact camp. 'But I've learned—and it's been a terrible education—that there are very few angels in the Temple of Science.'

'We may look back on this episode and feel more than a little embarrassed,' adds Raup. 'It's been exhilarating, but it's also been very upsetting. A lot of friendships have broken down as the result of this debate.'

Is this really the way science is supposed to work? Are rancor and personal animosity inevitable in the spread of new ideas? To some degree, the answer is yes, says sociologist Dorothy Nelkin, a visiting professor at New York University. 'Good science can get rough,' she says. 'How rough depends on the personalities and the stakes that are involved. But deep controversies are generated over every area of hot research. Their presence is actually a sign of a healthy science.'

Still, the controversy that Alvarez started has been more heated than most. Its origins date back to 1980 when Alvarez, his paleontologist son Walter, and their colleagues at Berkeley first proposed the impact theory in *Science* ...

The apocalyptic vision captured the imagination of scientists (although only a minority of paleontologists) and the public alike. Walter Alvarez's talks on the hypothesis played to packed houses at annual meetings of the American Association for the Advancement of Science, and a painting of the menacing asteroid even made the cover of Time magazine. ...

From the outset, however, plenty of reputable scientists did not buy the various aspects of the hypothesis. A number of them invoke volcanoes as the cause of the iridium layer and other peculiarities of the K/T boundary, while others accept the reality of the asteroid impact but not its cataclysmic biological and climatological consequences. But these opponents had a tough row to hoe because of the theory's enormous popularity and the influence of Luis Alvarez, its leading proponent.

Was Alvarez to blame for the acrimonious debate that followed? Some scientists have suggested, for example, that his entry into the fray was an example of the so-called 'Nobel Syndrome,' a foot-in-mouth condition that results from external and internal pressures on Nobel laureates to comment on a wide variety of matters that are outside their areas of expertise.

'They come to be defined as contemporary seers and sages,' says Harriet Zuckerman, a Columbia University sociologist who has written about the syndrome. 'But this was clearly not a case of it. Alvarez, even though a physicist, made himself as qualified as anyone else. He had the energy and imagination to start a new line of research.'

Of course, Luis Alvarez's caustic comments about paleontologists didn't exactly pour oil on troubled waters. 'Dad made some remarks that he shouldn't have made,' admits son Walter, adding that the 'shoot-out' was caused by far more than a few 'uncharitable comments. It's the result of a clash of scientific cultures.'

For one thing, Alvarez's foray into paleontology aroused a certain amount of jealousy, admits Yale paleontologist Leo Hickey, an expert on K/T plant communities who accepts the reality of the impact but does not see it as a world--wide 'angel of death.' Although the traditionally 'low-tech' world of paleontology had often been invaded by 'higher-tech' machines and methods, this newest assault by a pushy Nobelist—and an outsider, at that—who wielded the most advanced technology, denigrated paleontological researchers, and declared the K/T debate over, 'was bound to cause some resentment,' says Hickey.

In addition, the controversy was fueled by fundamental differences between scientific disciplines. Alvarez followers tend to view data with a kind of precision that is part and parcel of chemistry and physics, but which is alien to the tradition of paleontology, explains Hickey. 'One of Luie's points was that if we see a trend in physics— say, the coincidence of an extinction and a spike of iridium—the conclusion is incontrovertible: the two events are linked,' Hickey says. 'But the essence of the paleontological argument is that however soft and wishy-washy it sounds, biotic systems are extremely complex. They often respond to environmental stress in a nonlinear and unpredictable fashion. It's the height of hubris to say that changes in the record are caused by specific spikes.'

The impactors have an easy response to this charge. 'There's no reason that a simple answer can be right,' says Raup.

Another problem is that many paleontologists have wrapped entire scientific careers around the assumption that earth history can be explained by ongoing, earthly processes. The suggestion that the evolution of life was greatly influenced by sudden extraterrestrial events, therefore, smacks of heresy. 'Remember, a lot of science is really a belief system,' Nelkin explains. The inevitable result, says William Glen, a geologist currently at work on a book about the impact controversy, is an almost religious clash between two fundamental theories about how the planet and its life evolved.

And so, the consensus among those who have observed the K/T imbroglio is that the high level of acrimony was probably unavoidable. Hot, contentious areas of science, the observers explain, inevitably arouse strong passions and stimulate pitched battles. 'Unpleasant as it may be, controversy is important for the refinement of scientific ideas,' says sociologist Zuckerman. 'They rarely get straightened out without it.'

Zuckerman points to past battles that make the asteroid debate seem like a pillow fight. The debate over the theory of continental drift, for example, raged on for years, prompting all sorts of wild accusations, and leaving Alfred Wegener, who originally proposed the idea, a bitter, broken man. ...

Perhaps Hickey said it best, a few months before Alvarez's death. 'While I decry the nasty, bigoted, opinionated, cantankerous, crusty old bastard Luis Alvarez, he really has helped open up the thinking in paleontology.' Hickey said. 'His is a major contribution.'

(Fellman, 1988).

Appendix 8. Evolving Peer Review Processes.

Open peer review is a significant recent movement, with many journals now publishing review comments by reviewers, and revealing reviewer and author identities to each other in the review process. The following example of open peer review policy is from a stable of journals owned by *Science Direct International*, which includes *Physical Review and Research International*. The paper referred to here is original, heterodox, and interesting in its own right. The three reviewer's comments are reproduced below.

Physical Review & Research International, ISSN: 2231-1815, Vol.: 3, Issue.: 4 (Oct-Dec 2012) [134]

Research Paper

Quantum Gravity and the Holographic Mass
Nassim Haramein*

Director of Research, Hawaii Institute for Unified Physics, P.O. Box 1440, Kilauea, HI 96754.

General Comment

1. Up to 6th August, 2012, all SDI journals followed strict double blind fold review policy to ensure neutral evaluation. During this review process identity of both the authors and reviewers are kept hidden to ensure unbiased evaluation. More information is available in this *link*.

2. We have migrated to transparent and toughest 'Advanced OPEN peer review' system (Detailed general information is available in this *link*). Identity of the authors and reviewers will be revealed to each other during this review process. This transparent process will help to eradicate any possible malicious/purposeful interference by any person (publishing staff, reviewer, editor, author, etc) during peer review. As a result of this unique system all reviewers will get their due recognition and respect, once their names are published with the papers (Example *Link*). If reviewers do not want to reveal their identities, we will honour that request. In that case only the review reports will be published as 'anonymous reviewer report'.

3. Additionally 'Advanced OPEN peer review' greatly helps in 'continuity and advancement of science'. We strongly believe that all the files related to peer review of a manuscript are valuable and hold an important place in the continuity and advancement of science. If publishers publish the peer review reports along with published papers, this process can result in savings of thousands of hours of future authors during experiments, manuscript preparation, etc. by minimizing the

---

[134] See: http://www.sciencedomain.org/review-history/1298

common errors after reading these previously published peer review reports. Therefore, as per our new official policy update, if the manuscript is published, all peer review reports will be available to the readers. All files (like original manuscript, comments of the reviewers, revised manuscript, and editorial comment (if any)) related to the peer review, will be available in "Review history" link along with the published paper (Example *Link*).

4. Additionally we believe that one of the main objectives of peer review system is 'to improve the quality of a candidate manuscript'. Normally we try to publish the 'average marks (out of 1_o)' a manuscript received at initial peer review stage and at final publication stage to record its history of improvement during peer review. This process further increases the transparency. It is more important to record honestly the 'strength and weakness of a manuscript' than claiming that 'our peer review system is perfect'. Therefore, these transparent processes (i.e. publication of review history files and scores of a particular manuscript) additionally give a clear idea of the strength and weakness of a published paper to the readers, which enhances the chances of proper use of the result of a research (and or reduces the chances of misuse of the weakness of the findings of the paper). Thus this transparent process may prove to be highly beneficial for the society in long run.

This paper is another illustration of difficulty evaluating heterodox work. The three reviewers' comments, in full below, are completely inconsistent with each other.

Reviewer 1. "The author defined the holographic mass and applied it to the black hole and the proton. There are some interesting coincidences, e.g. similarity with strong force and Yukawa potential in short range in the study. However, There are still some questions to be answered: 1)The meaning of holographic principle(HP) is definitely not that used in the paper. Moreover, the area in Eq.(4) should be surface area. Generally, HP is no problem when used in black holes, but it is problematic when used in the weak gravitational field. How to guarantee its validity when used in the system of protons? 2) The black hole mass and proton mass were obtained from different formula (Eq. $_9$ and Eq. 24), although they were explained in holographic principle. Why? 3) The author described a system in section5 to use the gravitational interaction to explain the strong force. Where could we see the evidence of gluon? Or does the gluon derived from the evolution of graviton? 4) For the system consisted of two protons, there indeed are many coincidences. But if they cannot be extended into other systems, e.g. three protons etc., the phenomena described in the paper is not enough to support their conclusion.

In a word, the conclusions obtained in the paper are highly implicative in the physical mechanism. However, it is interesting to understand further these coincidences found by the author. **The paper is not proper to be published in the present form.**"

Reviewer 2. "This manuscript contains trivial mathematics and algebraic manipulations expresed as more important than they are. The manuscript is filled with nonsensical sentences without proper definitions etc. If there is an explanation, why the product of proton charge and Schwarzschild radii is about equal to the square of Planck length, it not given in this manuscript, where is explained nothing. **This text should not be published.**"

Reviewer 3. "The Paper is well-motivated, well-written and contains interesting results. All calculations are correct and the presentation is also suitable. **I recommend publication of this paper**. The only thing that is necessary to be considered before publication is a reference to the Planck scale physics. This can be done by referring to [arXiv:hep-th/0508078] and [arXiv:1205.0158]."

The published paper has a few paragraphs added to the draft, for clarification requested by the first reviewer. I add that to me, this paper is of significant interest. I believe the 'coincidences' observed by Nassim Haramein are of fundamental importance. If they are not coincidence but *law-like relationships*, then they reveal a profound interconnection in nature, ignored in mainstream physics. They are closely related to similar 'coincidences' I have tried to explain in a quite different framework, in (Holster (c) 2014/15). E.g. Haramein's dimensionless ratio (eq. 22, p. 277): $\eta = A_p/A_{lC} = 4.716551 \times 10^{40}$ (the number of Plank areas A_{lC} on the surface of a proton A_p, using the empirical value of the proton charge ratio), is intimately related in my view to the fundamental 'large dimensionless ratio' which in my version is: $D = hc/G(m_e^{1/3} m_p^{2/3})^2 = 1.5953 \times 10^{41}$. If the review process is dominated by orthodox physicists like Reviewer 2 above, investigation of such relationships could never be published.

Appendix 9. Predatory publishers: Beall's List.

The "ScienceDirect International" journal *Physical Review & Research International* used in the previous exhibit is a newcomer on the block, competing against *Physical Review* and other established physics journals. As one of around twenty SDI journals, it has been criticised as a *predatory journal* by some industry watchdogs. *"Beall's List of Predatory Publishers"* is probably the best known of these. The extract below is reproduced from Beall's 2015 website. This certainly reveals the escalating magnitude of this problem in the science and academic journal industry – where there are now some 25,000 indexed journals.

What is not clear to me is the demarcation of such judgements. How biased is this 'watchdog list' and others towards supporting the traditional journal industry, now dominated by a tiny number of corporate publishers owning thousands of journals. They usually charge substantial subscriptions, and for non-subscribers, they charge large one-off fees to access individual papers (typically $20-$30, much more than most online *e-books*, with no payment made back to authors), and they demand full commercial ownership of material. They could be accused of being *predatory monopolies.* The traditional journal industry has become a *profit-driven marketplace,* intent on sustaining academic power.

We must ask: how can *new, open access journals* enter this publishing space, and market themselves as alternative outlets for research work, *without* being accused of being 'predatory'? In reference to Beall's final 'criterion' below, why *should* 'alternative journals' be contrained to advertising 'impact factors' *controlled by their rivals who hold monopolistic or olargistic power?* The SDI journals I have looked at do not seem to me to be predatory (although they vary in quality). This situation is really highly confusing. What we can say is that *science journal authemticity is now in serious trouble,* and the industry is now unstable, changing rapidly with new players in the journal market, and widespread discontent among research authors.

Beall's List of Predatory Publishers 2o15.[135]

by Jeffrey Beall, January 2, 2o15 Each year at this time I formally release my updated list of predatory publishers. Because the list is now very large, and because I now publish four, continuously-updated lists, this year's release does not include the actual lists but instead includes statistical and explanatory data about the lists and links to them. **Potential, possible, or probable predatory scholarly open-access publishers**: This year, 2o15, marks the fifth annual release of this list, which is also continuously updated. The list this year includes 6o3 publishers, an increase of 216 over 2o14.

Publishers	
Year	Number of publishers
2011	18
2012	23
2013	225
2014	477
2015	693

Number of predatory publishers, 2o11-2o15

Potential, possible, or probable predatory scholarly open-access journals: This year, 2o15, marks the third annual release of this list, which is also continuously updated. The list this year includes 5o7 journals, an increase of 2o4 over 2o14.

Standalone Journals	
Year	Number of journals
2013	126
2014	303
2015	507

Number of predatory, standalone journals, 2o13-2o15

Two New Lists: Misleading metrics and Hijacked journals This year, I started two new lists that track two new areas of questionable practices related to open-access journals. The **Misleading metrics** list includes companies that "calculate" and publish counterfeit impact factors (or some similar measure) to publishers, metrics the publishers then use in their websites and spam email to trick scholars into thinking their journals have legitimate impact factors. The **Hijacked journals** list includes journals for which someone has created a counterfeit website, stealing the journal's identity and soliciting articles submissions using the author-pays model (gold open-access).

---

[135] http://scholarlyoa.com/2o15/o1/o2/bealls-list-of-predatory-publishers-2o15/

Misleading metrics: 26 (list debuted in March, 2o14)

Hijacked journals: 3o (list started in May, 2o14)

Here are links to current edition of each list:

- Predatory Publishers
- Predatory standalone journals
- Misleading metrics companies
- Hijacked journals

Updated Criteria: With the help and advice of others, I have updated the criteria that I use when evaluating publishers and standalone journals for possible inclusion on my list. The criteria document is called *Criteria for Determining Predatory Open-Access Publishers*, 3rd edition. Most of the document remains unchanged. The only external documents the criteria now refers to are two published by the Committee on Publication Ethics (COPE). The document has also been reformatted for easier reading, and some new criteria have been added:

- There is little or no geographical diversity among the editorial board members, especially for journals that claim to be international in scope or coverage.
- The editorial board engages in gender bias (i.e., exclusion of any female members).
- The publisher does not allow search engines to crawl the published content, preventing the content from being indexed in academic indexes.
- The publisher copy-proofs (locks) their PDFs, thus making it harder to check for plagiarism.
- There is little or no geographic diversity among the authors of articles in one or more of the publisher's journals, an indication the journal has become an easy outlet for authors from one country or region to get scholarly publications.

This criterion has been updated:

- In its spam email or on its website, the publisher falsely claims one or more of its journals have actual (Thomson-Reuters) impact factors, or advertises impact factors assigned by fake "impact factor" services, or it uses some made up measure (e.g. view factor), feigning/claiming an exaggerated international standing.

Appendix 10. Open Access Journals: paying for the privilege.

Researchers increasingly look to *free open access internet* journals. With such a proliferation of journals on the market, it encourages readership of your paper if it can be accessed immediately on the internet for free. OA journals usually charge author fees called *article processing charges* (APC) for open access publication. For example, Isaac Newton's old stamping ground, *The Royal Society*, runs a *society journal* (among other commercial functions), with the following open access fees:[136]

THE ROYAL SOCIETY

Venue hire

Home   Fellows   Events   Grants, Schemes & Awards   Topics & policy   Journals   Collections   About us

Article processing charges

Submission of an article is free, but if it is accepted for publication, the authors are asked to pay a fee to have their article made open access immediately upon publication. The charges are:

	GBP Pound sterling	USD United States dollar	EUR Euro
Royal Society Open Science	Free	Free	Free
Open Biology	£1200	$2100	€1680
All other journals	£1700	$2975	€2380

Three thousand US dollars seems like a lot of money to a poore scholar to get a short article (maximum 20 pages – extra charge for more pages) plonked on an internet site! This is an expensive example, but it is for high-ranked journals. More typical is about $US1,000, but for lower-ranked journals. What about the competitive prolific modern author, publishing say 200 papers over ten years? That is likely to cost $200,000 – or much more, if they want to buy their way into high-ranked journals.

What is peculiar in this industry is that *the researcher who does all the work then has to pay large amounts of money to get that work published.* Of course research-

---

[136] See: https://royalsociety.org/journals/authors/open-access/

ers can generally only afford this by getting grants for publication – locking most independent scholars financially out of the tournament. This is part of the circle of power and money in the modern research-publication business. (And we may wonder: what is the difference with vanity publishing? Or *infomercials* perhaps?) OA publishing is increasing rapidly, with some 10,000 scholarly journals estimated at present. For a recent survey of OA scholarly journals done in 2014, see Morrison *at alia* (2014). They found that about a quarter of OA journals had APCs:

"As of May 2014, the Directory of Open Access Journals (DOAJ) listed close to ten thousand fully open access, peer reviewed, scholarly journals. Most of these journals do not charge article processing charges (APCs). This article reports the results of a survey of the 2567 journals, or 26% of journals listed in DOAJ, that do have APCs based on a sample of 1432 of these journals. Results indicate a volatile sector that would make future APCs difficult to predict for budgeting purposes. (Morrison *et alia,* 2014, P.1.)

About 60% of APC journals were commercial:

"The largest group of publishers using APCs, by far, were clearly commercial in nature, a total of 1246 journals, or approximately 1567 (61% of the total) after adjusting for sampling. The second largest group was universities, with 36 titles or about 276 after adjusting for sampling factor, about 11% of the total. Other categories of publisher types may be too small to draw any conclusions. It is important to note that the categories of mixed publisher types such as commercial/society partnerships is likely understated as commercial publishers' title lists were not examined in detail to identify these mixed types. Ibid, P. 11.

Fees varied widely:

"The article processing amounts per se also suggest a business sector in a very volatile state. Prices ranged from $0 to $4114 USD; even with the zero factor removed the price range was $1 to $4114. The mean, median and mode were quite different. With the zero APCs, the mean was $964, median $800 and mode $0. With the $0 APCs removed, the mean was $1221, median $1145 and the mode $800. ... When the mean or average is higher than both the median and the mode, this suggests that the average is skewed by a small number of relatively high APCs. This makes it very difficult to suggest that there is, at the present time, an effective average APC usable for budgeting purposes. The analysis of APC by publisher types shows a wide range and noticeable differences in mean, median and mode for every publisher type. Ibid, P. 11.

Appendix 11. ORCID: Meta-organisation of researchers.

Meta-organization of *researchers* is now big IT business too. *The Royal Society* journals in the previous example now require ORCIDs for authors. This is a unique ID system for identifying researchers *and their positions*. The following marketing blurb is given by Phi Hurst, who makes it sound like a helpful and innocent convenience to researchers. But the real intent is to *censor individuals who are not members of approved institutions* (essentially academic or government institutions) from even being allowed to submit papers to journals! It is not even enough to have a PhD or other qualification. You now increasingly have to qualify by *occupation* to submit to many preprint sites and journals.

From January you'll need an ORCID
7 December 2015 by <u>Phil Hurst</u>

Mandating the use of Open Researcher and Contributor ID (ORCID) by submitting authors.

ORCID provides a unique identifier for all researchers that can be linked to their different research works and activities across multiple platforms. It also serves to distinguish authors with similar names and simplify searching of publications databases (such as PubMed, Scopus, etc.) to avoid retrieving articles by authors with similar names. Close to 1.8 million researchers have already created their ORCID iD. The service is non-profit and community driven.

From 1 January 2016, we will require the submitting author to provide an ORCID identifier as part of the manuscript submission process. Details of the practicalities will be provided shortly, but don't worry, it takes less than a minute to register for an ORCID iD and it's free of charge. A recent survey carried out by ORCID indicates strong support in the research community for mandates by publishers, funders, and universities.

Benefits of creating an ORCID iD

1. It's a time saver

ORCID provides a unique digital identifier, which distinguishes you from other researchers with similar names and also brings together any variant spellings of your name. It supports authentication across multiple journals, search engines and services, allowing you a single sign-on.

Up to now, each time you apply for a new grant, or register for a service (such as ResearchGate), you have had to re-enter all the same information in each system. Not only is this time-consuming, it also allows scope for errors. Once you have created an ORCID identifier and connected it with your publications, grants, and affiliations, your details will automatically be entered when using any compatible system.

2. It provides cross-platform compatibility

More and more systems are building in ORCID support, which means you can now easily onnect your ORCID iD to your existing publications via interfaces with systems such as Crossref, EuropePMC, or Scopus. Furthermore publishers' systems are now able to automatically update your ORCID record as you publish each new article, saving you the trouble of updating it manually. A number of systems, including university profiles, library repositories, funder reporting, pull data from ORCID, meaning you will spend less time manually updating in these as well.

3. Your privacy is protected

When you sign up for an ORCID iD you retain complete control and can set privacy settings on information in your record. For example, you can make some or all of your research works public whilst keeping other information (such as email addresses) private. You control which organizations have access to your record and which can add information.

4. It helps to build reputations

Using an ORCID identifier helps to build your reputation by making research works much more visible to potential collaborators, funders and employers. Publication repositories are using ORCID iDs to enable search, and universities are using them to support updates of local profile systems. ORCID also enables connections with, datasets, grant applications, reviewer reports, etc., to build a complete picture of your research contribution.

We believe that ORCID is beneficial to researchers and good for science as a whole. We hope that other societies, publishers, and funders will follow suit, and encourage them to contact ORCID to discuss this further.

With *millions* of researchers publishing papers – comparable to the population of a small country like NZ or Singapore or Scotland - organizing researcher identities and credentials has certainly become a serious business – a serious *marketing and control business*. Previously it was filtered primarily through *formals qualifications*, along with membership of diverse professional Societies; although these were never (officially) considered prerequisites for being allowed to submit an article until recent years. This now evolves into a Universal Global Identification System: a force for orderliness and standardization; and for centralized control and surveillance. In the future, it will certainly be used to regulate behavior; to exclude individuals without suitable *professional group membership;* to censor dissenters and enemies of the gatekeepers; and any supposed 'misdemeanors' will stay on your record forever. Can you think of any potential drawbacks to such a system?

Appendix 12. Nicky Hager: politically modified science.

In 2002, NZ politics was rocked by a scandal involving the illegal importation and cultivation of GM corn seeds, and a corporate and government cover up, revealed by Nicky Hager (2002) in a monograph, *Seeds of Distrust*. Hager is NZ's leading investigative journalist (actually almost the only one), with degrees in physics and philosophy, and author of half a dozen well-researched books. The GE scandal is a case study of state-corporate deceit and scientific ineptitude at the highest level of government. In Chapter 5, "Politically modified science", Hager gives a succinct and insightful account of how essentially meaningless science is to those in the circles of power. This is the best NZ case study of its kind, and although this particular scandal is over, the larger controversy over GM is ongoing, and it is well worth reading for its more general lessons. I have reproduced p. 49-55 in Chapter 5 below.

Nicky Hager (2002) *Seeds of Distrust.*

CHAPTER 5

POLITICALLY MODIFIED SCIENCE

We are brought up to think of science as solid and incorruptible. In public debates, scientific facts are presented as being like rocks in the sea of conflicting opinions and beliefs. It is an attractive idea and it should be true. Disappointingly, it does not always work that way.

Producers of CFCs denied they were harming the earth's ozone layer long after the damage was evident. Scientists paid by oil companies and car manufacturers have repeatedly questioned the scientific evidence for global warming. Money can buy scientists and 'independent' scientific studies that have discovered smoking is harmless, that logging does not harm rainforests and that – another hugely lucrative business – genetic engineering is thoroughly controlled and safe.

It pays to remind ourselves over and over of the corruptibility of science (or, more precisely, scientific claims). For science can only help with important human questions when its insights are sensibly integrated with those from other spheres, notably ethics and the consideration of social responsibility.

It is very damaging to society when the public start to feel that they cannot trust people in positions of responsibility. Whether it is scientists, businesspeople, lawyers, public servants or politicians, we rely on these people to perform their roles in society ethically and responsibly. They are not just working for themselves or the people

p.49

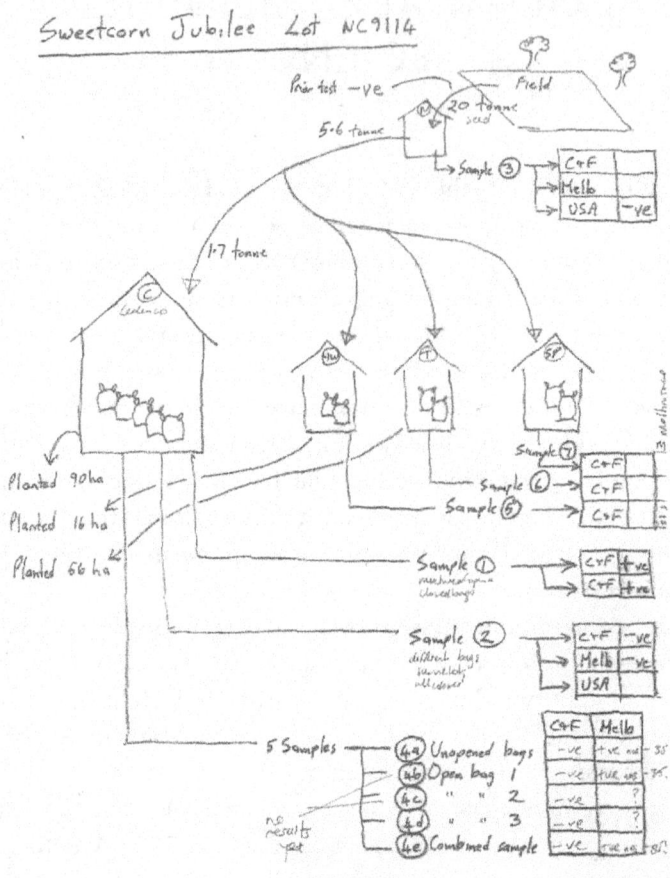

This hand-drawn diagram, by an official involved in the issue, records the various GE test results from different parts of the batch of Novartis seeds. With some tests positive and others negative, it was concluded that the GE seeds were probably spread unevenly through the batch.

p.50 This is as scientifically technical as it gets for the NZ Government Officials; but beyond their comprehension.

who pay them; they are part of what makes democratic society possible. An important part of the cohesion of society is about the bond of trust (or lack of it) between ordinary people and people in these positions.

In the case of the contaminated corn crops, the businesspeople, PR consultants, public servants and politicians involved had one possible scientific conclusion that suited them and one that did not. The conclusion that suited them was, as we know, that once the scientists had checked more carefully, there turned out to have been no 'reliable evidence' of GM contamination in the Lot NC9114 sweet corn seeds after all.

This chapter picks through the evidence to explain why this was not true. Yet the people who were responsible to the public to make good decisions on these issues worked together to present the opposite 'fact' to the Cabinet and ensured that no one could contradict them, or provide different advice, by keeping the entire issue very secret. There were no hearings or submissions to test the evidence – as happens in all other cases relating to GE decisions – and so the convenient idea that the contaminated crops might not be contaminated was repeated between the officials until it felt true.

When the full 20 tonne batch of Lot NC9114 sweet corn seeds was tested by Novartis in the United States in August 2000, the test result came back negative. A few weeks later, when 5.6 tonnes of the batch arrived in New Zealand and 1.7 tonnes went to Cedenco Food Ltd, Cedenco sent samples to the Crown Research Institute Crop and Food laboratory at Lincoln for testing. Those were the first positive results. Later Cedenco retested its batch and again got positive results. Heinz Wattie, Talley's and Seed Distributors apparently also did tests of their batches, but their results are not known. The known results were summarised in a hand-drawn diagram, shown opposite.

On 24 November, by which time the follow-up test results were known, the head of ERMA, Bas Walker, advised Marian Hobbs to 'Bear in mind that in this case [Lot NC9114], there are already several

p.51

POLITICALLY CONTAMINATED SCIENCE

1 November 2000 - First positive GE test result of Cedenco seeds by Crop and Food laboratory.

24 November 2000 - Novartis, the US seed supplier, revealed that 'further extensive testing' had also produced positive results of GE contamination, *Report from industry meeting.*

24 November 2000 - 'Bear in mind that in this case, there are already several positive tests for contamination which can hardly be ignored.' *Bas Walker, advice to Marian Hobbs.*

29 November 2000 - 'Some GM maize seed has been released in New Zealand.' *Ministry for the Environment Cabinet committee briefing paper.*

30 November 2000 - 'In essence some of the tests have shown the presence of a genetic sequence which is strongly indicative of a genetic modification...' *Ministry for the Environment confidential question and answer sheet on the incident.*

5 December 2000 - 'With a considerable degree of confidence it can be concluded the Novartis sweet corn Lot NC9114 has less than 0.5% GM contamination and hence, judged by that standard, does not contain a new organism.' *Donald Hannah, ERMA Manager of Science and Research.*

8 December 2000 - 'If genetically modified material was present at all, it was unlikely to be more than 0.04% of this.' *Ministry for the Environment confidential question and answer sheet on the incident (later draft).*

positive tests for contamination which can hardly be ignored.'[26] A Ministry for the Environment briefing on the results said that 'In essence some of the tests have shown the presence of a genetic sequence which is strongly indicative of a genetic modification....'.[27]

At this stage of the handling of the issue, when officials knew Helen Clark wanted the contaminated crops destroyed, they were happy to say that the seeds were contaminated.

But then the politics shifted. Once the Government accepted the industry's proposition of a contamination threshold, the issue

p.52

306

8 December 2000 - 'There is no reliable evidence of GM contamination.... If present at all, there is unlikely to be reliably discoverable GM contamination present in the particular [Lot NC9114] shipment.' And, 'A detailed evaluation has suggested that there is negligible contamination.' *Cabinet Paper signed by Marian Hobbs.*

15 December 2000 - 'If genetically modified material was present at all in that shipment, the amount present was below the level that could be reliably detected.' *Ministry for the Environment confidential question and answer sheet on the incident (final draft).*

18 December 2000 - 'No evidence of GE crops growing in New Zealand.' *Key points for Minister.*

19 December 2000 - A 'detailed evaluation concluded that with a high degree of confidence that, if present at all, the GM material was at levels below that which can be reliably detected.' *Marian Hobbs' media statement.*

28 February 2001 - 'Tests could not confirm whether or not GM material was present.' *Government letter to Royal Commission.*

Note how the 'scientific' conclusions metamorphosed over a six week period. You can observe the exact point, at the end of November, when the politics changed and it became important to play down and then deny the results. It is a study in the vulnerability of scientific fact to politics.

was not whether or not there was contamination but whether it was below the newly adopted threshold and could be ignored (i.e. the seeds thus 'deemed' to be GE-free). Without any new testing, a review of the test results was conducted by ERMA, which became the basis of subsequent written advice from the other agencies and the Ministers' offices.

The terms of reference for the review, which was done by ERMA's Manager of Science and Research, Donald Hannah, were to decide whether the contamination level in the sweet corn seeds was above

p.53

or below the new and arbitrary 0.5% level. He concluded: 'With a considerable degree of confidence it can be concluded the Novartis sweet corn Lot NC9114 has less than 0.5% GM contamination and hence, judged by that standard, does not contain a new organism.'[28]

This conclusion, combined with the 0.5% standard, had dealt neatly with the problem of the contaminated crops. It shows how flawed the threshold approach is. Judged by the new contamination standard, the crops were by definition 'GE-free' (even though they were not) and so they no longer had to be destroyed. Just add a drop of Policy Pragmatism and those difficult stains disappear!

But in case some Ministers, and later the public, felt uneasy about this sleight of hand, it became desirable to reinterpret the scientific findings. Dr Hannah's careful report was misinterpreted by other officials until they felt justified misleading Ministers by telling them that there turned out to be no reliable evidence of contamination of Lot NC9114 after all. The whole sorry progression of 'facts' about the contamination is set out in the box on pages 52 and 53.

According to Hannah's report, there were 12 separate samples taken of Lot NC9114 and sent for GE testing. Some of the results from the other companies had not been made available for his report. Of the data available to him, there was a mixture of positive and negative results. This is consistent with quite a low concentration of contamination. For instance, for three of the samples, the Genescan laboratory in Melbourne got positive results while the Crop and Food laboratory at Lincoln got negative results. It turns out that Genescan tested samples 100 times bigger than Crop and Food. Also, Hannah suggested, the contamination was probably not evenly distributed in the seed batch.

The test method used, called PCR, involves washing the sample to remove seed chemicals, grinding up the seeds and then extracting the DNA from sub-samples for testing. Each sub-sample has four PCR tests run on it so that the results can be compared. The PCR test kits can accurately detect GE contamination down to at least 0.1%

p.54

concentration. The report notes that it is much easier to say with certainty that there is some level of contamination in a set of samples than to say exactly what the level is in any one sample.

Dr Hannah made an estimate of the likely level of contamination by considering how many of the samples from the seed batch had 0.1% or more contamination (i.e. some level of contamination) and how many samples showed no contamination at the 0.1% level. He concluded that the overall level of contamination in the seed batch was about 0.04%. Contamination levels always sound small, but when spread across tens of millions of seeds (as in this case) it amounts to many thousands of GE plants scattered through the normal crops. Using his conservative estimate of 0.04% contamination, this amounted to about 15,000 GE sweet corn plants.

It is worth noting, in comparison, that the US Starlink contamination, which led to the US$100 million recall of corn-based products from shops and a sharp drop in US corn export prices, was estimated by the US Environmental Protection Agency (EPA) to involve a similar contamination level of 0.14%.[29] These EPA results were, by chance, released in the same week as the results of the testing of the New Zealand corn contamination.

Hannah cautiously considered the possibility that the positive results had been wrong (prefacing some comments with 'if there is contamination...'). He discussed the possibility that the test results were the result of lab error or dirty samples (giving 'false positives'). However, he found that results from unopened bags of seeds gave the same results as from opened (and so possibly dirty) bags. Similarly, he found that lab error was unlikely because the labs ran identical tests on 'controls' (seeds known to be GE-free) and these registered negative while the Lot NC9114 seeds registered positive. He also concluded that the likely contaminant was the genetically engineered 'Bt1' sweet corn.[30]

I have shown Hannah's report, containing the test findings, to genetic engineering specialists who say that the evidence of

p.55

Appendix 13. Rupert Sheldrake: ten dogmas of science.

In *The Science Delusion*, Rupert Sheldrake dissects "ten dogmas of science" that he holds are taken uncritically as facts of science, but that he rejects, and argues should be regarded sceptically:

1. That nature is mechanical.
2. That matter is unconscious.
3. The laws of nature are fixed.
4. The totally amount of matter and energy are always the same.
5. That nature is purposeless.
6. Biological inheritance is material.
7. That memories are stored as material traces.
8. The mind is in the brain.
9. Telepathy and other psychic phenomena are illusory.
10. Mechanistic medicine is the only kind that really works.

He says: *"The 'science delusion' is the uncritical belief in these dogmas, treating them not as beliefs but as truths... Science is much more fun, much more interesting, much more free, when we turn these dogmas into questions."* The primary question to ask is this:

- Is it scientifically reasonable to question these beliefs – or are they indeed so definitive of science that questioning them means rejecting science itself – rejecting scientific method or established scientific truth?

To my mind, these 'dogmas' are all eminently questionable, and they reveal *metaphysical assumptions* underpinning a certain paradigm of science – viz. the materialist, reductionist, Positivist paradigm. We can ask whether these dogmas have clear scientific explanations or decisive proofs. If so, what are they? I don't think they are robust myself, indeed I think most are probably false. However their truth or falsity is a secondary question that people may disagree about. The primary question is whether scientists should be allowed to question them – allowed to disagree about these beliefs, allowed to propose alternative beliefs.

Rupert Sheldrake is an iconic example in our time of a heterodox scientist, a former leading biologist who pursued unorthodox studies and was savaged by the Scientific Establishment. He is best known for investigations of telepathy, his theory of morphic resonance, and for challenging various 'dogmas of science', as above.

Attacks on Sheldrake began when Sir John Madox, long-time editor of *Nature*, wrote in his notorious 'book-burning' editorial in 1981, about Sheldrake's first book, *A New Science of Life*, that *"This infuriating tract... is the best candidate for burning there has been for many years."* In a BBC interview in 1994, Madox said: *"Sheldrake is putting forward magic instead of science, and that can be condemned in exactly the language that the Pope used to condemn Galileo, and for the same reason. It is heresy."*

Sheldrake has been subject to years of Wikipedia attacks by anonymous 'science debunkers' and 'guerrilla sceptics', constantly sabotaging entries to describe him derisively as a 'pseudo-scientist'. Perhaps because of his popularity outside science, he appears to be adopted as *Public Science Enemy #1* for years by the scientistic propagandists. Yet Sheldrake's views are much more philosophically sophisticated than his detractors, and scientifically original, insightful, challenging and well-argued. There is no doubt he has a finely developed sense of science. He believes in the values of science, and criticises the modern *scientistic* trend of treating science as a closed belief system, rather than an open method of inquiry. He is invariably polite and reasoned in debates. He emphasises values of tolerance, compassion and spirituality, and is much loved by his supporters.

The scientistic ideologists like Madox are surely wrong in wanting to burn books and condemn scientists as heretics for asking such questions. If the answers to questions Sheldrake raises are really are so conclusive, then surely they can give us the *evidence* without much trouble (instead of ranting against the person)? But of course they cannot: the evidence is really very complex.

The scientistic belief system is not really based on evidence at all. Rather, it is like the value system of *political conservatives* of a certain vintage – a bundle of interrelated feelings against 'liberals' or 'progressives' or 'radicals', that adds up to a moralistic paradigm. People with the wrong dress, hair style, musical taste, education, occupation, attitude, origin, are *bad, untrustworthy, immoral, despicable, inferior.* (In the US context, *un-American.*)

It may be difficult for outsiders to understand why Sheldrake draws such vitriolic abuse from the science establishment. But for those who have worked in conservative sciences, there is little mystery. The emotion springs from a personality type well known in business and politics: competitive, aggressive, egocentric, elitist. Madox and his cronies hate Sheldrake for the same reason Nixon and his cronies hate hippies. Because they *think wrong.* These judgements are visceral, not rational. They are expressions of a political personality, not a scientific one.

The question we may ask is: *do we want to do science under the liberal, tolerant, philosophical values of a Rupert Sheldrake, or under the antagonistic, technocratic, tyrannical values of a Sir John Madox?*

REFERENCES

Alvarez, LW, Alvarez, W, Asaro, F, and Michel, HV (1980). "Extraterrestrial cause for the Cretaceous–Tertiary extinction". *Science* 208 (4448): 1095–1108.

Bannister, Robert. 1991. *Sociology and Scientism: The American Quest for Objectivity, 1880–1940.* The University of North Carolina Press.

de Beauregard, Olivia Costa. 1980. *Time: The Physical Magnitude.* Riedel.

de Beauregard, Olivia Costa. 1980. "CPT Invariance and Interpretation of Quantum Mechanics". *Found.Phys.* 10 (1980) 7/8, pp. 513-531.

Benton, Michael J. 1997. "Dusk of the dinosaurs." *Scientific American.* Sep 97. Vol. 277 Issue 3.

Björk, Bo-Christer; Roos, Annikki and Lauri, Mari (2009). "Scientific journal publishing: yearly volume and open access availability." Information Research, Vol. 14, no. 1, March, 2009. http://www.informationr.net/ir/14-1/paper391.html

Bronowski, J. 1978. *The Common Sense of Science.* Harvard University Press.

Bryson, Bill. 2003. *A Short History of Nearly Everything.* Black Swan.

Callender, C. 2000. "Is Time Handed in a Quantum World?". *Proc.Arist.Soc,* 121 (2000) pp 247-269.

Chomsky, Noam. 1965. *Aspects of the theory of syntax.* Cambridge, Massachusetts: MIT Press.

Corredoira, Martín López. 2013. *The Twilight of the Scientific Age.* Brown Walker Press.

Devine, S. 2003. "A systems look at the science reforms". *New Zealand Science Review. Vol 60 (70-74).*

Duzí, Marie; Jespersen, Bjorn and Materna, Pavel. 2010. *Procedural Semantics for Hyperintensional Logic. Foundations and Applications of Transparent Intensional Logic.* Springer.

Edmeades, Doug. 2004. "Is the commercial model appropriate for science?" *New Zealand Science Review. Vol 61 (3-4).*

Edmeades, Doug. 2015. "McScience diet hard is to swollow". *NZ Farmer,* June 15, 2015

Einstein, Albert. 1918. "Temple of Science". Address at the Physical Society, Berlin, for Max Planck's 60th birthday.

Einstein, Albert. 1954. *Ideas and Opinions.* Crown Publishers (1982).

Einstein, Albert. 1920. "Aether and the Theory of Relativity". Address delivered on May 5th, 1920, at the University of Leyden, Germany. Copy sourced from:
http://www.aetherometry.com/Electronic_Publications/Science/Einstein's_aether_and_relativity

Faull, Richard L. M., *et alia.* 2007. "Human Neuroblasts Migrate to the Olfactory Bulb via a Lateral Ventricular Extension." Science 2 March 2007: 1243-1249.

Fellman, Bruce. 1988. "Shootout at the K/T boundary". *The Scientist. Oct 3. 1988. http://www.the-scientist.com/?articles.view/articleNo/9830/title/ Shootout-At-The-K-T-Boundary/*

Feyeraband, Paul. 1975. *Against Method.* Verso.

Gleick, James. 1987. *Chaos.* Cardinal.

Hager, Nicky. 2002. *Seeds of Distrust. The story of a GE cover-up.* Craig Potton Publishing.

Hankins, Thomas L. 1985. *Science and the Enlightenment.* Cambridge.

Haramein, Nassim. 2012. " Quantum Gravity and the Holographic Mass." *Physical Review & Research International.* ISSN: 2231-1815. Vol 3, Issue 4 (Oct-Dec).

Healey, R. 1981. "Statistical Theories, Quantum Mechanics and the Directedness of Time". (1981), pp.99-127. In *Reduction, Time and Reality,* ed. R. Healey. (Cambridge: Cambridge University Press. 1981).

Holster, A. 2003 (a). "The criterion for time symmetry of probabilistic theories and the reversibility of quantum mechanics". New Journal of Physics. 5. http://stacks.iop.org/1367-2630/5/130.

Holster, A. 2003 (b). "Proper Time, Coordinate Systems, Lorentz Transformations." *Internet Encyclopedia of Philosophy.* http://www.iep.utm.edu/proper-t/

Lee, T.D. 1988. *Symmetries, Asymmetries, and the World of Particles.* Washington University Press.

López Corredoira ,Martín. 2013. *The Twilight of the Scientific Age.* Brown Walker Press.

McCall, Storrs. 1994. *A Model of the Universe.* Oxford.

McCall, Storrs. 1976. "Objective time flow." Phil.Sci. 43. 337-362.

Meyer, Stephen. 2009. *Signature in the Cell. DNA and the evidence for intelligent design.* Harper Collins.

Misner, C.W., Thorne, K.S. and Wheeler, J.A., 1973. *Gravitation.* Freeman.

Mlodinow, Leonard and Chopra, Deepak. 2011. War of the World Views.

Morrison, Heather, Salhab, J. Calvé-Genest A. and Horava, T. 2014. "Open Access Article Processing Charges: DOAJ Survey." *Publications* 2015, *3,* 1-16. www.mdpi.com/journal/publications

Oddie, Graham. 1986. *Likeness to Truth.* D.Reidel.

Pollack, Gerald. 2013. *The Fourth Phase of Water.* Ebner and Sons. Seattle.

Robertson, D.S., Lewis, W.M., Sheehan, P.M. & Toon, O.B. 2013. "K/Pg extinction: re-evaluation of the heat/fire hypothesis". Journal of Geophysical Research: Biogeosciences.

Roy, R., Rao, M. L. and J. Kanzius. 2008. "Observations of polarised RF radiation catalysis of dissociation of $H_2O–NaCl$ solutions". *Materials Research Innovations* 2008 VOL 12 NO 1.

Sheldrake, Rupert. 2012. *The Science Delusion.* Hodder & Stoughton.

Spivak, Michael. 1979. *A Comprehensible Introduction to Differential Geometry.* Publish or Perish.

Thrussell, David. 2015. "Magician of the Gods. An interview with Graham Hancock." *New Dawn, No 153.* Nov-Dec 2015. pp.17-22.

Tichy, Pavel. 2004. *Pavel Tichy's Collected Papers in Logic and Philosophy.* 2004. Ed. V Svoboda, B. Jesperson, C. Cheyne. Otago-Praha, Otago UP, Filosofia.

Tichy, Pavel. 1988. *The Foundations of Frege's Logic.* De Gruyter.

Tichy, Pavel. 1984. "Subjunctive Conditionals: Two Parameters vs. Three". Philosophical Studies, **45,** 147-179. Reprinted in (Tichy, 2004).

Tichy, Pavel. 1974. "On Popper's definitions of verisimilitude." Brit.J.PhilSci., **25,** 155-188. Reprinted in (Tichy, 2004). *p.155-160.*

Watanabe, Satosi. 1955. "Symmetry of Physical Laws. Part 3. Prediction and Retrodiction." *Rev.Mod.Phys.* **27** (1) (1955) pp 179-186.

Watanabe, Satosi. 1965. "Conditional Probability in Physics". *Suppl.Prog.Theor.Phys. (Kyoto) Extra Number* (1965) pp 135-167.

Woit, Peter. 2006. *Not Even Wrong.* Jonathon Cape.

Ziman, John. 1984. *Introduction to Science Studies. The Philosophical and Social Aspects of Science and Technology.* Cambridge University Press.

Ziman, John. 1994. *Prometheus Bound: Science in a dynamic steady state.* Cambridge University Press.

Wolbach, W. S., I. Gilmour, E. Anders, C. J. Orth, and R. R. Brooks. 1988. Global fire at the Cretaceous-Tertiary boundary, Nature, 334, 665–669.

WEB REFERENCES

Atkinson, Nancy. 2015.   It Looks Like an Asteroid Strike Can't Cause a Worldwide, Dinosaur-Killing Firestorm.
http://www.universetoday.com/118633/it-looks-like-an-asteroid-strike-cant-cause-a-worldwide-dinosaur-killing-firestorm/

Beall. 2015. Beall's List of Predatory Publishers 2015.
http://scholarlyoa.com/2015/01/02/bealls-list-of-predatory-publishers-2015/

Faull, Richard. 2013. "Rethinking the brain". *TED Talks.*
www.youtube.com/watch?v=NT_Z6kUL0Vw

Institute for Venture Science.
http://www.theinstituteforventurescience.net/

Manifesto for a Post Materialist Science.
http://opensciences.org/about/manifesto-for-a-post-materialist-science

Open Science Movement.
http://opensciences.org/ https://en.wikipedia.org/wiki/Open_science

TIL Homepage *http://til.phil.muni.cz/index.php*

Willis, Dick. 2015. "AgResearch has lost its way." Sep 25, 2015.
http://www.nzherald.co.nz/science/news/article.cfm?c_id=82&objectid=11518965

WEB REFERENCES HOLSTER

These are free preprints and papers on the Internet by the author.

Holster, A.

(a) 2015 "Effects of Radio Frequency Water Treatment on Revival of Wilted Flowers." (Ref to be added).

(b) 2015 "Water: The Mystery". *http://philpapers.org/archive/HOLWTM*

(c) 2014/15 "A Geometric Model of the Universe with Time Flow". *http://philpapers.org/rec/HOLAGM*

(d) 2014 "How to Analyse Retrodictive Probabilities in Inference to the Best Explanation." *http://philpapers.org/rec/HOLHTA*

(e) 2014. "The Aethereal Universe". *http://philpapers.org/rec/HOLTAU*

(f) 2014. "Extrinsic and Intrinsic Curvature". *http://philpapers.org/rec/HOLD-5*

(g) 2014 (2003). "Geometric model of gravity, counterfactual solar mass, and the Pioneer anomalies." *http://philpapers.org/rec/HOLGMO*

(h) 2014. "Principles of physical time directionality and fallacies of the conventional philosophy." *http://philpapers.org/rec/HOLPOP-2*

(i) 2014. *The Time Flow Manifesto.* (Preprint book chapters).
"The Time Flow Manifesto: Introduction."
http://philpapers.org/rec/HOLTTF
"Chapter 1: Concepts of Time Direction."
http://philpapers.org/rec/HOLTTF-5
"Chapter 2: Time Symmetry in Physics."
http://philpapers.org/rec/HOLTTF-6
"Chapter 3: Reversibility in Physics."
http://philpapers.org/rec/HOLTTF-7
"Chapter 4: Metaphysical Time Flow."
http://philpapers.org/rec/HOLTTF-2
"Chapter 5: Time Flow Physics."
http://philpapers.org/rec/HOLTTF-3
"Chapter 6: Philosophical Issues."
http://philpapers.org/rec/HOLTTF-4

(j) 2004. "Time Flow Physics: Introduction to a unified theory based on time flow." *http://philsci-archive.pitt.edu/1641/*

(k) 2003. "An Introduction to Pavel Tichy and Transparent Intensional Logic." *http://philsci-archive.pitt.edu/1479/*

(l) 2003. "The time reversal invariance of classical electromagnetic theory: Albert versus Malament." *http://philsci-archive.pitt.edu/1475/*

(m) 2003. "The Quantum Mechanical Time Reversal Operator." *http://philsci-archive.pitt.edu/1449/*

(n) 2003. "The incompleteness of extensional object languages of physics and time reversal. Part 1". *http://philsci-archive.pitt.edu/1451/*

(o) 2003. " ... Part 2". *http://philsci-archive.pitt.edu/1452/*

(p) 1990. *Time Flow and Irreversibility in a Probabilistic Universe. (PhD Thesis) http://muir.massey.ac.nz/bitstream/handle/10179/3159/02_whole.pdf?sequence=1*

(q) 2007. US Patent 7979449 B2. "System and method for representing, organizing, storing and retrieving information." (2007/2008/2011). *https://www.google.com/patents/US7979449*

www.ingramcontent.com/pod-product-compliance
Lightning Source LLC
Chambersburg PA
CBHW071357170526
45165CB00001B/85